MODELING THE METABOLIC AND PHYSIOLOGIC ACTIVITIES OF MICROORGANISMS

MODELING THE METABOLIC AND PHYSIOLOGIC ACTIVITIES OF MICROORGANISMS

Edited by

Christon J. Hurst
U.S. Environmental Protection Agency
Cincinnati, Ohio

JOHN WILEY & SONS, INC.
New York • Chichester • Brisbane • Toronto • Singapore

In recognition of the importance of preserving what has been
written, it is a policy of John Wiley & Sons, Inc., to have books
of enduring value published in the United States printed on
acid-free paper, and we exert our best effort to that end.

Copyright © 1992 by John Wiley & Sons, Inc.

All rights reserved. Published simultaneously in Canada.

Reproduction or translation of any part of this work
beyond that permitted by Section 107 or 108 of the
1976 United States Copyright Act without the permission
of the copyright owner is unlawful. Requests for
permission or further information should be addessed to
the Permissions Department, John Wiley & Sons, Inc.

Library of Congress Cataloging-in-Publication Data

Modeling the metabolic and physiologic activities of microorganisms
  edited by Christon J. Hurst.
        p.   cm.
    ISBN 0-471-54271-7 (cloth)
      1. Microbial metabolism—Mathematical models.   2. Microorganisms-
  -Physiology—Mathematical models.   I. Hurst, Christon J.
  QR88.M63   1992
  576'.11'011—dc20   92-3482
                                                                CIP

Printed in the United States of America

10 9 8 7 6 5 4 3 2 1

For Pei-Fung and our children Allen and Rachel

PREFACE

Microorganisms were the first chemical engineers, and their metabolic activity provided the modifications of Earth's environment that necessarily preceded evolution of oxygen-dependent plants and animals. Microorganisms remain the cornerstone of carbon and nitrogen recycling, which sustains the life on Earth. We humans have learned to harness the metabolic and physiologic activities of microorganisms for the purpose of biosynthesizing compounds of interest such as antibiotics, which are beneficial to our continued survival, and for the biotransformation and biodegradation of municipal and industrial waste compounds, which are of our own creation. This harnessing of microbial activity constitutes the basis of microbial biotechnology. Our ability to model the metabolic and physiologic activities of microorganisms helps us in understanding, predicting, and directing the activities of those organisms.

The U.S. Environmental Protection Agency was not involved with the editing of this book.

CHRISTON J. HURST

Cincinnati, Ohio
March 1992

CONTRIBUTORS

J. N. BOYER, Department of Biology and Institute for Coastal and Marine Resources, East Carolina University, Greenville, North Carolina 27858

W. MICHAEL CHILDRESS, Center for Biosystems Modeling, Department of Industrial Engineering, Texas A&M University, College Station, Texas 77843

R. R. CHRISTIAN, Department of Biology and Institute for Coastal and Marine Resources, East Carolina University, Greenville, North Carolina 27858

ROBERT M. CLARK, U.S. Environmental Protection Agency, Cincinnati, Ohio 45268

TOMAS B. CO, Department of Chemical Engineering, Michigan Technological University, Houghton, Michigan 49931

RICHARD B. COFFIN, U.S. Environmental Protection Agency, Environmental Research Laboratory, Gulf Breeze, Florida 32561

JOHN P. CONNOLLY, Environmental Engineering & Science Program, Manhattan College, Riverdale, New York 10471

ROSEANNE M. FORD, Department of Chemical Engineering, Thornton Hall, University of Virginia, Charlottesville, Virginia 22903-2442

ANTHONY F. GAUDY, JR., 9379 Fox Hollow Lane, Brooksville, Florida 34613

M. R. GRAY, Department of Chemical Engineering, University of Alberta, Edmonton, Alberta T6G 2G6, Canada

DAVIS W. HUBBARD, Department of Chemical Engineering, Michigan Technological University, Houghton, Michigan 49931

CHRISTON J. HURST, U.S. Environmental Protection Agency, Cincinnati, Ohio 45268

ROBIN E. LANDECK, Environmental Engineering & Science Program, Manhattan College, Riverdale, New York 10471

P. K. NAMDEV, Department of Chemical Engineering, University of Alberta, Edmonton, Alberta T6G 2G6, Canada

W. M. RIZZO, Department of Biology and Institute for Coastal and Marine Resources, East Carolina University, Grenvile, North Carolina 27858

STEVEN K. SCHMIDT, Department of Environmental, Population and Organismic Biology, Campus Box 334, University of Colorado, Boulder, Colorado 80309

PETER J. H. SHARPE, Center for Biosystems Modeling, Department of Industrial Engineering, Texas A&M University, College Station, Texas 77843

D. W. STANLEY, Department of Biology and Institute for Coastal and Marine Resources, East Carolina University, Greenville, North Carolina 27858

D. W. S. WESTLAKE, Department of Microbiology, University of Alberta, Edmonton, Alberta T6G 2G6, Canada

DEANNA K. WILD, National Institute for Occupational Safety and Health, Cincinnati, Ohio 45226

CONTENTS

MODELING OF BIODEGRADATION

1 Introduction to Modeling the Metabolic and Physiologic Activities of Microorganisms Using Wastewater Treatment as an Example   1
Anthony F. Gaudy, Jr.

2 Models for Studying the Population Ecology of Microorganisms in Natural Systems   31
Steven K. Schmidt

3 A Compartment Model Approach to Bacterial Population Genetics and Biodegradation   61
W. Michael Childress and Peter J. H. Sharpe

MODELING OF INDIVIDUAL CELLULAR FUNCTIONS

4 Modeling the Growth of Cellulase-Producing Ogranisms   89
Davis W. Hubbard and Tomas B. Co

5 Kinetic Model of Antibiotic Production by *Streptomyces clavuligerus*   115
P. K. Namdev, M. R. Gray, and D. W. S. Westlake

6 Comparing the Accuracy of Equation Formats for Modeling Microbial Population Decay Rates   149
Christon J. Hurst, Deanna K. Wild, and Robert M. Clark

7 Mathematical Modeling and Quantitative Characterization of Bacterial Motility and Chemotaxis   177
Roseanne M. Ford

MODELING OF ENVIRONMENTAL CYCLES

8 Network Analysis of Nitrogen Cycling in an Estuary   217
R. R. Christian, J. N. Boyer, D. W. Stanley, and W. M. Rizzo

9 Modeling Carbon Utilization by Bacteria in Natural Water Systems   249
John P. Connolly, Richard B. Coffin, and Robin E. Landeck

INDEX   279

MODELING THE METABOLIC AND PHYSIOLOGIC ACTIVITIES OF MICROORGANISMS

1 Introduction to Modeling the Metabolic and Physiologic Activities of Microorganisms Using Wastewater Treatment as an Example

ANTHONY F. GAUDY, JR.
H. Rodney Sharp Professor Emeritus, Department of Civil Engineering, University of Delaware

1. Introduction 1
2. Modeling aerobic decay 6
3. Modeling aerobic microbial growth 9
   3.1. Kinetics of growth 9
      3.1.1. Accumulation and deaccumulation of biomass 9
      3.1.2. Dependence of μ on S 12
4. Conversion of respiration to growth and substrate removal 15
5. Reactor engineering considerations 16
   5.1. Prediction of effluent substrate and reactor biomass concentrations ... 16
   5.2. Effects of cell recycle 19
   5.3. Prediction of excess biomass X_W 22
6. Future potential and status of the model 22
   6.1 Effects of influent substrate concentration S_i 23
   6.2. Effects of changing feed rate 23
   6.3. Effects of changing feed composition 25
   6.4. Role of oxidative assimilation 26
7. Concluding statements 27
   References 28

1. INTRODUCTION

Writing the introductory chapter to a text such as this is a much different task than introducing a text one has written. In the latter case, the writer is thoroughly familiar with the contents and with what one is attempting to do for the user of the text. And one can afford the feeling that if readers do not bother to study the introduction carefully, there is ample time to hit the mark in the ensuing chapters. Not so in the present case, and an attempt is made in this chapter to make an input, which, while making a suitable introduction and foundation for the subjects that follow, also makes an individual contribution to the general subject of modeling the metabolic and physiologic activities of microorganisms in the biological treatment of waste substrates.

A brief scanning of the Contents reveals that, while there is a broad range of subjects applicable to a wide array of biological processes, the subject matter is particularly apropos to the environmental sciences and pollution control. This observation is justification for stressing this important application of such work.

This chapter deals with some commentary, observations, conclusions, and recommendations on the general subject of modeling biological processes. It then evolves to more specific comments on the status of modeling and my experiences with regard to modeling aerobic biological purification of wastewaters and includes some speculation on possible ways to expand the approach to modeling. To some readers, the entire area of modeling biological processes may be unfamiliar. Thus the starting point in this chapter is quite fundamental. However, we are dealing with a rather complex physiological phenomenon and specialized information will necessarily be developed. The references cited herein should be freely consulted by the reader. The term "introduction" used for this chapter is one that might be applied to the entire text, since this area of work is rather new and all who work in it or who aim to learn something about it must realize that it is an undertaking that, for completion, requires continued amassing of knowledge of the mechanisms of the metabolism and physiology of microorganisms.

One must also realize the difficult task modelers of biological processes set for themselves. The life processes are so intricately controlled and affected by the environment in which they take place that oversimplification of mechanism becomes one of the major pitfalls blocking real success in this field. An author of much more power than myself put the situation into focus by declaring something to the effect that we ought not lower the levels of our aims simply to ensure the complacency of success. Herein lies one of the major hurdles modelers must consider. That is, how do we define success in modeling? Although this kind of work is best placed in the category of a scientific endeavor, its "raison d'être" is clearly practical; that is, it has an engineering purpose.

Models are devised in order to predict an outcome, or several outcomes, of the operation of the biological process being modeled. They are used for purposes of functional process design and to develop strategies for controlling processes. Therefore the tendency is to assume that, if the model does yield fairly good prediction of events under the conditions of interest for which it was conceived, it is a good and true model of the process. The trouble here is that this may be a gross oversimplification and the model may not suffice under new conditions and may not be extrapolatable beyond the numerical bounds for which it was developed.

Yet, because of its success as a predictor of events under certain conditions, there is a tendency to declare the model to be "basic" and to be of "theoretical" value or to represent proven theory. This may lead to gross errors and one is reminded that the word theory may be more equivalent to hypothesis than to scientific truth. Such warnings are not generally needed for scientists who use models as attempts to improve and extend understanding of events. However, models are increasingly used as tools by on-line practicing professionals to control and design processes or to make managerial decisions, and these people may be neither inclined nor prepared to consider the limitations of the latest model of a particular process. It is because models are becoming such important practical tools that continued research aimed toward improving them is so vital to the field. There is indeed no room for complacency, even though we may tend to simplify in order to say we understand the process.

On the other hand, one cannot declare as unworthy or unusable a model that does a fair job of predicting outcomes, simply because it may contain some flaw(s) with respect to a physiological mechanism. Always before us is the practical need to solve problems by making quantitative predictions regarding the products of the biological process. Ideally, any biological model describes a known or proven mechanism. This description is one that predicts a result, for example, a quantity of product or an effect on a resource, under certain time and operational constraints. However, the more complete the knowledge of the mechanism, generally the more complicated the model becomes. These complications, if they can be handled mathematically with ease, may make the model prediction more precise but may militate against its use by those who need to use it.

In defense of enlightened compromise, it is certainly true that very accurate prediction of events can be attained using models that may be in error from a mechanistic standpoint, provided the model is devised after repeated and accurate measurements of the events being modeled. The calendar and the kinetic definition of time provide classic examples of very accurate prediction, which was originally based on a totally erroneous mechanistic model, that is, the notion that the Sun revolved around Earth. The Michaelis–Menten model describing the dependence of specific velocity of an enzyme reaction on substrate concentration resulted from repeated experimental observations and provides an example of a case wherein practical kinetic observations permitted the investigators to postulate biological mechanism; it was only through decades of research that proof of the postulated mechanism was obtained. And it must be realized that this led to more accelerated research on enzyme catalysis rather than to termination of research due to success afforded by the ideal wedding of mechanism with its predictive kinetic model. While models are devised and tailored for practical purposes, satisfaction of a particular practical purpose does not indicate a stopping place, because the mechanism purported to be modeled may be so faulty or incomplete as to negate use of the model for extrapolated purposes or new applications to which professionals in the field may wish to extend it.

It should also be noted that the practical need to tailor a metabolic and physiologic process for a particular application, while not affecting the basic process being modeled, may bring into play sociotechnological aspects that affect not only the specific uses of the model but also its acceptance by professionals. For example, there is great underlying sameness in modeling physiologic action regardless of whether the model is being employed in production of single-cell protein or of pharmaceuticals or in purification of wastewater. But the differences in the specific reasons for modeling may greatly affect the very terms in which the model is couched. In industrial aerobic fermentations (an often-used misnomer), one may select both substrate and microbial species, whereas in the field of environmental pollution control, substrates and the microbes that act on them represent heterogeneous mixtures. While microbes may be measurable in terms of biomass for either application, quantitative description of substrate(s) in the environmental field represents a considerable problem, which has itself been the subject of extensive investigation. In the wastewater pollution control field, growth of microorganisms is modeled in order to predict substrate removal (i.e., purification) so that the oxygen resource in the natural receiving body of water will not fall below that required to maintain suitable environmental conditions to sustain normal healthful conditions for the biota in the receiving body. The normal healthful conditions are usually determined by the uses (political decision) to which the receiving body (the surface water resource) has been allocated.

Growth, substrate removal, and the assimilation capacity of the stream are related and, to be usefully handled by modeling procedures, they need to be couched in terms readily identified with those employed by practicing professionals in design, operation, and regulation, who will live by the numbers the models produce. It is indeed a very serious area of activity in which mistakes are inevitable and forgivable but one in which propagation of mistakes is unforgivable, not only in terms of natural resources but perhaps in lives. Thus models ought to be continuously refined, simplified concepts altered when new information demands, and overcomplicated concepts pared down when possible. All this requires continuing inquiry and investigational effort. Fine-tuning of the models and demanding their use in environmental management are not simply desirable refinements warranted because of new knowledge, but a necessity due to increased human activity and demands placed on the life support system.

Before discussing the technological status of modeling the methodologies for the biological treatment of municipal and industrial wastewaters, elaboration of some of the unique characteristics of this field may aid in understanding the directions that such modeling efforts have taken. One of the unique considerations is that treatment of wastewaters is mandated by federal, state, and local laws. Producers of wastes do not generally treat them for profit. Polluters treat because a majority of the people have demanded legislation requiring treatment. While today's laws are justifiably rather tough on polluters, there is a demand that methodologies for purification of wastewaters be as economical as possible simply because those wise enough to demand treatment also pay for it, whether the waste is of municipal or industrial origin. One can see that this is an ideal arena for engineers because the tenets of the engineering profession demand economical solutions to problems through use of various technologies.

Before the laws were strengthened, there were, of course, technologies for treating wastewaters of various kinds. These methods were, in general, capable of doing the job as defined by the existing laws and regulations, and the general idea was that, if the technologies were only more generally applied, the problem of water pollution would be solved. Thus some cities that formerly collected wastewater and ran it directly to the receiving stream were forced to install a treatment process prior to discharge, and industrial waste producers were forced to treat prior to discharge either to a municipal sewer or to a natural receiving body. This is by no means ancient history and the clean-up process is still in progress, has already cost billions, and will cost billions more. The progress of time has also seen continued tightening of standards, and the pollution problems have become ever more complicated. Many of the older engineering practices cannot meet the requirements of today's laws. It is entirely too late to handle the economic problem by legislative undoing of the laws or by looking the other way regarding their enforcement because, during such a hiatus, the life support system deteriorates further and costs become greatly escalated when one inevitably comes to grips with the problem. Thus new methods must be devised or existing methodologies must be improved.

One of the generalities that is becoming a justifiable technological policy in the minds of many scientists and engineers and the public whom they serve is one that has become largely understood and generally embraced by the public who pay the bills. Nearly every young student is now routinely being made aware that the activities of our species are conducted as a part of the carbon–oxygen cycle and that we function largely in the decay leg of the cycle. They learn that this is, for the most part, a balanced cycle and its disturbance can adversely affect the environment (i.e., the life support system) and that we have now constitutionally required ourselves to maintain the balance. They learn that this decay side of the cycle is biological and that it largely deals with aerobic micro- and macroorganisms, from single-celled organisms all the way to themselves. Unless their interest is stirred to go further, their knowledge may taper off here, but people are now tuned in to aerobic decay, which is indeed the mechanism of successful biological treatment of wastewater. Since the people are involved, informed, and foot the bill, there is ample logic to the now general acceptance of aerobic biological treatment of wastewaters as a national policy. In addition to this important force, it has become more and more evident to pollution control technologists that biological treatment is, in the long run, the most economical type of treatment. This technology has been around for a long time and, although older methodology using biological treatment has failed to meet some of the increasingly stringent requirements, the general approach is not to replace it (there would not appear to be any replacement) but to improve it by enhancing its reliability and perhaps by extending the range of its application to detoxification of organics and to removal of inorganic toxicants.

Modeling offers a way to examine the factors affecting the outcome(s) of operation of a process and, since aerobic biological treatment of wastewaters is really nothing more than technological enhancement of the natural process of aerobic decay, it is a good basis on which to build.

2. MODELING AEROBIC DECAY

A simple model of the decay leg of the carbon–oxygen cycle is shown in Fig. 1.1. Organic matter (synthesized largely by the photosynthetic leg of the cycle, not shown) is metabolized by aerobic organisms (largely microorganisms). A portion of this organic food material is respired (i.e., oxidized) and dissipated to the environment as CO_2 and H_2O, which is returned to the supply of organic matter through photosynthesis. The remainder is used to produce biomass, which is recycled as food or carbon source for other organisms, and the process proceeds. Sometimes this forward reaction and recycle can go essentially to completion, in which case the decay leg is said to have provided total oxidation of the organic matter. Due to lack of oxygen in some locales, organic matter may be anaerobically converted into gas or placed under great pressure and, due to geologic processes, in time be converted to coal and oil, and not converted to CO_2 until extracted and then oxidized; that is, some organic matter may for a time be sequestered from aerobic decay.

In order to convert this model to one suitable for biological treatment of wastes,

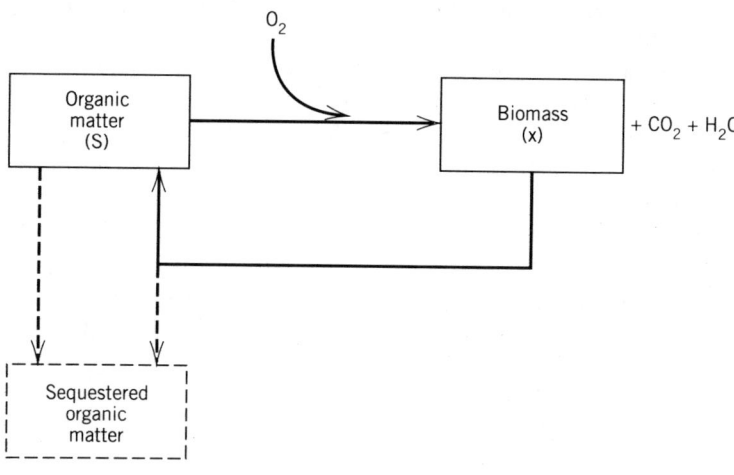

FIGURE 1.1. Global model of aerobic decay.

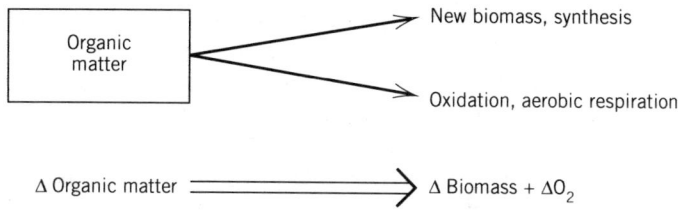

FIGURE 1.2. Partition of carbon source between respiration and synthesis.

each forward pass may be written in the form of a chemical equation, as seen in Fig. 1.2.

There are several questions to be asked of this simple model of biological treatment. Is it correct? Can one go from a macromodel known to provide an adequate gross description of a global process to an essentially microscale process in which a decrease in the amount of organic substrate in an isolated or local system is accounted for solely by that which has been channeled into synthesis of new biomass in the form of microorganisms and that which has been oxidized by them? Does the sum of O_2 used and biomass produced in the process equal the amount of organic matter that has been used? The answers to these questions are not obvious yeses. There are evident problems with such a simple model. For example, it is well known that organotrophic microorganisms fix a certain amount of single carbon compounds (i.e., CO_2); thus some of the organic carbon that has been oxidized may be reconverted to organic matter, lessening the validity of the equation. Also, if the original organic matter is strippable to some extent or if some portion is made so through production and elaboration of volatile metabolic intermediate compounds, this physical removal process could detract from the validity of the equation. In addition, it is obvious that the quantities in Fig. 1.2 must be converted to like terms in order to form an equation of quantitative value. Another consideration regarding potential use of this model (equation) is whether it, like other chemical equations, represents the conditions of the system at rest. That is, is it valid only after the reaction has gone to completion or, in any event, has ceased? Might it also be valid at any time during the reaction?

There are a number of ways one can quantitatively deal with the portioning of the organic substrate between synthesis and respiration so as to check its validity. If one knew the chemical formulas of the organic matter and the biomass that was grown on it, a "balanced equation" could be written. One could calculate the mass of substrate organic matter oxidized, sum this with the biomass produced, and compare to the mass of organic matter used up in the reaction. Alternatively, one could analyze for the carbon used up and compare it to the sum of carbon in the new growth and the CO_2 produced. Also, one could determine the oxygen equivalent of the substrate organic matter and compare this value with the O_2 equivalent of the cellular organic matter produced and the O_2 used in the reaction.

My first experience with using this model in an experiment goes back many years and the methodology was the first described above, that is, expressing respiration, synthesis, and substrate in terms of mass of organic substrate (Gaudy and Engelbrecht, 1963). The course of the reaction was followed over time and the results indicated that respiration and synthesis accounted quite well for substrate disappearance; recovery of material varied from 87 to 98% when one omitted the very early samples. Later experiments were designed to compare various ways of assessing the validity of the model for an isolated system. It was found that an energy balance using biological O_2 uptake and COD values of substrate and cells provided excellent recovery of substrate as well as methodological convenience for field application (Gaudy et al., 1964). The best way to write the fundamental equation was concluded to be as shown in Eq. (1) or in differential form [Eq. (1a)]:

$$\Delta COD_t = \Delta COD_{X_t} + \Delta O_{2,t}, \tag{1}$$

$$\frac{dS}{dt} = \frac{dX}{dt} + \frac{dO_2}{dt}. \tag{1a}$$

This basic equation for aerobic treatment simply states that the amount of waste substrate disappearing from the system during aerobic metabolism, ΔCOD_t, that is, $COD_i - COD_t$, during any time period is accounted for as the increase in COD in the biomass, ΔCOD_{X_t}, that is, $COD_{X_t} - COD_{X_i}$, and the amount of O_2 used during the same time period. If one initiates the system with only a small amount of biomass, there is no need to be concerned with making a subtraction for O_2 used by the seed biomass. If the reaction goes to completion, that is, the organisms that have been previously acclimated to the organic matter grow to some peak level, then cease to grow, the ΔCOD at this point in time provides a useful measure of the strength of the waste (Gaudy and Gaudy, 1972; Hiser and Busch, 1964; Symons et al., 1960).

Equation (1) provides an excellent method for checking the accuracy of experimental results. However, it does much more for the cause of modeling. It is not, in itself, a functional model, because it does not purport to describe how the process occurs or what factors affect utilization of the substrate, nor does it predict the time required to attain any degree of substrate utilization. It is simply a good bookkeeping system for the process of aerobic growth of biomass on organic substrate. Equation (1) provides two practical and fundamental requirements for modeling.

1. It provides a ground equation on which to base models describing aerobic microbial growth on substrates.
2. Since it truly does delineate a fundamental relationship between substrate, biomass production, and respiration, it may be used as a basis for calculating one term from another provided that more detailed relationships between the three factors can be illuminated.

The usefulness of the equation in both these regards is discussed below.

3. MODELING AEROBIC MICROBIAL GROWTH

In the pollution control field, the basic interest is in designing and operating a process that reduces the oxygen demand in the natural receiving stream. However, few models ever deal directly with this aspect other than a traditional and outdated attempt to measure the strength of influent organic matter, S_i, and effluent, S_e, solely in terms of 5-day or total carbonaceous biochemical oxygen demand, BOD. The concept of BOD is a very important one in the water pollution control field, but respiration, or O_2 uptake, which is what the BOD test measures, is not a good candidate for use in modeling the treatment process because BOD exertion can go on long after the original substrate has been used. Also, in systems containing predator organisms, largely protozoa, the end of the substrate removal phase is not always well delineated in such dilute systems as are necessarily employed under conditions of BOD testing (Bhatla and Gaudy, 1965; Gaudy, 1972). Under experimental conditions permitting more rapid microbial growth, respiration can ideally be related to either biomass production or substrate removal, as discussed in a later section. There would appear to be a fairly general consensus that either substrate S_i, as measured by either ΔCOD or ΔTOC, or biomass growth ΔX, expressed either as mass of cells or as their COD, is the best candidate for use in definitive modeling.

Some models are couched in terms of substrate removal. While use of this parameter moves us somewhat closer to the goal of treatment, it is not the natural mechanism and all models looking at substrate removal relate it to the real biological mechanism responsible for both substrate removal and O_2 uptake, that is, microbial growth. It seems sensible to focus on metabolism and growth as the subject for modeling because it gets to the heart of the mechanism of purification. Moreover, metabolic activities critical to growth, and thus to treatment, can be affected by many environmental factors, and these effects can be inserted into the equations in various ways. Also, in general, much more fundamental information is available on microbial growth than on substrate utilization. Thus substrate removal is rightfully considered as a consequence of growth, as is O_2 utilization.

Both substrate removal and oxygen uptake can be calculated for models focusing on biomass X. It should be remembered that oxygen dissolved in the waste solution is a vital reactant. Thus in modeling Eq. (1), the approach is to make sure that this reactant is present in sufficient concentration so that it exerts no control on the growth process. Much work has indicated that DO concentrations of 1–2 mg/L are more than ample to assure that the rate of metabolism is not limited by O_2 concentration.

3.1. Kinetics of Growth

3.1.1. Accumulation and Deaccumulation of Biomass

The growth term ΔX is kinetically defined in several ways, for example,

$$\frac{\Delta X}{\Delta t} = \frac{X_t - X_0}{\Delta t}, \tag{2}$$

and, in general, has been observed to be defined by

$$\frac{dX}{dt} \approx \frac{\Delta X}{\Delta t} = \mu X. \quad (3)$$

That is, the rate of growth, dX/dt, is proportional to the amount of biomass. The term specific growth rate, that is, the proportionality factor μ, or the change in biomass per unit of biomass in the system, is

$$\mu = \frac{dX}{dt}\frac{1}{X}. \quad (4)$$

Under conditions wherein μ is observed to have a constant numerical value, the biomass is said to be in an exponentially increasing phase of growth. In a once-fed batch system, in accord with this equation, concentration at any selected time can be calculated using the integrated form of Eq. (4):

$$X_t = X_0 e^{\mu t}. \quad (4a)$$

Microorganisms will accumulate in accord with Eq. (4a) provided the carbon source, S, is maintained in sufficient supply to keep them in exponential growth and the environment in which they are growing does not become deficient in other nutrients, for example, phosphorus, nitrogen, or DO. Eventually they would stop growing because of spatial considerations and loss of fluidity of the reactor system. Microbial populations do not continually accumulate according to Eq. (4a) in local, isolated systems because the food resource, S, becomes depleted. The growth curve assumes its typical S shape; X peaks and then may decrease because of autodigestion of the biomass.

The question of autodigestion is a very interesting one and was the subject of considerable debate in the basic field of microbiology several decades ago. That is, does metabolism of internal microbial substance (in pure cultures) or that plus active predatory feeding (in heterogeneous or mixed cultures) go on during active metabolism of the external carbon source? Much study indicates that decay or autodigestion does proceed in the presence of an exogenous carbon source, but the rate of decay is much lower than the maximum rate of synthesis (on the order of 10–100 times lower). Thus it is only measurable at slower growth rates, that is, when the substrate concentration nears exhaustion or is held to a low value by other means. There is, to my knowledge, no convincing evidence of the manifestation of decay during the exponential phase or, for that matter, in the declining growth phase in batch cultures. However, there is much evidence consistent with concurrent growth and decay in slowly growing continuous culture systems. Decay of biomass already synthesized is, as one would expect, the only source of metabolic carbon in the system after the original exogenous substrate has been exhausted. In continuous flow wastewater reactor systems, slow growth will be shown to be desirable, and it is well to modify Eq. (3) as shown below:

$$\frac{dX}{dt} = \mu X - k_d X \qquad (5)$$

The term k_d is considered to be a biokinetic constant and is designated as the specific decay rate. Like μ, its numerical value can vary depending on the environmental conditions as well as the physiological state of the biomass. Unlike μ, which in modeling will be shown to be the chief operating variable, k_d is assumed to be a system constant, the first thus far introduced. There are four more to be discussed to bring us to the current status of modeling.

From Eq. (1) it is apparent that there will always be less COD in the cells than there is in the carbon source, because some must be oxidized to provide energy needed to produce new cells, even neglecting cell decay. Thus one can say that the yield of biomass, in terms of COD, is always less than 1. Conventionally, one measures X in terms of mass, while substrate S is generally measured in terms of metabolizable COD. However, it should be noted that sometimes S is given in terms of weight and it is important to make note of the units used, because the values can be widely different according to the units of measurement. If one measured S in terms of the mass of carbon source metabolized and X in terms of mass, the yield of biomass could exceed unity, depending on the oxidation state of the substrate and the synthesized cells. For example, if the substrate consisted of 1000 mg/L of a hydrocarbon, the substrate COD may be 2.5 times this amount, that is, 2500 mg/L COD. Let us say that this amount of hydrocarbon permitted growth of 1250 mg/L of new biomass. The biomass yield, Y_t, would be calculated as 0.5 mg X/mg COD removed or 1.25 mg X/mg hydrocarbon removed. In this chapter and in most modeling work in this field, S is given in terms of oxygen equivalents, that is, chemical oxygen demand or COD, and X in terms of biomass, and the biomass yield Y_t is given by

$$Y_t = \frac{\Delta X}{\Delta S} = \frac{dX/dt}{dS/dt}. \qquad (6)$$

The rate of substrate utilization is thus related to growth by

$$\frac{dS}{dt} = \frac{dX}{dt}\frac{1}{Y_t}. \qquad (7)$$

The numerical value of Y_t has been found not to change during the entire course of the growth curve. In a batch experiment, Y_t remains constant throughout the exponential and declining phases of growth, that is, the entire range of the substrate removal phase. As the top of the growth curve is approached, it may decrease due to cell decay, but, with the accuracy of methods employed for measurement, there can be shown to be a constant ratio of ΔX to ΔS and Eqs. (6) and

12 INTRODUCTION TO MODELING MICROBIAL ACTIVITIES

(7) can be used to help describe the kinetics of the system. When decay becomes a factor, that is, after X peaks in a batch growth system, or when μ is held low by other system controls under continuous culture conditions, the cell yield decreases. This yield has been termed the observed cell yield Y_o and will be discussed later. Note that the yield obtained in batch experiments is termed the "true" cell yield, Y_t. It, like k_d, is a "constant" that must be determined for the particular system of substrate, cells, and environmental conditions under study. For a variety of compounds readily metabolized by microorganisms, yield values commonly range between 0.4 and 0.6, but values outside this range are occasionally observed.

3.1.2. Dependence of μ on S. Thus far, it is possible to say that the amount of cells produced per unit of substrate removed is constant during the substrate removal phase (i.e., Y_t is a constant) because the decay rate k_d of the biomass produced is negligible or so small as to be unmeasurable by the methods employed to observe the system. However, it is known that the specific rate of growth μ changes during this phase. Its value can be observed to be constant for a while (exponential phase). Then μ decreases during the declining phase and eventually becomes zero at the peak of the curve. All else being equal, this is ascribed to depletion of the carbon source S. Microbial kineticists have, for decades, sought to relate μ and S, that is, to describe or model this relationship. Many equations have been suggested, but the one that has been found to be highly adequate in most modeling situations is the rectangular hyperbolic equation suggested by Monod (1942):

$$\mu = \frac{\mu_{max} S}{K_s + S}. \tag{8}$$

The equation is plotted in Fig. 1.3. It is apparent that μ_{max} and K_s are two additional biokinetic "constants," μ_{max} being the upper value of μ at high values of S, while K_s, termed the saturation "constant," determines the shape of the curve as it approaches μ_{max}. K_s can be shown to be numerically equal to the substrate concentration at which μ is $0.5\mu_{max}$. In the early days, some of the most widely published data on several species of microorganisms and carbon compounds suggested that K_s was rather low, on the order of a few milligrams per liter. This was not true for all pure cultures and has been found not to be the general case for heterogeneous microbial populations, especially when COD is used as a measure of S. For many readily metabolized substrates, values can be expected to range between 50 and 150 mg/L COD. Values for μ_{max} of 0.4–0.6 h^{-1} are commonly observed (Benefield and Randall, 1980; Gaudy and Gaudy, 1988b; Grady and Lim, 1980). For some of the more difficultly metabolized manufactured compounds increasingly found in wastewaters in low concentrations, one may expect lower numerical values.

The Monod equation now appears in nearly every microbiology text, in modeling texts, and in all applied fields of microbial kinetics. It is an equation based on experimental observation and has been found to fit experimental data for sys-

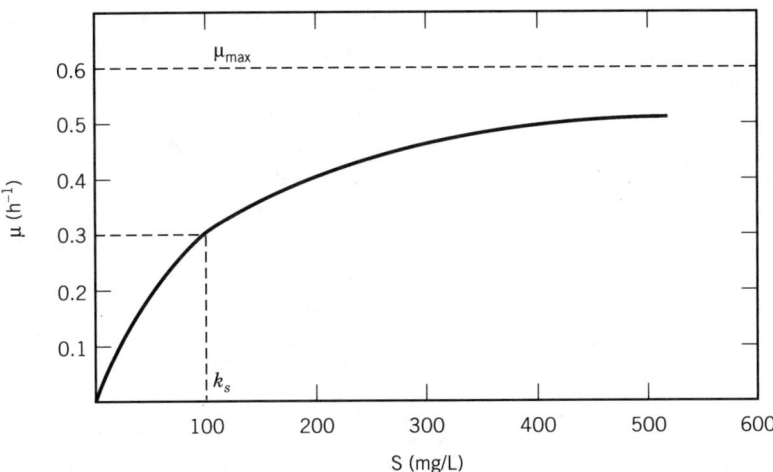

FIGURE 1.3. Plot of μ versus S for the Monod equation; $\mu_{max} = 0.6\ h^{-1}$, $K_s = 100$ mg/L.

tems consisting of single substrates and microbial species as well as heterogeneous systems of both S and X. Several processes other than microbial processes follow a similar curve, for example, adsorption of molecules onto surfaces. It has become of basic value in the field, not because it represents basic theory but because it has permitted fairly accurate and reliable approximation of the course of microbial growth on exogenous substrates, and it has practical use in modeling. This fact should encourage speculation as to theory; that is, it should cause scientists to hypothesize about mechanisms of growth, but it must be concluded that such efforts in the past have not resulted in convincing arguments. The equation is best looked on as a good, but empirical, modeling equation. When inserted into Eq. (5), it provides a model for growth:

$$\frac{dX}{dt} = \frac{\mu_{max} S}{K_s + S}(X) - k_d X. \tag{9}$$

Equations (7) and (9) can be solved simultaneously and integrated to produce the time course of S and X for a given initial concentration of biomass and substrate, X_0 and S_0, and is best handled by numerical integration. Two new biokinetic constants, μ_{max} and K_s, have been added, bringing the total to four. The major use of the equation is not in approximating the curve of S and X in batch systems but in describing growth in continuous culture devices, which will be described later. Before doing this, it is important to expand the model to include toxic or inhibitory substrates.

At times, the substrate or one of the chemical entities comprising the substrate (note that ΔCOD is used as a measure of S and generally represents a heterogeneous mixture of compounds) not only permits growth but is also an inhibitor of growth. That is, an increase in S increases μ at some concentrations, but at higher concentrations is observed to decrease μ; μ_{max} is not approached asymptotically as in the Monod equation. Much research using both single carbon sources and pure cultures as well as heterogeneous systems has shown that several forms of equations can be used to fit the experimental observations for growth on toxic substrates, but one that provides a fit equal to, or better than, most is a modification of the Monod model, the Haldane equation (D'Adamo et al., 1984; Hill and Robinson, 1975; Jones et al. 1973; Pawlowsky and Howell, 1973; Yang and Humphrey, 1975). Equation (10) for growth on inhibitory substrates introduces a new term, K_i, the inhibition constant.

$$\mu = \frac{\mu_{max} S}{K_s + S - S^2/K_i}. \tag{10}$$

Figure 1.4 shows a plot of μ versus S according to Eq. (10). When growth data behave in accordance with Eq. (10), μ_{max} is never approached. The highest value of μ observed, that is, the critical specific growth rate, termed μ^*, which is the peak in the plot of μ versus S, is given by

$$\mu^* = \frac{\mu_{max}}{1 + 2(K_s/K_i)^{0.5}}. \tag{11}$$

This value of μ occurs at an S value of S^*, given by

$$S^* = \sqrt{K_s K_i}. \tag{12}$$

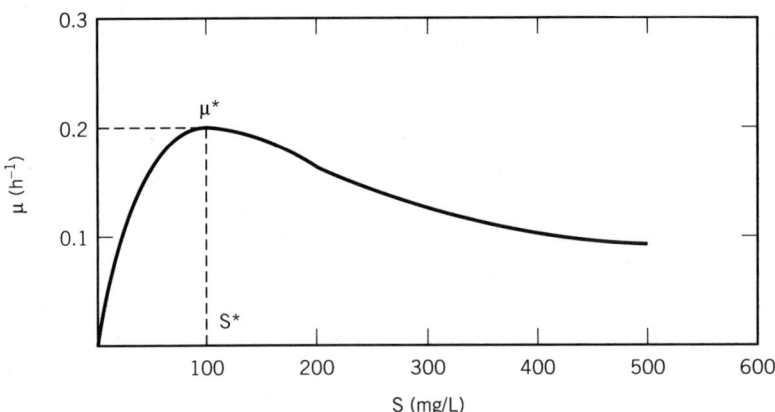

FIGURE 1.4. Plot of μ versus S for the Haldane equation; $\mu_{max} = 0.6 \text{ h}^{-1}$, $K_s = 100$ mg/L, $k_i = 100$ mg/L.

Equation (9) can be modified to represent growth on inhibitory carbon sources:

$$\frac{dX}{dt} = \left[\frac{\mu_{max} S}{K_s + S - S^2/K_i}\right] X - k_d X. \quad (13)$$

Equation (13) should be used instead of Eq. (9) when a plot of μ versus S provides a better fit to the Haldane equation than to the Monod equation (compare Figs. 1.3 and 1.4).

4. CONVERSION OF RESPIRATION TO GROWTH AND SUBSTRATE REMOVAL

Thus far, Eq. (1) has been used to relate dS/dt and dX/dt and the system has been modeled in accordance with growth of the microorganisms. It is important to examine the second use of Eq. (1). The equation can be used to supply missing information, provided one can arrive at appropriate conversion factors. The part of the equation yet to be discussed is O_2 uptake, or respiration. One would not usually need to calculate the O_2 utilization from growth data, but, due to the ease of obtaining the time course of O_2 uptake (respirometry), Eq. (1) can be used to determine growth and substrate removal from the respiration data. Equation (1) may be rewritten as follows:

$$\Delta COD_t = Y_t O_X \Delta COD_t + \Delta O_{2,t}. \quad (14)$$

The biomass term has been related to ΔCOD_t by the biomass yield, Y_t, and converted from mass to biomass COD using O_X, the unit COD of biomass, that is, COD/mg biomass. O_X may be assumed to be 1.42, in accordance with the empirical formula of Porges et al. (1953) for growing biomass. Although this is a reasonable assumption, the value can generally range from 1.2 to 1.6 and it is best to determine the unit cell COD experimentally (Bhatla and Gaudy, 1965).

The amount of COD removed at any time during the course of its removal is

$$\Delta COD_t = (X_t - X_0)/Y_t, \quad (15)$$

and substitution of Eq. (15) for ΔCOD_t in Eq. (14) allows calculation of X_t:

$$X_t = X_0 + (\Delta O_{2,t})/(1/Y_t - O_X) \quad (16)$$

Similarly, the COD concentration can be calculated from the O_2 uptake data:

$$COD_t = COD_0 - (\Delta O_{2,t})/(1 - O_X Y_t). \quad (17)$$

The term $1/Y_t - O_X$ can be combined into one conversion factor R, the amount of O_2 required to produce 1 mg of biomass:

$$R = 1/Y_t - O_X. \qquad (18)$$

Equations (16) and (17) may be rewritten as

$$X_t = X_0 + (\Delta O_{2,t})/R; \qquad (19)$$

$$COD_t = COD_0 - (\Delta O_{2,t})/RY_t. \qquad (20)$$

R may be directly determined by experimentation as a check on the value calculated from the experimentally determined values of Y_t and O_X.

These equations, (16) or (19) and (17) or (20), make it very easy to obtain the kinetic data on growth and substrate removal that are needed to evaluate the biokinetic constants. This approach has been employed to obtain numerical values for μ_{max}, K_s, and K_i from batch experiments in respirometer flasks. These values were then employed successfully to predict S and X in growth reactors (Gaudy et al., 1988, 1990). This basic respirometric concept in somewhat different form has also been employed by others to obtain kinetic constants (Grady et al., 1989; Zabriskie and Humphrey, 1978).

It is important to realize that the values of the biokinetic constants are unique for the substrate–biomass system examined and the environmental conditions, that is, temperature, pH, and so on. Since the reactor system being modeled is heterogeneous with respect to organic substrate and species comprising the biomass, there will be a range, rather than a single numerical value, for each "constant." Thus for any model that employs these constants to provide reasonable predictions, one must determine the numerical values periodically. Using respiration as a "surrogate" in obtaining the needed kinetic growth and substrate removal data greatly aids in facilitating the modeling strategy.

5. REACTOR ENGINEERING CONSIDERATIONS

5.1. Prediction of Effluent Substrate and Reactor Biomass Concentrations

Although Eqs. (7), (9), and (13) constitute a rather good model of growth provided one has equally good values for the five biokinetic constants and knowledge of how they may vary with environmental conditions, the equations alone do not suffice until they are inserted into an operational scheme peculiar to a specific process. The global process shown in Fig. 1.1 is a specific one involving the entire Earth. Some attempts have been made to deal with the general decay model [Fig. 1.1, Eq. (1)] on the global scale, but the reactor is so large that estimates of organic decay and O_2 utilization are only rough approximations and of course it is difficult to verify the model on a quantitative basis. There is much more chance of successfully predicting aerobic organic decay in isolated systems such as aerobic wastewater treatment plants, which can be subjected to engineering controls. Reducing the scale of the reactor to the size of an activated sludge tank or any other industrial growth reactor and imposing certain engineering restrictions permit one to expand the practical utility of the growth model.

Starting with a simple once-through reactor, such as represented in Fig. 1.5, it is possible to demonstrate an extremely useful and important fact germane to modeling. Into the reactor of volume V, a solution containing carbon source at some concentration, S_i, is pumped at a flow rate F; it is pumped out, or allowed to exit, at the same flow rate. If we stipulate that all parts of the reactor fluid are thoroughly and completely mixed so that the concentration of substrate S and biomass X are the same throughout volume V, then X and S in the reactor effluent are the same as in the reactor. Using nontoxic waste as an example, growth and decay of X in the reactor can be defined in accordance with Eq. (9) and the mass rate of change in biomass for this system is given by

$$\frac{dX}{dt} = VX\mu - VXk_d - FX. \tag{21}$$

The equation can be simplified by dividing through by V and assuming that the system will attain a steady state of growth; that is, dX/dt will approach zero:

$$0 = \mu - k_d - \frac{F}{V}. \tag{22}$$

The last term is recognized as the reciprocal of the mean hydraulic retention time $\bar{t}$ and is commonly termed the dilution rate D, which is thus equal to the net specific growth rate μ_n:

$$D = \mu - k_d = \mu_n, \tag{23}$$

$$\mu = D + k_d = \mu_n + k_d. \tag{24}$$

It can be said that the specific growth rate (μ or μ_n) is subject to hydraulic control; that is, it can be subjected to engineering control as well as being biologically

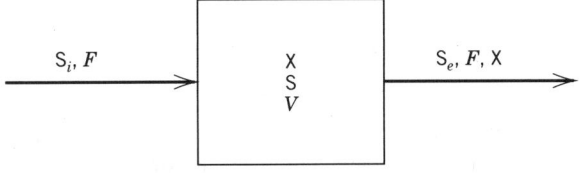

Note: $S = S_e$, $X = X_e$

FIGURE 1.5. Bioreactor of the once-through type (completely mixed).

determined in accordance with Eq. (9). Combination of this significant fact with the biological growth model provides a complete "theory" of continuous culture in completely mixed reactors and permits one to calculate the X and S as well as the biomass output for the system.

Expressing substrate utilization rate dS/dt in terms of growth, using Eqs. (3) and (7), and writing a simple mass balance in S for the reactor lead to

$$V\frac{dS}{dt} = FS_i - FS_e - \frac{VX\mu}{Y_t}. \tag{25}$$

Then solving Eqs. (21) and (25) simultaneously provides the model equations, which have become so widely studied and eventually accepted over the last 40 years.

$$X = \frac{Y_t D(S_i - S_e)}{D + k_d}, \tag{26}$$

$$S = \frac{K_s(D + k_d)}{\mu_{max} - (D + k_d)}. \tag{27}$$

These equations (without k_d) were independently derived by scientists in France (Monod, 1950) and the United States (Novick and Szilard, 1950) and since then have provided a basis for much modeling research on microbial growth. They comprise the so-called theory of continuous culture, and they have been shown to be usable predictive equations for heterogeneous populations of microorganisms (Gaudy et al., 1967; Ramanathan and Gaudy, 1969).

If one compares values for X and S at several values of D (i.e., μ_n), using Eqs. (26) and (27), it is found that X decreases and S increases as μ_n is increased, whereas both X and S decrease at very low values of μ_n. In between, X and S may remain fairly constant depending on the numerical values of K_s and k_d. At the lower values of D, X decreases because μ is approaching the value of k_d; that is, the net specific growth rate μ_n approaches zero. If one determines the biomass yield in the reactor at progressively lower values of μ, progressively lower values of yield will be observed. Values of cell yield observed in the continuous flow reactor, Y_o, can be obtained in accordance with

$$Y_o = \frac{X}{S_i - S_e}. \tag{28}$$

Determining the cell yield in a continuous flow reactor at progressively lower rates of dilution D is analogous to determining values of cell yield in batch cultures at progressively greater time intervals after the biomass curve has peaked and the system has entered the autodigestive phase. At the lower dilution rates, autodigestion decreases X in the continuous flow reactor and this provides one way to obtain

5. REACTOR ENGINEERING CONSIDERATIONS

a numerical value for k_d. Substituting Eq. (26) for X into Eq. (28) produces a relationship that is very helpful in determining k_d from experimental data:

$$Y_o = \frac{Y_t D}{D + k_d} = \frac{Y_t \mu_n}{\mu_n + k_d} = \frac{\mu_n Y_t}{\mu}. \tag{29}$$

5.2. Effects of Cell Recycle

Cell recycle has a considerable effect on the equations for X and S. Figure 1.6 is similar to Fig. 1.5 except for the addition of a cell separation device and provision for recycle of concentrated cells at flow F_R; the recycle flow ratio F_R/F is designated as α. The concentration of cells in the recycle flow is X_R, and the ratio X_R/X is designated as C, the recycle cell concentration ratio. It is assumed that no substrate removal takes place in the separator.

Writing a mass balance for change in X around the bioreactor,

$$V\frac{dX}{dt} = VX\mu - VXk_d - F(1 + \alpha)X + \alpha F X_R, \tag{30}$$

and allowing that a steady state in X will prevail, an equation comparable to Eq. (24) is obtained:

$$\mu = D(1 + \alpha - \alpha X_R/X) + k_d \tag{31}$$

or

$$\mu = D(1 + \alpha - \alpha C) + k_d. \tag{32}$$

In terms of the net specific growth rate,

$$\mu_n = D(1 + \alpha - \alpha X_R/X). \tag{33}$$

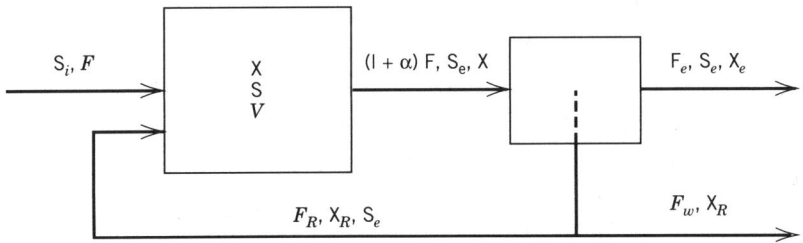

Note: $F_R = \alpha F$; $X_R = CX$, $F_w + F_e = F$

FIGURE 1.6. Bioreactor with recycle of biomass.

The mass balance for substrate may be written as

$$V\frac{dS}{dt} = FS_i + \alpha FS_e - \mu XV/Y_t - F(1 + \alpha)S_e. \tag{34}$$

Setting the values of dS/dt and dX/dt at zero and solving Eqs. (30) and (34) simultaneously leads to Eqs. (35) and (36) for X and S in the steady state:

$$X = \frac{Y_t D(S_i - S_e)}{D(1 + \alpha - \alpha C) + k_d}, \tag{35}$$

$$S_e = \frac{K_s[D(1 + \alpha - \alpha C) + k_d]}{\mu_{max} - [D(1 + \alpha - \alpha C) + k_d]}. \tag{36}$$

This is a rather uncomplicated model in which four biokinetic constants and three engineering parameters, D, α, and C, and the waste strength determine biomass in the reactor and the effluent quality. Equations (35) and (36) were presented by Herbert (1961) without the inclusion of k_d. In my laboratory, it was found that, although the once-through equations of Monod (1950) and Novick and Szilard (1950) were adequate for describing heterogeneous populations of microorganisms, Eqs. (35) and (36) did not apply because a steady state was not attained. This resulted from the engineering constraints involved in using C as an operational constant. In a heterogeneous microbial system, C does not remain constant because in such systems random, small but significant changes in cell yield occur. If one attempts to hold C constant by varying X_R in proportion to X, the system is forced further away from the steady state ($dX/dt = 0$). It was possible to provide fairly practical means to maintain X_R, rather than C, as an operational constant. This led to development of new equations:

$$X = \frac{Y_t(S_i - S_e) + \alpha X_R}{1 + \alpha + k_d/D}, \tag{37}$$

$$S_e = \frac{-b \pm \sqrt{b^2 - 4ac}}{2a}, \tag{38}$$

$$a = [(1 + \alpha)D + k_d] - \mu_{max},$$

$$b = \mu_{max}(S_i + \alpha X_R/Y_t) + [(1 + \alpha)D + k_d](K_s - S_i),$$

$$c = S_i[(1 + \alpha)D + k_d]K_s.$$

Experimentation showed that this modification to the operation of the system and its resulting effect on the equations enable one to fulfill the model's requirement of the steady state, and very good estimates of S and X were possible using values

5. REACTOR ENGINEERING CONSIDERATIONS

of the biokinetic constants determined from separate batch experiments on the system (Gaudy and Srinivasaraghavan, 1974; Gaudy et al., 1977; Ramanathan and Gaudy, 1969, 1971; Srinivasaraghavan and Gaudy, 1975).

This approach has also been examined using toxic substrates by inserting the Haldane relationship, yielding the model equations given below (Gaudy and Rozich, 1982; Rozich and Gaudy, 1984; Rozich et al., 1983):

$$X = \frac{Y_t(S_i - S_e) + \alpha X_R}{1 + \alpha + k_d/D}, \tag{37}$$

$$0 = aS_e^3 + bS_e^2 + cS_e + d, \tag{39}$$

$$a = \frac{(1 + \alpha)D + k_d}{K_i},$$

$$b = [(1 + \alpha)D + k_d](1 - S_i/K_i) - \mu_{max},$$

$$c = \mu_{max}(S_i + \alpha X_R/Y_t) + [(1 + \alpha)D + k_d](K_s - S_i),$$

$$d = -S_i[(1 + \alpha)D + k_d]K_s.$$

As with the model for nontoxic carbon sources, experimentation has shown that Eqs. (37) and (39) provide rather good predictions of S and X, using independently determined values of the biokinetic constants (D'Adamo et al., 1984; Gaudy et al., 1989; Rozich and Gaudy, 1985; Rozich et al., 1983, 1985). In the space available, full exposition of the effect of the variables on X, S, and X_w (waste sludge) is not practical and it must be left to the reader to examine the ways in which the equations behave as one changes the waste strength S_i and the parameters D (or $\bar{t}$), α, and X_R and inserts varying values for μ_{max}, K_s, Y_t, and k_d (and K_i if the waste is shown to be toxic).

Some brief comments may be helpful here. First, the applicable model equations are determined by examination of plots of μ versus S for batch experiments (see Figs. 1.3 and 1.4). One can see from Eqs. (8) and (10) that if K_i is large, the Haldane equation approaches the Monod equation. Thus Eq. (39) becomes Eq. (38); that is, the model for toxic substrates grades into that for nontoxic substrates. One can see from Eq. (33) that μ_n can be controlled by engineering factors, D (or $\bar{t}$), α, and X_R, and, from the general shape of the Monod relationship and the Haldane relationship (up to μ^*), that the higher the μ_n, the higher will be S. Thus any biokinetic and engineering factors that tend to increase μ will allow more substrate to exit the reactor. Since the aim of biological treatment is to produce an effluent with low S_e, wisdom indicates using a μ_n as low as is practical; to lower μ_n, one increases α, X_R, or $\bar{t}$ within economic limits. One can see from Eqs. (38) and (39) for S_e that all three controllable engineering parameters and the five biokinetic constants, along with the waste strength, have interrelated effects on S_e and this comes about because they have an effect on the specific growth rate μ.

5.3. Prediction of Excess Biomass X_w

Considering that the excess biomass consists of the sum of that in the separator underflow at concentration X_R and the overflow at concentration X_e (see Fig. 1.6), a mass balance for the separator shows that the excess biomass (mass/time) can be given by

$$X_w = VXD(1 + \alpha - \alpha X_R/X). \tag{40}$$

Combining Eqs. (40) and (33), we find

$$X_w = VX\mu_n. \tag{41}$$

Equation (41) points out the existence of a useful identity:

$$\mu_n = \frac{X_w}{VX} = \frac{1}{\theta_c}. \tag{42}$$

The term θ_c is recognized as the mean sludge replacement time (sludge age) in water pollution control parlance. Equations (41) and (42) show that lowering the specific growth rate μ_n (i.e., increasing θ_c) will result in lessening the amount of excess biomass produced. Thus slow growth improves effluent quality by reducing S_e and also produces less excess biomass to be further treated for ultimate disposal. Equations (37), (38), and (39) provide a rather good working model for the complete metabolic process of growth and substrate removal for toxic or nontoxic carbon sources, reducing its description to five biokinetic constants, input substrate concentration, and three engineering control variables. While it may be considered to be an oversimplification of the natural global process depicted in Fig. 1.1, it is made usable as a good predictor of events because of the engineering constraints and controls placed on an isolated reactor system. Simplification is a desirable characteristic if a model is to be useful in practice of the pollution control profession. Since it deals with growth rather grossly, it may be considered by some microbial physiologists to be too simple to satisfy their needs; however, it is generally considered a good foundation for further research. The model is just now finding its way into engineering practice, and it would seem that it need not, and cannot, be further simplified for engineering utility.

6. FUTURE POTENTIAL AND STATUS OF THE MODEL

While the model has been found to predict S, X, and X_w rather well, it is by no means flawless and there is a need for continuing investigation. For one thing, the increasing demand for pollution control, as times goes by, and the increased expectations in regard to process efficiency and effluent quality control require continuing investigation, as does the increasing complexity of wastewaters.

6. FUTURE POTENTIAL AND STATUS OF THE MODEL

6.1. Effects of Influent Substrate Concentration S_i

As things stand now, the theory of continuous culture, as expressed by Monod and by Novick and Szilard and embodied in reactor Eqs. (26) and (27), states that effluent concentration S_e is independent of S_i. Analysis of the equations shows that, if Y_t remains constant, an increase in S_i simply causes an increase in X and S_e is unaffected, because μ remains unchanged. The same is true for the equations of Herbert [see Eqs. (35) and (36)] because C is a constant. Thus the essence of the theory of continuous culture is that only those things that affect μ affect S_e. However, if we examine Eqs. (38) and (39), which employ X_R as a control variable, rather than C, it is seen that S_i, the waste loading, in addition to the biokinetic constants and the three engineering parameters, affects the value of S_e. Study of these equations indicates that an increase in S_i causes an increase in S_e. This, however, is not out of line with the original theory because it can also be shown that, when one employs X_R as an engineering control, an increase in S_i causes μ to increase.

There are experimental data indicating that an increase in S_i to a completely mixed reactor leads to a slightly greater increase in S_e than that predicted by Eqs. (38) and (39). This effect can also be experimentally observed for once-through reactor systems. These observations have some conceptual ramifications because they go against the basic theory or, in any event, may weaken the confidence with which one uses the model (Grady et al., 1972; Manickam and Gaudy, 1980). The cause is generally ascribed to production and elaboration of biodegradable and nonbiodegradable microbial products. These may consist of metabolic intermediates leaked into solution during growth or may be products of microbial decay. At specific growth rates slow enough to be comparable to those usually found in activated sludge processes, and using high concentrations of defined and easily biodegradable carbon sources, the increase in effluent COD was found to be in the range of 10-15 mg/L COD per 1000 mg/L increase in feed COD (Manickam and Gaudy, 1980). However, in studies employing municipal sewage as feed, there was no increase in effluent COD for influent COD values ranging between 125 and 375 mg/L (Gaudy et al., 1987). Thus, in a practical sense, the effect seems too small to negate the utility of the theory of continuous culture in modeling activated sludge processes. However, the effect would seem to warrant the attention of microbial physiologists interested in employing this simple kinetic model as a basic tool in studying species interrelationships and production and utilization of organic metabolic compounds produced from the original organic substrates.

6.2. Effects of Changing Feed Rate

It has been pointed out that the reactor equations deal with the "steady state." Monod's original thinking led him to reason that small changes in X and S would be self-correcting because if, for example, S were to rise slightly so would μ, in accordance with Eq. (8), thus returning S to its former value. In general, the equations have been shown to be less accurate predictors of values of X and S in the transient state between the old and a new steady state caused by rather large changes in S externally imposed on the system, for example, shock loadings in S_i and/or F. Both analytical and experimental responses to such perturbances have been the subject of attention by many researchers in the pure science and engineering fields. Exposition of this area of investigation would require much space and only one or two findings will be delineated here because they serve to illuminate some interesting physiological aspects of the growth model.

If one changes S_i from a lower to a higher value, the expected response is an increase in μ, and thus in X, until some new steady concentration of S is attained. The question is: Can one use the model equations in the time-dependent state to predict the transient concentration of S_e and X as the new steady state is approached? Experiments with continuous flow reactors using heterogeneous populations of microorganisms but well-defined feed media have shown that the actual increase in μ as S increased due to a step-up in S_i lagged the increasing value of μ computed using the model equations. As the system responded and the value of S in the reactor began to decrease, the actual (experimentally observed) value of μ again lagged behind the value computed from the equation (Storer and Gaudy, 1969). This finding was taken as experimental evidence for the existence of "growth rate hysteresis," postulated originally by Perret (1960).

This inertial effect is also believed to play a role in permitting exponential growth in batch systems at S_0 values lower than those needed to generate μ_{max} (Gaudy and Gaudy, 1988a; Gaudy et al., 1971, 197.). These findings show that simply integrating the steady-state equations does not provide precise prediction of the transient state and may serve as a criticism of the models. More correctly, they should provide a starting point for investigation of and delineation of more sophisticated and, to be sure, more complicated dynamic models. However, it must be stated that the predicted transient plots of X and S were not widely divergent from the observed response and, as a practical matter, very large changes in S_i (e.g., threefold) were needed in order to provide significant leakage of substrate during the transient (Krishnan and Gaudy, 1976; Manickam and Gaudy, 1979, 1985; Rozich and Gaudy, 1985; Storer and Gaudy, 1969).

The fact that some refinement in the current model may be desirable to predict transient states does not militate against practical use of the theory of continuous culture. Imposition of cyclical, rather than simple, step increases in loading provides evidence that the self-correcting property of the system strives to attain and retain a steady output condition regardless of input oscillations. When a reactor was subjected to a continued 12-h cycle, S_i = 500 ⇌ 1500 mg/L COD, the soluble COD (S_e) rose and fell slightly in a cyclical manner from approximately 10 to 20 mg/L. As expected, the biological solids concentration in the reactor, X, and in the clarifier effluent, X_e, fluctuated somewhat more. However, the amplitude of the fluctuations in S_e and X_e gradually decreased and after the 10th day of such operation the effluent values steadied even though the feed concentration continued in the threefold cycle in S_i (Saleh and Gaudy, 1978). This finding of a dampening in output fluctuation in response to unsteady input feed concentration not only offers an interesting basis for further investigation of physiological and ecological response but should provide some assurance to practicing engineers of the inherent steadiness of the bioreactor system and its representation by Eqs. (.7) and (.8) or (.9).

It should be pointed out that the total mass rate of feed to a bioreactor consists of F, the hydraulic factor, and S_i, the influent substrate concentration. One can combine various changes in F and S_i while maintaining the mass loading constant. However, it has been found that similar mass loadings do not elicit similar responses. A change in F, that is, a hydraulic shock loading, appears to cause a more deleterious response than does a change in S_i (Manickam and Gaudy, 1985).

6.3. Effects of Changing Feed Composition

Another of the complications in modeling natural biological systems is the fact that S_i usually consists of many carbon sources and the composition of the feed may be subject to significant qualitative changes. It is well known that some carbon sources can interfere with the metabolism of others in that they may be selectively metabolized at the expense of others or may interfere with the rate at which others are metabolized (Gaudy, 1962; Gaudy and Gaudy, 1980a; Komolrit and Gaudy, 1966; Monod, 1947). Appearance in the reactor effluent of specific feed compounds due to interference with their rate of metabolism is related to the specific growth rate and/or age of the biomass in the system as well as to the relative concentrations of competing substrates. Faster growth in continuous flow reactors or use of young cells in batch reactors tends to enhance the severity of the effect, that is, may even lead to sequential removal of the carbon sources, whereas in more slowly growing microbial populations the effect is attenuated. That is, the rates of utilization of certain compounds may be retarded, but seldom is total blockage manifested in more slowly growing systems. Thus, while a continuous flow reactor system being fed a multicomponent carbon source may leak only minute concentrations of specific feed compounds at a relatively slow specific growth rate (i.e., dilution rate), it may selectively leak significantly higher concentrations during a transient increase in feed flow rate F or during hydraulic shock load. For example, when a system consisting of equal parts of glucose, glycerol, and galactose was subjected to a step change in D from $\frac{1}{6}$ to $\frac{1}{2}$ h^{-1}, there was a rather serious transient increase in effluent COD. Analysis for the specific feed compounds showed that galactose accounted for approximately half of the effluent COD, with glycerol and glucose accounting for decreasing portions in that order (Gaudy and Gaudy, 1980b).

Selective metabolic blockage of metabolism of compounds in the feed to a biological system can occur under steady growth conditions as well. For example, with substrate consisting of 500 mg/L each of glucose, sorbitol, and propionic acid, little or no amounts of these compounds were found in the effluent at low dilution rates, but at progressively higher D values the increased steady-state concentration of effluent COD contained larger and larger proportions of the individual feed compounds in the following order of concentrations, high to low: propionic acid, sorbitol, and glucose (Gaudy and Gaudy, 1980c).

Whether it will be possible to gain enough understanding of these phenomena to devise new models or to modify the existing growth model to obtain adequate prediction of the course of such events cannot now be foreseen, but it seems important that modelers turn their attention to such phenomena. In the future, the study and incorporation into models of aspects of metabolic control mechanisms triggered by substrates in the waste may be essential to adequate control of treatment processes. Such considerations may become necessary as wastewaters become increasingly complex and variable. Changing concentrations of various toxic substances in waste may also require consideration of control mechanisms in relation to these compounds if the scope of biological treatment is to be safely extended to include mixed substrates containing toxic or inhibitory compounds. Based on the current state of knowledge of control mechanisms, their inclusion in kinetic models would appear to be a very advanced consideration, but it is consistent with the philosophy of modeling in that it precedes complete mechanistic or physiological information on a process. How to approach the inclusion of metabolic controls in the model is an area for future research.

6.4. Role of Oxidative Assimilation

Another approach to more physiologically satisfactory modeling of the growth process may be to look in more detail at the biomass and the sequence of processing carbon sources from entry to the cell through production of the new cell. Presently, the process is dealt with quite simply without any quantitative consideration of the metabolic processing of the carbon source. That is, substrate enters the cell and new cells are produced. No consideration is given to how this occurs, only to the effect that the concentration S has on the specific rate at which new biomass is produced. However, under some circumstances, growth may occur with the involvement of two separate, but interrelated, physiological functions. The first metabolic step in such cases does not involve production of the nucleic acids and proteins necessary for replication of the cell, but involves the production from the original exogenous carbon of nonnitrogenous carbohydrate and, in some species, lipid products. In the absence of a source of nitrogen, these compounds accumulate in the cell, whereas in the presence of a nitrogen source they serve as a common store of carbon from which the compounds necessary for replication are synthesized. The first function, oxidative assimilation, may be the rate-determining step that provides the specificity of kinetic values for each substrate. The question of how one approaches experimental evaluation of such a compartmentalized model I cannot answer.

There are data that show that synthesis of carbohydrate stores precedes synthesis of proteinaceous cellular constituents, especially when biomass concentration is high in relation to feed substrate concentration, a condition typical of most activated sludge processes (Krishnan and Gaudy, 1968). It is possible that the process of oxidative assimilation of exogenous carbon to common storage products, which are then drawn upon to synthesize the proteins and nucleic acids requisite to replication, may proceed in the same way in sparsely seeded systems as well, but this is not readily observable with current means of measuring cell composition. It has also been shown that the two functions can be physically separated (i.e., accomplished sequentially in different reactors) and still produce a viable self-sustaining system capable of purifying a feed medium under either batch (Komolrit et al., 1967) or continuous flow steady-state conditions (Gaudy and Gaudy, 1988c; Gaudy et al., 1968, 1969).

7. CONCLUDING STATEMENTS

From the preceding discussion, it is hoped that the reader will have gained an appreciation of the technological approaches and the practical goals and pitfalls of biological modeling, in particular, for the process of biological treatment of wastewaters. Modeling of the process, as it stands now, is based largely on the theory of continuous culture as described by Monod and by Novick and Szilard. Various modifications have been made, largely of an engineering nature, in order to enhance practical application and engender conformation to the physical boundaries of the reactor system.

Some flaws have been pointed out as well as some avenues for further investigation, because while either the basic equations or the more complete engineering equations have use in today's practice of biological engineering, they have perhaps even more usefulness as a basic framework for future study. It must be remembered that the original goal of the basic scientists was to describe and control the growth of a single microbial species on a single carbon source. In the environmental engineering field, much experimentation showed that the approach could be extended to a heterogeneous mixture of microbes, then to a heterogeneous mixture of substrates as well. It is assurance of such commonality of application that enhances use of the concept for wastewater treatment. It is because of the wide scope of application as well as naturally occurring changes in wastewater composition and in ecological composition of the biomass that the four or five required biokinetic "constants" are not completely constant. Also, it should be remembered that their numerical values are affected by environmental conditions such as temperature and pH. Thus to make optimum practical use of the model equations in predicting and/or controlling process performance, one must be concerned with methodology for determining the numerical values of the growth constants on a routine or semiroutine basis at the treatment plant. Here the universality and veracity of Eq. (1) make it possible, through equations such as (16) and (17), to obtain such values with relative ease. While refinement of the engineering equations through continued investigation, perhaps including those lines of approach herein discussed, is warranted, experience has shown that the model can be used with confidence.

REFERENCES

Benefield, L. D., and Randall, C. W. (1980), *Biological Process Design for Wastewater Treatment*, Prentice-Hall, Englewood Cliffs, NJ, pp. 216–217.

Bhatla, M. N., and Gaudy, A. F. Jr. (1965), *Biotechnol. Bioeng.*, **7**, .87–404.

D'Adamo, P. C., Rozich, A. F., and Gaudy, A. F. Jr. (1984), *Biotechnol. Bioeng.*, **26**, .97–402.

Gaudy, A. F. Jr. (1962), *Appl. Microbiol.*, **10**, 264–271.

Gaudy, A. F. Jr. (1972), in *Water Pollution Microbiology* (R. Mitchell, Ed.), Wiley, New York, pp. .05–..2.

Gaudy, A. F. Jr., and Engelbrecht, R. S. (196.), in *Advances in Biological Waste Treatment* (J. McCabe and W. W. Eckenfelder, Eds.), Vol. 1, Pergamon Press, New York, pp. 11–26.

Gaudy, A. F. Jr., and Gaudy, E. T. (1972), *Ind. Wastes Eng.*, **9**, .0–.4.

Gaudy, A. F. Jr., and Gaudy, E. T. (1980a), *Microbiology for Environmental Scientists and Engineers*, McGraw-Hill, New York, pp. 599–614.

Gaudy, A. F. Jr., and Gaudy, E. T. (1980b), *Microbiology for Environmental Scientists and Engineers*, McGraw-Hill, New York, pp. 611–612.

Gaudy, A. F. Jr., and Gaudy, E. T. (1980c), *Microbiology for Environmental Scientists and Engineers*, McGraw-Hill, New York, pp. 612–614.

Gaudy, A. F. Jr., and Gaudy, E. T. (1988a), *Elements of Bioenvironmental Engineering*, Engineering Press, San Jose, CA, pp. 240–252.

Gaudy, A. F. Jr., and Gaudy, E. T. (1988b), *Elements of Bioenvironmental Engineering*, Engineering Press, San Jose, CA, p. 249.

Gaudy, A. F. Jr., and Gaudy, E. T. (1988c), *Elements of Bioenvironmental Engineering*, Engineering Press, San Jose, CA, pp. 451–461.

Gaudy, A. F. Jr., and Rozich, A. F. (1982), *Civil Eng. Practicing and Design Eng.*, **2**, 55–70.

Gaudy, A. F. Jr., and Srinivasaraghavan, R. (1974), *Biotechnol. Bioeng.*, **16**, 723–738.

Gaudy, A. F. Jr., Bhatla, M. N., and Gaudy, E. T. (1964), *Appl. Microbiol.*, **12**, 254–260.

Gaudy, A. F. Jr., Ramanathan, M., and Rao, B. S. (1967), *Biotechnol. Bioeng.*, **9**, 387–411.

Gaudy, A. F. Jr., Goel, K. C., and Gaudy, E. T. (1968), *Appl. Microbiol.*, **16**, 1358–1363.

Gaudy, A. F. Jr., Goel, K. C., and Freedman, A. J. (1969), in *Advances in Water Pollution Research*, Pergamon Press, New York, pp. 613–623.

Gaudy, A. F. Jr., Obayashi, A. W., and Gaudy, E. T. (1971), *Appl. Microbiol.*, **22**, 1041–1047.

Gaudy, A. F. Jr., Yang, P. Y., Bustamante, R., and Gaudy, E. T. (1973), *Biotechnol. Bioeng.*, **15**, 589–596.

Gaudy, A. F. Jr., Srinivasaraghavan, R., and Saleh, M. (1977), *J. Environ. Eng. Div.*, *ASCE*, **103**, 71–84.

Gaudy, A. F. Jr., Manickam, T. S., Chen, Y. K., D'Adamo, P. C., and Reddy, M. P.

(1987), *Proceeding, 41st Annual Industrial Waste Conference*, vol. 41, Lewis Publishers, Chelsea, MI, pp. 48–58.

Gaudy, A. F. Jr., Rozich, A. F., Garniewski, S., Moran, N. R., and Ekambaram, A. (1988), *Proceedings, 42nd Annual Industrial Waste Conference*, vol. 42, Lewis Publishers, Chelsea, MI, pp. 573–584.

Gaudy, A. F. Jr., Ekambaram, A., and Rozich, A. F. (1989), *Proceedings, 43rd Annual Industrial Waste Conference*, vol. 43, Lewis Publishers, Chelsea, MI, pp. 35–44.

Gaudy, A. F. Jr., Ekambaram, A., Rozich, A. F., and Colvin, R. J. (1990), *Proceedings, 44th Annual Industrial Waste Conference*, vol. 44, Lewis Publishers, Chelsea, MI, pp. 393–403.

Grady, C. P. L. Jr., and Lim, H. C. (1980), *Biological Wastewater Treatment*, Marcel Dekker, New York, p. 327.

Grady, C. P. L. Jr., Harlow, L. J., and Riesing, R. R. (1972), *Biotechnol. Bioeng.*, **14**, 391–410.

Grady, C. P. L. Jr., Dang, J. S., Harvey, D. M., Jobbagy, A., and Wang, X. L. (1989), *Water Sci. Technol.*, **21**, 957–968.

Herbert, D. (1961), in *Continuous Culture of Microorganisms*, Science Monograph No. 12, Macmillan, New York, pp. 21–53.

Hill, G. A., and Robinson, C. W. (1975), *Biotechnol. Bioeng.*, **17**, 1599–1615.

Hiser, L. L. and Busch, A. W. (1964), *J. Water Pollut. Control Fed.*, **36**, 505–516.

Jones, G. L., Jansen, F., and McKay, A. S. (1973), *J. Gen. Microbiol.*, **74**, 139–168.

Komolrit, K., and Gaudy, A. F. Jr. (1966), *J. Water Pollut. Control Fed.*, **38**, 1259–1272.

Komolrit, K., Goel, K. C., and Gaudy, A. F. Jr. (1967), *J. Water Pollut. Control Fed.*, **39**, 251–266.

Krishnan, P., and Gaudy, A. F. Jr. (1968), *J. Water Pollut. Control Fed.*, **40**, 495–510.

Krishnan, P., and Gaudy, A. F. Jr. (1976), *J. Water Pollut. Control Fed.*, **48**, 906–919.

Manickam, T. S., and Gaudy, A. F. Jr. (1979), *J. Water Pollut. Control Fed.*, **51**, 2033–2042.

Manickam, T. S., and Gaudy, A. F. Jr. (1980), *Proceedings, 34th Annual Industrial Waste Conference*, vol. 34, Lewis Publishers, Chelsea, MI, pp. 854–867.

Manickam, T. S., and Gaudy, A. F. Jr. (1985), *J. Water Pollut. Control Fed.*, **57**, 241–252.

Monod, J. (1942), *Recherches sur la Croissance des Cultures Bacteriennes*, Hermann et Cie, Paris, pp. 1–211.

Monod, J. (1947), *Growth*, **11**, 223–289.

Monod, J. (1950), *Ann. Inst. Pasteur*, **79**, 390–410.

Novick, A., and Szilard, L. (1950), *Science*, **112**, 715–716.

Pawlowsky, V., and Howell, J. A. (1973), *Biotechnol. Bioeng.*, **15**, 889–896.

Perret, C. J. (1960), *J. Gen. Microbiol.*, **22**, 589–617.

Porges, N., Jasewicz, L., and Hoover, S. R. (1953), *J. Appl. Microbiol.*, **1**, 262–270.

Ramanathan, M., and Gaudy, A. F. Jr. (1969), *Biotechnol. Bioeng.*, **11**, 207–237.

Ramanathan, M., and Gaudy, A. F. Jr. (1971), *Biotechnol. Bioeng.*, **13**, 125–145.

Rozich, A. F., and Gaudy, A. F. Jr. (1984), *J. Environ. Eng. Div., ASCE*, **110**, 562–572.

Rozich, A. F., and Gaudy, A. F. Jr. (1985), *J. Water Pollut. Control Fed.*, **57,** 795–804.

Rozich, A. F., Gaudy, A. F. Jr., and D'Adamo, P. C. (1983), *Water Res.*, **17,** 1453–1466.

Rozich, A. F., Gaudy, A. F. Jr., and D'Adamo, P. C. (1985), *Water Res.*, **19,** 481–490.

Saleh, M. M., and Gaudy, A. F. Jr. (1978), *J. Water Pollut. Control Fed.*, **50,** 764–774.

Srinivasaraghavan, R., and Gaudy, A. F. Jr., (1975), *J. Water Pollut. Control Fed.*, **47,** 1946–1960.

Storer, F. F., and Gaudy, A. F. Jr. (1969), *Environ. Sci. Technol.*, **3,** 143–149.

Symons, J. M., McKinney, R. E., and Hassis, H. H. (1960), *J. Water Pollut. Control Fed.*, **32,** 841–852.

Yang, R. D. and Humphrey, A. E. (1975), *Biotechnol. Bioeng.*, **17,** 1211–1235.

Zabriskie, D. W., and Humphrey, A. E. (1978), *Am. Inst. Chem. Eng. J.*, **24,** 138–146.

2 Models for Studying the Population Ecology of Microorganisms in Natural Systems

STEVEN K. SCHMIDT
Department of Environmental, Population and Organismic Biology,
University of Colorado

1. Introduction... 31
2. Models of population dynamics 33
   2.1. The logistic equation 33
   2.2. The Monod equation... 35
   2.3. Blackman kinetics.. 35
   2.4. Modifications of the Monod equation............................... 37
       2.4.1. Maintenance energy 37
       2.4.2. Inhibitory substrates 37
   2.5. Dual-substrate models 38
3. Models of growth-related microbial activity 38
   3.1. Model relatedness 40
4. Parameter estimation procedures 41
5. Model validation... 42
   5.1. The dangers of curve fitting 42
       5.1.1. Validation of the Monod equation 43
       5.1.2. A failure of the Monod equation 43
       5.1.3. Validation of a dual-substrate model 47
6. Studies of populations in natural systems............................... 47
   6.1. Population dynamics in soil 47
   6.2. Simulation models of soil population dynamics............ 49
   6.3. Simple models of soil population dynamics 50
   6.4. Dual-substrate kinetics in soil 53
7. Conclusions.. 55
   References 56

1. INTRODUCTION

The aim of this chapter is to demonstrate how simple models of population dynamics can be used to study microbial populations in natural and human-made systems. The first part of this chapter is devoted to a description of the models, emphasizing similarities between the models and how they have been used in the past. The other sections deal with parameter estimation procedures, model validation, and finally examples of how the models have been used to study actual microbial populations.

Models have long been used in the fields of population biology and microbiology but the recent increase in the number and complexity of models has outpaced the actual application of models to real-world problems. Although many complex models have been proposed, in many studies researchers are looking for answers to relatively simple questions, such as: (1) How many of a given type of organism are present in the system? (2) What is the affinity of these organisms for a given substrate in the system? (3) How fast is a given population of organisms growing or dying in the system? The answers to these types of questions are also of interest to policy-makers and researchers studying the fate of toxic chemicals in the environment (Boethling, 1985; Button, 1985; Lewis and Gattie, 1991; Paris and Rogers, 1986). In most cases, supercomputers and complex simulation models are not needed to answer the types of questions listed above. Rather, in many cases, simple and biologically interpretable models, used in combination with relevant microbial activity measurements, are all that are needed.

Two main criteria were applied to decide which models are usable to answer the types of questions discussed above:

1. *Simplicity*. The model should contain the minimum number of parameters necessary to describe the observed results. If a model with fewer parameters explains as much of the variance as a model with more parameters, then the former is the preferred model (Gilpin and Ayala, 1973).
2. *Reality*. All the parameters of the model should have biological meaning and be measurable independently of the model estimate. If any parameter of a model cannot be independently determined, then the model should be considered invalid (Salmon and Bazin, 1988). Unvalidated models may be useful for stimulating thought processes and organizing ideas (Levin, 1982), but they are of little use for testing hypotheses about real populations of organisms.

These general guidelines will obviously limit the scope of this chapter to simple models that contain few but realistic parameters. Complex simulation models that attempt to describe the intricate functioning of a system and detailed structured models of microbial physiological processes will not be discussed. Rather, the models discussed are those that incorporate only the supposed controlling variables and parameters for a system. Examples are given to demonstrate how such models can be used to obtain estimates of parameters for specific microbial populations in a variety of systems.

2. MODELS OF POPULATION DYNAMICS

In this section models of microbial population dynamics are discussed with an eye toward whether they fulfill the two criteria discussed in the previous section. In addition, a historical perspective is taken because some models have been subject to various biological interpretations and an awareness of these interpretations can lead to unexpected insights into the functioning of biological systems. Historical records also reveal that some models have arisen via several independent routes, perhaps indicating that such models have general applicability.

The study of population dynamics has a long history in which many models have been proposed. Malthus (1798) is generally cited as the starting point for theoretical discussions of population dynamics. He hypothesized that a population grows at a rate that is the difference between the birth rate and death rate of the population. If it is assumed that birth and death rates can be combined into one net rate, then Malthus' model is equivalent to the equation for an exponentially increasing population and can be written as

$$\frac{dN}{dt} = \mu_{max} N, \tag{1}$$

which upon integration yields

$$N = N_0 e^{\mu_{max} t} \tag{2}$$

in which N is the population density at time t, N_0 is the initial population density, and μ_{max} is the maximum specific growth rate. In the ecological literature (Hutchinson, 1978), μ_{max} is usually represented by r (the intrinsic rate of increase) and is sometimes called the Malthusian parameter (Christiansen and Fenchel, 1977). Basically, Eq. (2) describes the infinite growth of a population if μ_{max} is greater than zero and an infinitely decreasing population if μ_{max} is less than zero. Obviously neither scenario is probable and thus model (2) is of limited utility as a general model for describing population growth in nature. Equation (2) is useful, however, for describing short-term population growth in the absence of extrinsic limiting factors. Examples are the initial growth of microorganisms in a previously unexploited environment (e.g., a rotting carcass) or the growth of the human population on Earth over the last few hundred years.

2.1. The Logistic Equation

One of the first attempts to model the population growth of microorganisms was that of McKendrick and Pai (1911), who believed that the rate of bacterial growth was proportional to the initial number of organisms in the environment and the concentration of "food" in that environment. From these assumptions they derived an equation that was essentially identical to the "logistique" equation proposed by Verhulst (1838) in his studies of human populations and the "autocatalytic" equation of Robertson (1923). Pearl (1922) rediscovered the work of Verhulst and restored the title of logistic to the equation used by previous workers. Pearl was a tireless promoter of the logistic equation and even claimed that it was a "law" of population growth (Kingsland, 1982, 1985). Some of the claims of Pearl have been discounted, but the general utility of the logistic equation as a starting point for population modeling is still widely accepted in ecology. The equation is the core of many complex models of population dynamics including models of competitive and predatory interactions (Hutchinson, 1978; Whittaker, 1975). The logistic equation describes the linear relationship between increasing population density and decreasing growth rate and can be written as

$$\mu = \mu_{max}(1 - N/N_{max}), \tag{3}$$

where N and μ_{max} are as defined above, μ is the specific growth rate, and N_{max} is the maximum population density achievable in the environment. In the ecological literature (Hutchinson, 1978), N_{max} is usually represented by K, the carrying capacity of the environment. In the microbiological literature the growth rate μ is usually defined as the number of new organisms produced per initial number of organisms per unit time, or

$$\mu = \frac{dN}{dt}\left(\frac{1}{N}\right). \tag{4}$$

One problem with the logistic equation is that it is an empirical equation that does not specify what controls the rate of population growth. It does specify that growth rate decreases as population density approaches some maximum (N_{max}), but the factors that determine the level of N_{max} are not explicit in the model. Thus the logistic model can be used no matter what the controlling factors are in a given system. This can be an advantage for curve fitting, but it limits the usefulness of the logistic equation as a tool for understanding microbial population dynamics in natural systems. For as yet unknown reasons, the logistic equation often accurately describes the pattern of filamentous fungal growth both in soil (Couteaudier and Steinberg, 1990; Scow et al., 1990a) and liquid culture (Bull et al., 1975; Scow et al., 1990a).

Several examples of how the mechanistic basis of the logistic equation have been interpreted can be taken from previous studies. McKendrick and Pai (1911) studied batch cultures of *Escherichia coli* in which the concentration of substrate initially present was the factor that they thought was determining the size of the final population. Gause (1934) later hypothesized that the amount of toxic waste products, such as ethanol, formed by yeast could limit population size. In both cases the logistic equation was used to describe the data because both studies dealt with populations whose growth decelerated as it approached some maximum. Other factors that have been postulated to determine N_{max} are space limitations and other density-dependent factors (Hutchinson, 1978).

Another problem with the logistic equation is that in its integrated form it describes only symmetrical growth curves, whereas in reality many growth processes are not symmetrical. The logistic equation can be slightly modified to fit asymmetric curves (Carlson, 1913; Gilpin and Ayala, 1973):

$$\mu = \mu_{max}[1 - (N/N_{max})^c], \tag{5}$$

where N, N_{max}, and μ_{max} are as previously defined and c is a fitting parameter that gives an inflection point at $N > N_{max}/2$ when $c > 1$ and $N < N_{max}/2$ when $c < 1$. Although Eq. (5) can describe asymmetrical as well as symmetrical growth curves, it still suffers from most of the drawbacks of the logistic equation. In addition, a satisfactory biological explanation of the meaning of the parameter c is not intuitively obvious and was not given by the formulators of this model (Ayala et al., 1973; Gilpin and Ayala, 1973).

2.2. The Monod Equation

A more mechanistic model of substrate-concentration limited population growth was introduced by Monod (1949). The Monod equation describes the relationship between the concentration of a limiting nutrient and the growth rate of a population of microorganisms. The Monod equation is analogous to the Michaelis–Menten model for enzyme kinetics and can be presented as

$$\mu = \mu_{max} S/(K_s + S), \tag{6}$$

where S is the concentration of the limiting substrate, μ_{max} is the maximum specific growth rate, and K_s is the half-saturation constant for microbial growth.

The main advantage of the Monod equation over the logistic equation is that it embodies a mechanistic explanation for the slowdown of population growth in response to dwindling concentrations of the limiting nutrient. The Monod equation is also more flexible than the logistic equation in that it can describe symmetrical and asymmetrical growth curves without the addition of undefined fitting parameters to the model.

2.3. Blackman Kinetics

Koch (1971, 1985) has asserted that Blackman kinetics are more applicable to the growth of *E. coli* than is the Monod equation. Koch derived the Blackman model from the writings and a diagram of Blackman (1905). The Blackman model(s) can be written as follows (Bader, 1982):

$$\mu = \mu_{max} \quad \text{for } S \geq 2K_s, \quad (7a)$$

$$\mu = \mu_{max} S/(2K_s) \quad \text{for } S < 2K_s, \quad (7b)$$

where μ, K_s, and μ_{max} are as defined above.

Blackman was a plant physiologist who speculated that the relationship between growth rate and the abundance of a limiting factor is linear up to a point where another factor becomes limiting. At this transition point there would be an abrupt flattening in the growth rate versus substrate concentration relationship (Fig. 2.1). This is the same as saying that growth kinetics are discontinuous and jump directly from first-order to zero-order with respect to substrate concentration. In contrast, the Monod equation predicts a gradual change from first-order to zero-order kinetics (Fig. 2.1). Evidence for the Blackman model comes from curve fits to data sets that purportedly show a flat rather than an asymptotic response at high substrate concentrations (Dabes et al., 1973; Koch, 1982, 1985). Disadvantages of the Blackman model are that it actually consists of two models and that it does not account for mixed-order growth kinetics often observed in bacterial cultures and samples from natural systems. Thus the present evidence for the Blackman model is unconvincing, but hopefully future work will further clarify its applicability to the relationship between growth rate and substrate concentration.

A natural compromise between Monod and Blackman kinetics would be a model that allows a gradual slowdown in growth rate as S approaches $2K_s$ followed by a horizontal flattening of the μ versus S curve. Several such models have been pro-

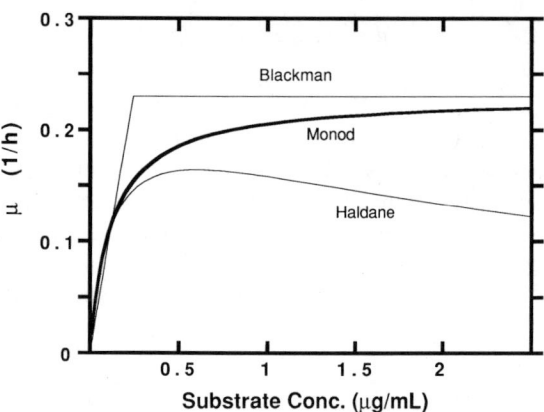

FIGURE 2.1. Growth rate versus substrate concentration plots for three basic models of substrate-limited growth. The curves were generated using parameter estimates for the growth of a *Rhodococcus* sp. utilizing 2,4-dinitrophenol as its sole carbon, nitrogen, and energy source (Hess et al., 1990). The parameter values used to generate the curves were $K_s = 0.1 \ \mu g/mL$, $\mu_{max} = 0.23 \ h^{-1}$, and $K_i = 3 \ \mu g/mL$.

posed (Dabes et al., 1973; Koch, 1985), but they usually contain more parameters than the Monod equation and there has not been enough experimental work done to validate them or to justify their use in place of the Monod equation. Other workers have also reported deviations from Monod kinetics at high concentrations of noninhibitory substrates (Shehata and Marr, 1971).

2.4. Modifications of the Monod Equation

2.4.1. Maintenance Energy.
Deviations from the predictions of the Monod equation have been noted at very low concentrations of limiting nutrients (Pirt, 1975; Schmidt et al., 1985a). At very low substrate concentrations the endogenous metabolism of microorganisms can constitute a major drain on the energy needed to sustain growth. To account for this metabolic drain, models such as the Monod equation can be modified in the following way:

$$\mu = \mu_{max} S/(K_s + S) - a, \qquad (8)$$

where a is the "specific maintenance rate," which has the same units as μ and expresses the rate of energy consumption as a result of non-growth-related metabolic processes (Neidhardt et al., 1990; Tempest and Neijssel, 1984). A more commonly used parameter is the "maintenance coefficient" (Pirt, 1975), which is related to the specific maintenance rate by

$$m = a/Y, \qquad (9)$$

where Y is the yield coefficient with units of grams of biomass produced per gram of substrate consumed. Thus the units of m are grams of substrate consumed per gram of cells per hour. Typical values of a for $E.\ coli$ are 0.02–0.03 h^{-1} (Neidhardt et al., 1990). Values of a and m for natural microbial populations are up to three orders of magnitude lower than those for laboratory strains of $E.\ coli$ (Smith, 1989; Williams, 1985) and thus deviations from the Monod relationship at low S values in nature are probably less than those observed in laboratory cultures.

2.4.2. Inhibitory Substrates.
Another class of models that describes deviations from classical Monod kinetics are models of microbial growth on inhibitory substrates. One model that has been used in a number of recent studies (D'Adamo et al., 1984; Klecka and Maier, 1985; Sokol, 1987; Suflita et al., 1983) is a modification of the Monod equation known as the "Haldane" equation (Edwards, 1970):

$$\mu = \mu_{max} S/[(K_s + S)(1 + S/K_i)]. \qquad (10)$$

A special form of Eq. (10) has also been used by D'Adamo et al. (1984):

$$\mu = \mu_{max} S/[K_s + S + (S^2/K_i)], \qquad (11)$$

where S, K_s, and μ_{max} are as defined above and K_i is the inhibition constant defined as the highest concentration of substrate at which $\mu = \mu_{max}/2$. It should be noted that in order to distinguish K_s from K_i the definition of K_s can be restated as: the lowest concentration of substrate at which $\mu = \mu_{max}/2$ (Andrews, 1968).

Haldane (1930) modified the Michaelis–Menten equation to obtain the original "Haldane" equation and only after the advent of the Monod equation were equations like (10) and (11) developed. The Haldane equation and a number of variations on its theme have been used to model microbial population dynamics in response to a wide array of toxic substances, including ammonia, benzoate, nitrite, phenol, and pentachlorophenol (Boon and Laudelout, 1962; D'Adamo et al., 1984; Edwards, 1970; Klecka and Maier, 1985). A useful feature of the Haldane equation is that it reduces back to the Monod equation at noninhibitory substrate concentrations (S $\ll K_i$). A comparison of the relationship between μ and substrate concentration for the Haldane, Monod, and Blackman models is shown in Fig. 2.1.

2.5. Dual-Substrate Models

Most of the models discussed so far are only applicable if it is known that the concentration of one substrate is limiting the rate of microbial growth. It is possible, however, that in some instances two or more substrates can simultaneously limit either the extent or rate of microbial growth. Theoretical operational limits under which this could occur have been proposed (Bader, 1978, 1982), but experimental evidence for dual-substrate limited growth is scant (Lee et al., 1984). In addition, the models proposed to describe dual-substrate limited growth are often too complicated to be easily verified in the laboratory, let alone in nature. It is very difficult to determine kinetic constants for two substrates that are being simultaneously metabolized especially if there are interactive effects.

Other dual-substrate models have been used to model diauxic growth (Kompala et al., 1984), substrate inhibition (Papanastasiou and Maier, 1982), secondary substrate utilization (LaPat-Polasko et al., 1984; Rittmann and McCarty, 1980; Schmidt et al., 1985b), and cometabolism (Schmidt et al., 1985b). Unlike the models of Bader (1982), the assumption of simultaneous limitation by more than one nutrient is not part of these models. Examples of how dual-substrate models have been used to study natural microbial populations are given in Sections 5.1.3 and 6.4.

3. MODELS OF GROWTH-RELATED MICROBIAL ACTIVITY

In this section it is demonstrated how some of the simple models described above can be modified and used to estimate values for parameters such as μ_{max} and N_0 from substrate depletion or product formation data. In other words, it is shown how one can use these models to study microbial population dynamics without having to directly count organisms. This approach is advantageous because in most habitats it is extremely difficult to directly count numbers of individuals in specific microbial populations or guilds (Mills and Bell, 1986). In many cases substrate disappearance or product formation kinetics are an indirect measure of the activity of specific microbial groups in the system. By using properly formulated models, it is possible to use such kinetic data to estimate parameters of microbial population dynamics (Robinson and Tiedje, 1983; Schmidt et al., 1985b; and Simkins and Alexander, 1984).

As an example, the Monod equation can be converted from the growth form [Eq. (6)] to the substrate disappearance form. To do this it is first necessary to rearrange the Monod equation by substituting Eq. (4) into Eq. (6) to give

$$\frac{dN}{dt} = \frac{\mu_{max} SN}{(K_s + S)}. \tag{12}$$

In order to convert population growth into substrate disappearance, a mass-balance equation is also needed:

$$S_0 + qN_0 = S + qN, \tag{13}$$

in which q is the inverse yield $(1/Y)$ given that Y is the yield coefficient defined as grams of biomass produced per gram of substrate metabolized, N_0 is the initial population density and S_0 is the initial substrate concentration. If q is assumed not to vary with time (Simkins and Alexander, 1984), then qN_0 and qN can be replaced with X_0 and X, respectively, and Eq. (13) can be rewritten as

$$S_0 + X_0 = S + X, \tag{14}$$

where X_0 and X are the concentration of substrate needed to produce a population density of N_0 and N, respectively. It also follows that the Monod equation can now be expressed as

$$\frac{dX}{dt} = \frac{\mu_{max} SX}{(K_s + S)}. \tag{15}$$

Equation (14) and its derivative form can then be solved for X and dX/dt and substituted into Eq. (15) to yield a differential equation that expresses the depletion of substrate in terms of Monod growth kinetics:

$$\frac{-dS}{dt} = \frac{\mu_{max} S(S_0 + X_0 - S)}{(K_s + S)}. \tag{16}$$

Integration of Eq. (16) yields the following equation (Simkins and Alexander, 1984):

$$K_s \ln (S/S_0) = (S_0 + X_0 + K_s) \ln (X/X_0) - (S_0 + X_0)\mu_{max} t, \tag{17}$$

which can be used to fit substrate depletion curves caused by populations of microorganisms growing according to Monod kinetics.

Using the general approach presented above, other models of population dynamics can be converted to their respective substrate depletion forms (Brunner and Focht, 1984; Hess et al., 1990; Robinson and Tiedje, 1983; Schmidt et al., 1985b; Simkins and Alexander, 1984, 1985). In some cases a population growth model can be converted into a number of different substrate depletion forms depending on how the model is parameterized. For example, several different substrate depletion forms of the logistic equation have been derived. Simkins and Alexander (1984) presented the following integral form of the logistic equation:

$$S = (S_0 + X_0)/[1 + (X_0/S_0) \exp(k_4(S_0 + X_0)t)], \qquad (18)$$

in which k_4 is equal to μ_{max}/K_s and all other parameters are as described above. Equation (18) has proved useful for estimating parameters such as X_0 but it cannot be used to directly obtain values for other useful parameters such as μ_{max}. Therefore Hess et al. (1990) reparameterized the logistic equation so that estimates of μ_{max} could be obtained from symmetrical substrate depletion curves. The integrated form of the model of Hess et al. (1990) is

$$S = (S_0 + X_0)[(S_0 e^{-\mu_{max}t})/(X_0 + S_0 e^{-\mu_{max}t})]. \qquad (19)$$

Equations (18) and (19) can be used interchangeably and both yield the same curves of substrate depletion when the same parameter values are used in each equation. Equation (18) can be used to estimate K_s if μ_{max} is known, whereas Eq. (19) can be used in studies in which an estimate of μ_{max} is needed (Hess et al., 1990). As with the logistic equation, several different substrate depletion forms of the Monod equation have also been derived (Robinson and Tiedje, 1983; Simkins and Alexander, 1984, 1985).

Under some circumstances it is easier to measure the accumulation of a by-product of population growth, such as CO_2, CH_4, and fermentation products, than it is to measure the actual compound under study. In such cases, data on the accumulation of the product can be fit using modified forms of models of microbial population dynamics. Such an approach is especially useful in soil and other systems in which it is often difficult to measure either substrate depletion or population density. Approaches for expressing models in the product accumulation form have been given elsewhere (Brunner and Focht, 1984; Nelson et al., 1982; Robinson, 1985; Scow et al., 1986; Smith, 1989). Another approach that works equally well is to convert product accumulation data into substrate disappearance form before analyzing it using nonlinear regression techniques (Scow et al., 1986, 1989).

3.1. Model Relatedness

One advantage of using the Monod equation as a starting point for modeling is that many simpler models [e.g., Eqs. (1), (7a), and (18)] are mathematically related to it. In addition, more complex models of population dynamics can be constructed from the Monod equation by adding more parameters to it [e.g., Eqs. (8), (10), and (11)]. This ability to add and subtract parameters to and from the Monod equation gives one the flexibility to use the simplest form of the equation necessary to adequately describe a given data set. This approach allows one to eliminate superfluous parameters from the model and to determine which parameters are important for describing the data. Using mathematically related models also facilitates statistical comparisons between model fits to a given data set.

An example of how related forms of the same general model can be used to describe data sets of varying complexity was presented by Simkins and Alexander (1984). They showed that the substrate disappearance form of the logistic equation can be viewed as a limiting case of the Monod equation when $S \ll K_s$. Under these conditions growth rate is first-order with respect to S and growth curves are usually symmetrical. This means that at low substrate concentrations the parameter K_s is not needed to model the data and thus a model that includes the parameter K_s should not be used. Conversely, at very high concentrations of S_0 ($S_0 \gg K_s$) the Monod equation reduces to the exponential growth equation [Eqs. (1) and (7a)] and the rate of growth is zero-order with respect to S and is essentially equal to μ_{max}. The actual Monod equation is thus only really useful for situations in which S_0 is in the vicinity of K_s (Simkins and Alexander, 1984, 1985) or more precisely when S_0 is from 5 to 15 times higher than K_s (Schmidt, 1988; Schmidt et al., 1987).

4. PARAMETER ESTIMATION PROCEDURES

One of the biggest problems facing modelers of biological processes is that of parameter estimation. Without reliable methods for estimating parameter values, there is no way to determine the usefulness or validity of a given model. To avoid confusion about definitions of parameters and variables, consider again the Monod equation in is simplest form:

$$\mu = \mu_{max} S/(K_s + S). \tag{20}$$

As written, μ_{max} and K_s are the parameters to be estimated and are constants for a given set of conditions. The dependent variable is μ, and S is the independent variable.

Parameter estimation procedures utilize statistical and numerical techniques to analyze data in order to obtain reasonable estimates of model parameters. A common and long used method to estimate parameter values in linear regression. In linear regression, data are graphed with the independent variable on the x axis and the dependent variable on the y axis. The line determined by the linear regression procedure is chosen to minimize the sum of squares of the vertical distances between the data points and the line. In other words, the goal of most regression procedures is to minimize the function

$$\text{RSS} = \text{sum}[(y_\text{obs} - y_\text{pred})^2], \qquad (21)$$

where RSS stands for the residual sum of squares, y_obs are the observed values for the dependent variable, and y_pred represent the values for the dependent variable predicted by the model being fit to the data.

Although linear regression procedures are simple and easy to interpret, most natural processes are not linear and linearization can only vaguely approximate many natural phenomena. In addition, statisticians have repeatedly pointed out that parameter estimates from linearized forms of many models are inherently unreliable (Richter and Söndgerath, 1990). This unreliability is mostly due to the unpredictability of error transformation and invalid assumptions about error distribution and occurrence in association with the variables of model equations (Robinson, 1985). Other problems and drawbacks to using linearized forms of nonlinear models are discussed in detail elsewhere (Motulsky and Ransnas, 1987: Richter and Söndgerath, 1990; Robinson, 1985, 1986).

With the recent explosion in computer technology, there is no longer a need to linearize complex data sets in order to analyze them. Nonlinear regression techniques have come into widespread use and many packaged programs that can perform nonlinear regression are now available (Richter and Söndgerath, 1990). The goal of any nonlinear regression routine is to manipulate parameter values in order to minimize the difference between the model estimate and the actual values of the dependent variable of the model equation. The general approach is the same as described above for linear regression [Eq. (21)] except that parameter values that minimize the RSS must be obtained iteratively, usually using gradient techniques that mathematically determine the best path to be taken to minimize the RSS. A variety of algorithms can be used to perform nonlinear regression and these are described in detail elsewhere (Bard, 1974; Dennis and Schnabel, 1983; Simkins and Alexander, 1984).

5. MODEL VALIDATION

5.1. The Dangers of Curve Fitting

One danger of the curve-fitting approach outlined above is that nonlinear sets of data can be fit by an almost infinite number of curves described by an equal number of models. The universe of models that adequately describe a given data set can be narrowed somewhat by choosing models with parameters that make biological sense. Hutchinson (1978) has emphasized this point with respect to ecological modeling: "we choose the logistic because it is general and because it makes simple minded biological sense." Even models that make biological sense, however, can fit a given data set without accurately representing the mechanistic basis underlying the observed data. One way to partially alleviate this problem is to

5. MODEL VALIDATION

determine whether the environmental or experimental conditions correspond to the theoretical assumption that went into formulating the model (Schmidt et al., 1987). One can also evaluate whether the parameter estimates from the model are in line with accepted norms for microbial growth under similar circumstances (Robinson, 1986). Other criteria for determining "the model of best fit" to a data set are presented in detail elsewhere (Motulsky and Ransnas, 1987; Richter and Söndgerath, 1990; Robinson, 1985, 1986; Simkins and Alexander, 1984).

Given the vast number of possible models that can be used to fit a given data set, one should always beware of placing too much emphasis on finding the "model of best fit." Without some sort of independent validations of model parameter estimates, there is really no way of knowing that a model accurately represents a realistic description of the underlying processes being studied. The best way to verify whether one is using the right model is to compare the parameter estimates from the model fit to independent measurements of the same parameters. Thus before trying to apply nonlinear models of population dynamics to data from environmental samples or natural systems, it is important to first test each model under laboratory conditions where most variables can be controlled or monitored (Schmidt et al., 1985b; Van de Werf and Verstraete, 1987b). For example, the data presented in Figs. 2.2 and 2.3 are from a series of laboratory experiments done to test the applicability of a number of models for estimating parameters of population growth from substrate disappearance data (Schmidt et al., 1985b, 1987).

5.1.1. Validation of the Monod Equation. The curve presented in Fig. 2.2a represents the mineralization of phenol by a growing culture of a *Pseudomonas* sp. Of the 12 models tested, the Monod model best fit these data and the parameter estimates from this fit are compared to independent estimates of the same parameters in Table 2.1. Figure 2.2b shows a plot of the residual differences between the model curve and the actual data or $y_{obs} - y_{pred}$ from Eq. (21). The residual plot is presented to magnify the differences between the model fit and the actual data so that deviations between the model fit and the data are made more obvious (Suflita et al., 1987). Note that the model fit the data very closely and that no drastic systematic deviations between model and data can be detected in Figs. 2.2a and 2.2b. In this example, independent parameter estimates were close enough to those estimated using the Monod equation that we can be fairly confident that the Monod model is indeed the model that best describes the data presented in Fig. 2.2a.

5.1.2. A Failure of the Monod Equation. In the preceding example, the model of best fit to the data set also proved to give an adequate description of the biology of the system under study. In other cases, however, the model of best fit to a data set can lead to a completely erroneous picture of how microbial populations are behaving. As stated before, this is because a given curve can be described by many different mathematical expressions and it is impossible to say that the model of best fit to a data set is necessarily the best model to describe the data.

44 POPULATION ECOLOGY OF MICROORGANISMS

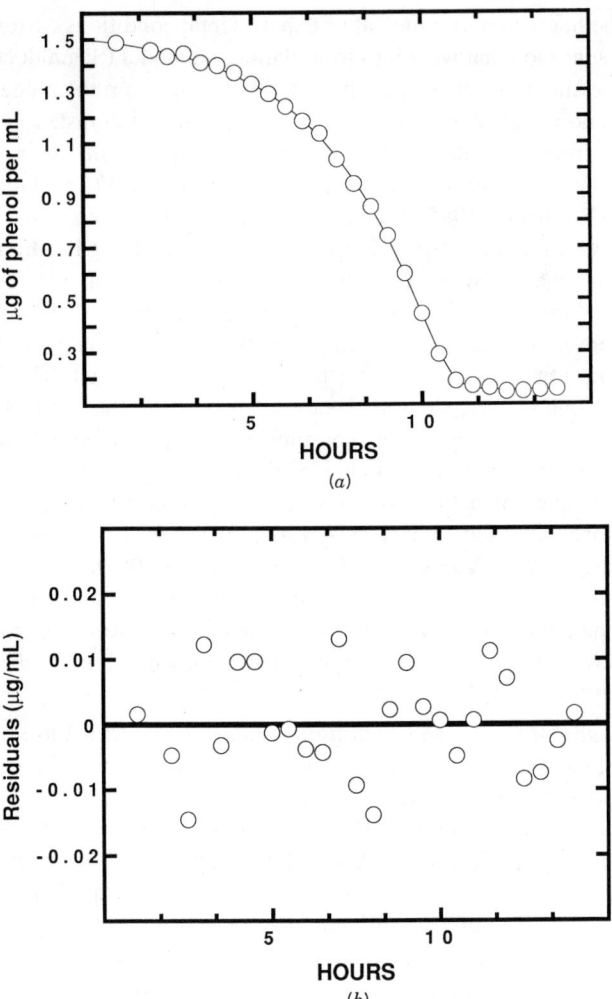

FIGURE 2.2. (*a*) Mineralization of 1.5 µg of phenol per mL by a growing strain of *Pseudomonas* sp. The data, represented by the open circles, were best fit by an integrated form of the Monod model [Eq. (17)] represented by the drawn line. Parameter estimates from the model fit are compared to independent estimates of the same parameters in Table 2.1. (*b*) Residual differences between the model fit and the actual data points. The figure was drawn using data from Schmidt et al (1987).

As an example of how relying on only curve fitting can lead to the wrong interpretation of a data set, consider the data in Fig. 2.3. This mineralization curve is of the same general shape as the curve in Fig. 2.2*a* and corresponds to the general shape of the integrated form of many different models, including the Monod equation, the asymmetrical logistic equation (Gilpin and Ayala, 1973), and models IV and V of Schmidt et al. (1985b). Using standard nonlinear regression pro-

TABLE 2.1. Parameter Estimates from the Fit of the Monod Equation to the Data Presented in Fig. 2.2a Compared to Independent Estimates of the Same Parameters[a]

Source of Estimate	μ_{max} (h^{-1})	N_0 (cells/mL)	S_0 (μg/mL)	K_s (μg/mL)
Monod equation	0.37 ± 0.01	$1.4 \pm 0.14 \times 10^5$	1.5 ± 0.004	0.10 ± 0.015
Independent measurements	$0.33\ (r = 0.99)$	$1.0 \pm 0.3 \times 10^5$	1.5	ND

[a]Independent estimates of μ_{max} and N_0 were obtained from plate count data (Schmidt et al., 1987) and S_0 represents the amount of phenol added at the beginning of the experiment.

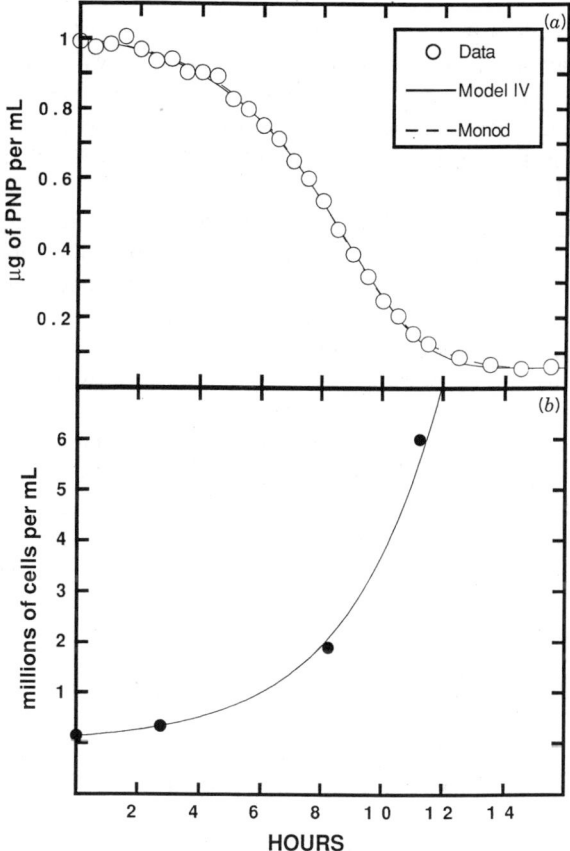

FIGURE 2.3. (a) Mineralization of 1.0 μg of para-nitrophenol (PNP) per mL in the presence of 20 μg of glucose per mL by growing cells of a *Pseudomonas* sp. The data were best fit by the integrated form of the Monod equation but model IV of Schmidt et al. (1985b) also fit the data well and gave more realistic parameter estimates (Table 2.2). (b) Actual cell growth data from the same microcosm from which the mineralization curve was taken. The independent estimate of μ in Table 2.2 was obtained by fitting an exponential growth function to these data. Curves were redrawn using data from Schmidt et al. (1987).

cedures and several criteria for determining the curve of best fit, it was determined that of the 12 models tested, the Monod equation gave the best fit to the data (Schmidt et al., 1987). The parameter estimates from the Monod equation were of reasonable magnitude and the error estimates were less than are normally observed for the integrated Monod equation. Thus, if only standard curve-fitting criteria are applied, it could reasonably be assumed that the Monod equation best describes the population dynamics of the organisms responsible for the curve presented in Fig. 2.3a.

Attempts to validate the applicability of the Monod equation to this data set, however, revealed that it is not appropriate for describing the data. Figure 2.3b shows simultaneous independent measurements of population growth in the same microcosm from which the data depicted in Fig. 2.3a were taken. By fitting the simple exponential growth function [Eq. (2)] to these data, a μ_{max} value of 0.33 h^{-1} was obtained. In comparison, the nonlinear regression estimate of μ_{max} from fitting the Monod equation to the mineralization curve was 0.76 h^{-1} (Table 2.2). In addition, other measurements demonstrated that the factor limiting population growth in this flask was not para-nitrophenol (PNP) but rather glucose, which had also been added in the incubation medium (Schmidt et al., 1987). This *Pseudomonas* sp. could simultaneously mineralize PNP and glucose and the presence of glucose stimulated the growth and thus the rate of PNP degradation by this organism. Thus even though the Monod equation was the model of best fit to the data, it is not a valid model for use in this particular case. The underlying assumptions of the Monod equation were not met by environmental conditions and at least one of the parameter estimates from the Monod fit is not even close to an independent measure of that parameter (Table 2.2).

This example points out a general problem inherent in studies that depend only on curve-fitting procedures to study populations and activity of microorganisms. Shapes of growths or activity curves can be caused by a myriad of factors and unless something is known about the limiting factors of a system it should not be assumed that the model of best fit is the best model to describe the data.

TABLE 2.2. Comparison of Parameter Estimates for μ_{max} from Two Models Fit to the Curve of PNP-Mineralization Depicted in Fig. 2.3a and an Independent Estimate of μ_{max} from the Growth Data Shown in Fig. 2.3b

Source of Estimate	μ_{max} (h^{-1})
Monod	0.76 ± 0.14
Model IV	0.39 ± 0.007
Growth curve	0.33 ($r = 0.99$)

5.1.3. Validation of a Dual-Substrate Model. Of the other models that gave reasonable fits to the data shown in Fig. 2.3*a*, model IV of Schmidt et al. (1985b) best approximated the actual experimental conditions and gave a reasonably accurate estimate of μ_{max} (Table 2.2). Model IV can be represented in its substrate dissappearance form as

$$S = S_0 \exp[-(k_1/\mu_{max})(e^{\mu_{max}t} - 1)], \qquad (22)$$

where k_1 is the pseudo-first-order rate constant and is equal to $V_{max}X_0/K_m$, where V_{max} is the maximum specific mineralization rate, K_m is the half-saturation constant for substrate mineralization by nongrowing cells, and all other variables and parameters are as previously described. The theoretical basis of Eq. (22) is grounded in observations that cells can metabolize one substrate while their rate of growth is being controlled by the concentration of anothter substrate or some other environmental factor (Harder and Dijkhuizen, 1982; Law and Button, 1977; Schmidt and Alexander, 1985; Schmidt et al., 1985b). Equation (22) is only one of a family of models that were developed for systems in which the substrate being measured is not the growth limiting substrate (Schmidt et al., 1985b). Such situations may be common under natural conditions, especially if the substrate of interest is not essential for growth or when the concentration is either very high or very low (Anderson et al., 1990; Hess et al., 1990; Law and Button, 1977; Schmidt et al., 1985b, 1987; Scow et al., 1989; 1990b, Simkins et al., 1986).

Further validation of Eq. (22) (model IV) was obtained by Schmidt et al. (1985b). Figure 2.4 shows the mineralization of a trace level of glucose by a nongrowing population of *Salmonella typhimurium* LT_2 compared to the mineralization of the same concentration of glucose by a population of *S. typhimurium* growing on arabinose. The concentration of glucose used, 0.4 ng/ml, was too low to support the growth of *S. typhimurium* and growth-related kinetics were only observed when these organisms were growing at the expense of other substrates (Schmidt et al., 1985b; Schmidt and Alexander, 1985). As with the previous example, however, the Monod equation fit the glucose mineralization data better than did Eq. (22) even though the Monod equation was biologically inappropriate in this case.

6. STUDIES OF POPULATION IN NATURAL SYSTEMS

In this section examples are presented from studies in which some of the models presented above have been used to help understand microbial population dynamics in soils from a variety of natural systems.

6.1. Population Dynamics in Soil

Studying microbial population growth and activity in soil is difficult because soil is a heterogeneous habitat with a large and very sorptive surface area that can

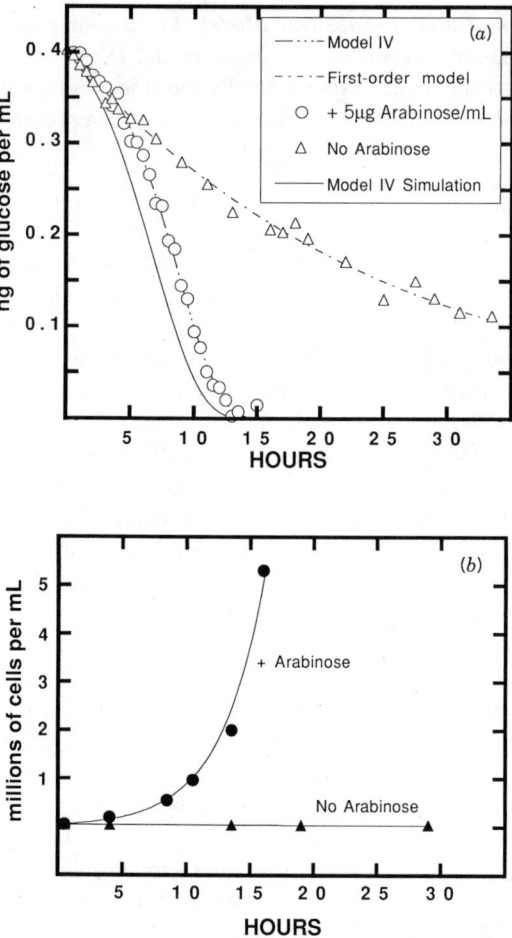

FIGURE 2.4. (a) Mineralization of 0.4 ng of glucose per mL by nongrowing cells of *Salmonella typhimurium* LT$_2$ or by cells growing on 5 µg of arabinose per mL and simultaneously mineralizing 0.4 ng of glucose per mL. The data for glucose mineralization in the absence of arabinose, shown by the open triangles, were best fit by the pseudo-first-order model (Schmidt et al., 1985b) and the data for glucose mineralization in the presence of arabinose, shown by the open circles, were best fit by mdoel IV [Eq. (22)]. In addition to curve fits to the data, a simulated curve generated using Eq. (22) is also shown for the mineralization of 0.4 ng of glucose in the presence of 5 µg of arabinose per mL. Parameter values used to create this simulation were $\mu = 0.29$, from the growth data shown in Fig. 2.4b, and $k_1 = 0.039$, from fitting the pseudo-first-order model to the glucose mineralization curve in the absence of arabinose. (b) Actual cell growth data (filled circles and triangles) from the same microcosm from which the mineralization curves were taken. The independent estimate of μ used above was obtained by fitting an exponential growth function (drawn line) to the data for cell growth in the presence of arabinose. Curves were redrawn using data from Schmidt et al. (1985b).

interfere with many methods for measuring microbial population levels. As a result, there have been many attempts to indirectly determine total microbial numbers and activity using a variety of approaches. Methods that have been used include ATP measurements (Jenkinson and Ladd, 1981; Sparling et al., 1981), soil fumigation (Soulas et al., 1984), and measurements of the respiratory response of soils to added substrates (Anderson and Domsch, 1978; Smith et al., 1985).

Given the difficulties in measuring microbial populations in soil, modeling approaches have proved to be quite useful for gaining insight into the population biology of soil microorganisms. Modeling of microbial population dynamics and activity in soil has ranged from the simple, such as the power-rate model of Hamaker (1972), to very complex simulation models (Smith, 1982; Soulas, 1982). Most of the more complex models and even some of the simpler ones remain to be validated and thus are of limited use at the present time (Alexander and Scow, 1989). The most successful attempts at modeling population dynamics in soil have been with models of the physiological response of soils to various added substrates. The basic premise of this approach is that the size and growth rate of specific populations can be determined by analyzing curves of substrate depletion or product formation using appropriate forms of simple models of population dynamics. Some of the more informative attempts at modeling population dynamics in soil are presented in the following sections.

6.2. Simulation Models of Soil Population Dynamics

In general, any of the models discussed so far can be used as simulation models. Simulation modeling entails putting reasonable (preferably measured) values for parameters into a model and then observing how close the model prediction comes to an actual data set. As an example, Nelson et al. (1982) developed a simple simulation model of microbial population dynamics in soil. They modeled the by-product formation and population dynamics of parathion-degrading bacteria in response to varying concentrations of parathion added to soil. Their model for product formation was a modification of the model of Williams (1973) and can be restated as follows:

$$\frac{dP}{dt} = \frac{1}{Y}\frac{dN}{dt} + m(N - N_0), \tag{23}$$

where P is the concentration of hydrolysis product formed, and m, Y, and N are as defined previously. To relate product formation to initial parathion concentration (S) they used the relationship

$$\frac{dP}{dt} = -c\frac{dS}{dt}, \tag{24}$$

where c is the recovery factor, which varied from 0.6 to 0.8 in their experimental work. To express bacterial growth and death rates in relation to S, they used a modification of the Monod equation:

$$\mu = \mu_{max}\, S/(K_s + S) - mY[1 - (N_0/N)]. \tag{25}$$

Equation (25) was used to model both growth and death of the parathion degrading population because when $N = N_0$, Eq. (25) reduces to the Monod equation and when parathion is exhausted (S = 0), Eq. (25) reduces to

$$\mu = -mY[1 - (N_0/N)], \tag{25}$$

which expresses the maintenance related decay of the population. Using Eqs. (23), (24), and (25) in their finite difference forms, measured and estimated values of the parameters of Eqs. (23), (24), and (25) and a time step of 0.1 day, Nelson et al. (1982) successfully simulated the relationship between microbial population dynamics and parathion concentration. They were less successful in modeling the dynamics of product formation, presumably because their model lacked a parameter to account for a lag period noted in the actual measurements of product formation (Nelson et al., 1982). Despite some failings of the model of Nelson et al. (1982), their approach was very instructive in terms of defining some of the most important parameters for successfully simulating a rather complex data set.

More complex simulation models than that of Nelson et al. (1982) have been developed (Hunt et al., 1984; Paustian and Schnürer, 1987; Smith, 1982; Soulas, 1982; Soulas and Lagacherie, 1990; Stroo et al., 1989) for studying a variety of processes and interactions in soil. In general, such models are too complex and contain too many parameters to be easily used in studies of short-term microbial population dynamics. These models are useful, however, for understanding long-term population trends and complex relationships and processes such as nutrient cycling (Hunt et al., 1984; Smith, 1982), food webs (Hunt et al., 1984), and plant–microorganism interactions (Smith, 1982).

6.3. Simple Models of Soil Population Dynamics

One problem with simulation modeling is that it is usually difficult to obtain reasonable estimates of all the parameters of the model. Thus, in many cases, a curve-fitting approach, using simple models of population dynamics, can be a more powerful tool for understanding the behavior of microbial populations than using complex simulation models. In this section several examples of how models of population dynamics can be used to determine parameters of microbial population dynamics from curves of microbial activity measurements are discussed.

Schmidt and Gier (1989, 1990) conducted intensive studies of the population dynamics of toxic-chemical mineralizing microorganisms in two soils (soils 1 and 2). By modeling the response of the soil populations (Fig. 2.5) to various concen-

6. STUDIES OF POPULATION IN NATURAL SYSTEMS 51

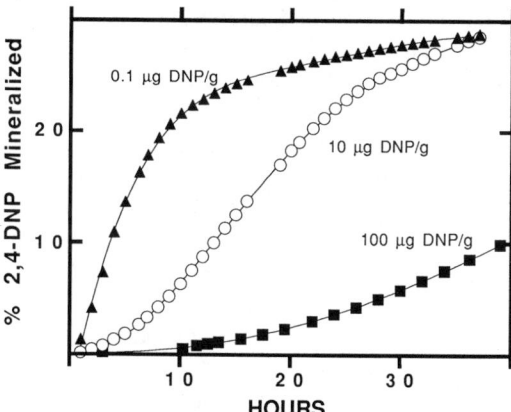

FIGURE 2.5. Mineralization of three concentrations of 2,4-dinitrophenol (DNP) added to a forest soil. The curve of mineralization of 0.1 µg of DNP per gram of soil shows typical nongrowth kinetics and was best fit by the pseudo-first-order model (Schmidt et al., 1985b), whereas the other two curves are typical growth-related shapes and were best fit by the product formation form of Eq. (27). The symbols represent the actual data points. The drawn lines represent curve fits using the models. This figure was redrawn using data from Schmidt and Gier (1989).

trations of added 2,4-dinitrophenol (DNP) they were able to determine the size of the DNP-mineralizing populations in the two soils studied. By visual inspection of the shapes of the curves presented in Fig. 2.5, it is apparent that 0.1 µg of DNP per gram was not enough substrate to induce growth of the DNP-mineralizing population whereas, 10 µg/g did cause growth. The shape of the curve at 0.1 µg/g was concave down, indicating kinetics that are first-order and not growth related. In contrast, at DNP concentrations of 1 µg/g (data not shown) and higher (Fig. 2.5) the mineralization curves show an acceleration phase that indicates growth was occurring.

The model of best fit to each curve corroborated the intuitive interpretation of the curves presented in Fig. 2.5. The curve of 0.1 µg/g was best fit by the pseudo-first-order model (Schmidt et al., 1985b), whereas all other curves were fit by models incorporating parameters for population growth. Parameter estimates were partially validated using most-probable-number (MPN) estimates of the DNP-mineralizing population (Schmidt and Gier, 1989). For instance, nonlinear regression estimates of the initial population capable of mineralizing DNP ranged from 1.6×10^4 to 50×10^4 cells per gram of soil compared to an average MPN estimate of $3.3 \pm 1.6 \times 10^4$ cells per gram of soil 1. In the other soil studied (soil 2), nonlinear regression estimates averaged $1.35 \pm 0.01 \times 10^3$ cells per gram and MPN estimates averaged $2.5 \pm 1.7 \times 10^3$ cells per gram. Given the large potential error of the MPN estimates, the overall agreement between these two independent means of measuring population size was taken as qualitative validation of the modeling approach of Schmidt and Gier (1989).

Another interesting observation of Schmidt and Gier (1989) was that model estimates of population density and μ_{max} from one of the soils (soil 1) varied with concentration of added substrate. This was interpreted to mean that there existed more than one population, with different DNP affinities, of DNP-mineralizing organisms in this soil. The existence of more than one population with the ability to mineralize a given substrate gives rise to what has been termed multiphasic (Azam and Hodson, 1981) or multisystem (Lewis and Gattie, 1991) kinetics. Multiphasic kinetics can also be attributed to one population of organisms with multiple uptake or metabolic systems for a given substrate (Hwang et al., 1989; Lewis et al., 1985). Multiphasic kinetics have previously been invoked to explain deviations from saturating kinetics observed in a number of aquatic systems (Jones and Alexander, 1986).

The prediction that there were two coexisting populations mineralizing DNP in the soil studied by Schmidt and Gier (1989) was corroborated when two organisms with different substrate affinities were isolated from this soil (Table 2.3). In Table 2.3, parameter estimates for μ_{max} and K_s from soil incubations are compared to independent estimates of the same two parameters for pure cultures of the DNP-mineralizing bacteria isolated from the soil (Hess et al., 1990; Schmidt and Gier, 1990). Nonlinear regression estimates of μ_{max} were almost identical between soil incubations and pure cultures, whereas estimates of K_s were much higher in soil than in pure culture (Table 2.3). Possible reasons for the discrepancy between K_s estimates are discussed after a similar observation of Focht and Shelton (1987) is presented below.

Several other recent modeling studies have provided similar insights into the population dynamics of soil microorganisms (Moorman, 1990). Van de Werf and Verstraete (1987a,b) used a curve-fitting approach to estimate the population levels of microorganisms capable of metabolizing several different substrates in soil. They used the Monod equation with a maintenance coefficient to fit oxygen uptake and substrate depletion curves from soil incubations. They verified the validity of their approach using axenic and mixed liquid cultures of bacteria.

A similar approach to understanding microbial population dynamcis using

TABLE 2.3. Comparison Between the Values Obtained for the Parameters K_m, K_s, and μ_{max} from Pure Culture Studies Conducted in Liquid Media and Incubation Studies with Samples of the Soil from Which Both Bacteria Were Isolated

	K_m^a (μg/mL)	K_s^b (μg/mL)	μ_{max} (h^{-1})
Soil population A	0.19 ± 0.005	ND	0.08 ± 0.003
Janthinobacterium sp.	0.01 ± 0.002	ND	0.065 ± 0.014
Soil population B	ND	357 ± 118	0.35 ± 0.03
Rhodococcus sp.	2.3 ± 0.37	2.7 ± 1.08	0.31 ± 0.06

[a]K_m is the half-saturation constant from the Michaelis–Menten equation.
[b]K_s is the half-saturation constant from the Monod equation.

models was that used by Focht and Shelton (1987). They compared the growth kinetics of a 3-chlorobenzoate-degrading strain of *Pseudomonas alcaligenes* in soil and pure culture experiments. In the soil used, 3-chlorobenzoate mineralization was not detected unless it was inoculated with *P. alcaligenes*. *P. alcaligenes* was inoculated at different densities into the soil and the resulting 3-chlorobenzoate depletion curves were analyzed using the Monod equation. As with the studies of Schmidt and Gier (1989, 1990), good agreement was seen between parameter estimates for μ_{max} but not for K_s. K_s estimates were more than 30 times higher in soil incubations than pure culture studies. Possible reasons for the discrepancy between K_s estimates are discussed below.

Some generalizations about microbial population dynamics in soil can be made based on the modeling approaches of Focht and Shelton (1987) and Schmidt and Gier (1989, 1990). In both studies μ_{max} was a constant for a given species of bacteria utilizing a given limiting substrate, whether it was in the soil or in pure culture. That μ_{max} was the same in culture as in soil indicates that the basic growth physiology of the organism was not appreciably affected by the soil environment. The effect of the soil environment on K_s, on the other hand, was quite dramatic in both studies. Estimates of K_s from soil incubations apparently reflect environmental constraints on substrate availability in soil (Focht and Shelton, 1987; Scow et al., 1986). Lowered rates of diffusion due to increased diffusion path lengths in soil is one possible explanation of the effects of soil on values of K_s. Sorption of chemicals to inorganic or organic soil constituents would also tend to lower the rate of flux of substrate molecules to microbial cells and thus raise apparent values for K_s. Whatever the physical reason, the practical implication of this observation is that much more substrate is needed in soil for an organism to grow at μ_{max} than is needed in aquatic systems. This finding is one example of the type of unexpected information that can be obtained by applying simple models of population dynamics to studies of microbial activity in natural systems.

To more accurately model the effects of substrate availability on microbial populations in soil, the modeling approaches of Focht and Shelton (1987) and Schmidt and Gier (1989, 1990) can be modified to account for the slower rate of substrate flux in soil versus acquatic systems. Scow et al. (1986) have argued that a multicompartmental modeling approach is needed to account for the partitioning of substrate in different physical realms in soil. More recently, Scow and Hutson (unpublished data) have developed a process-based, deterministic model that describes the diffusion, sorption, and biodegradation of organic compounds in aggregated systems such as soil. Their model accurately simulated data describing biodegradation in defined experimental systems in which the substrate was limited in its availability to bacteria by diffusion and sorption (Scow and Alexander, unpublished data).

6.4. Dual-Substrate Kinetics in Soil

As pointed out above, the Monod equation is only applicable to situations in which the substrate of interest is the one limiting the extent and rate of microbial growth.

In some cases other nutrients or factors may be limiting the growth rate of the microbial population under study. Scow et al. (1989) studied the mineralization of PNP in soils at concentrations too low (5 ng/g) to support microbial growth. They carried out studies to determine the effects of supplemental carbon and inorganic nutrient additions on the kinetics of PNP mineralization. Simple organic compounds, such as amino acids, acetate, and glucose, and inorganic nutrients had no effect on the kinetics of PNP mineralization. In contrast, additions of phenol dramatically altered the kinetics of PNP mineralization from non-growth-shaped curves in the absence of phenol to sigmoidal curves of $^{14}CO_2$ accumulation when phenol was present (Fig. 2.6). One of the dual-substrate models (model I) of Schmidt et al. (1985) that incorporates parameters for population growth gave good fits to these curves. The integrated form of model I is

$$S = S_0[\phi(e^{\mu_{max}t} - 1) + 1]^{-k_2/\mu_{max}}, \quad (27)$$

where k_2 is the rate constant equal to $V_{max}X_{max}/K_m$, ϕ is equal to X_0/X_{max}, and all other variables and parameters are as previously described. Theoretically, Eq. (27) describes the pseudo-first-order mineralization of substrate by a population of organisms growing logistically when the growth limiting factor is not the concentration of substrate being measured.

Growth of the PNP-mineralizing population at the expense of phenol was verified for the highest concentration of phenol used. Most-probable-number (MPN) estimates indicated that 250 μg of phenol per gram of soil caused significant growth of the PNP-mineralizing population. In addition, estimates of μ_{max} for the PNP-

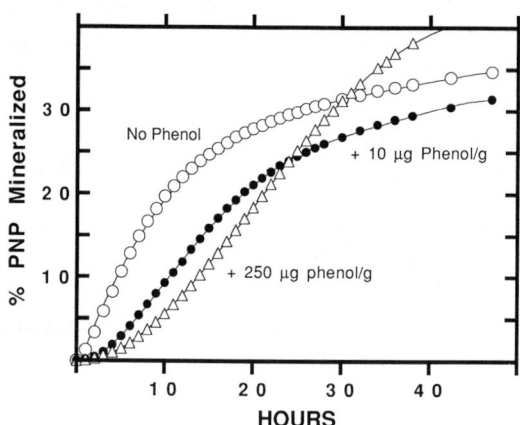

FIGURE 2.6 Mineralization of 5 ng of PNP in the presence of either 0, 10, or 250 μg of phenol per gram of a forest soil. The PNP-mineralization curve in the absence of phenol was best fit by the pseudo-first-order model (Schmidt et al., 1985b), whereas the other two curves were best fit by the product formation form of Eq. (27). The symbols represent the actual data points. The drawn lines represent curve fits using the models. This figure was redrawn using data from Scow et al. (1989).

mineralizing population (0.13 ± 0.004 h^{-1}) were almost identical to the estimates of μ_{max} (0.12 ± 0.01 h^{-1}) for the population simultaneously mineralizing 250 μg of phenol per gram. These results indicated that phenol was controlling the growth rate of the PNP-mineralizing population and that the PNP-mineralizing population was a subset of the phenol-mineralizing population in this soil.

Comparable verification was not obtained for the mineralization of PNP in the presence of 10 μg of phenol per gram of soil. MPN estimates did not indicate significant growth of the PNP-mineralizing population nor did growth rates from PNP- and phenol-mineralizing populations match. Thus although Eq. (27) best fit the data, it seems unlikely that this model adequately reflects population behavior in this case. An alternative explanation of the shape of the curve of PNP mineralization in the presence of 10 μg of phenol per gram is suggested by the substrate inhibition study of Schmidt et al. (1987). In this study it was shown that phenol acted as both a growth-supporting substance and as a competitive inhibitor of PNP uptake by a *Pseudomonas* sp. isolated from the soil studied by Scow et al. (1989). When phenol acted soley as a competitive inhibitor, in the absence of cell growth, the substrate depletion curves were growth shaped (sigmoidal) and were best fit by the logistic equation, even though no growth was detected (Schmidt et al., 1987). This again is an example of the failure of relying only on curve-fitting procedures to understand the behavior of microbial populations from a natural system.

7. CONCLUSIONS

The main conclusion of this chapter is that simple models of microbial population dynamics can be used as effective tools for studying the population dynamics and activity of microorganisms in natural systems if the proper precautions are observed. These precautions can be summarized as follows: (1) Use only models with biologically interpretable parameters. (2) Use the simplest possible model needed to describe the data. (3) Validate the model before applying it to natural systems. (4) Validate the model again using data from the system being studied.

The above precautions are meant only as guidelines for using models in studies of microbial activity or population dynamics. In some cases it is very difficult to validate a model using data obtained from the system under study. In such cases models can still be used but too much significance should not be placed on the parameter values obtained. As was shown with several examples (Sections 5 and 6.4), relying on curve-fitting procedures can lead to erroneous conclusions about microbial population dynamics in a given system. These considerations are especially important if information derived from the use of models is to be used to estimate the fate of a given organism or chemical in the environment.

Finally, it is important to realize that no model can completely explain the complexity of even the simplest of biological systems. The true value of models is that they formalize ideas into a manageable set of mathematical hypotheses that can then be tested. Modeling should thus be viewed as a dynamic process in which models that are either too complex or too simple are rejected in favor of models that contain the optimal number of controlling parameters to describe the system under study.

ACKNOWLEDGMENTS

I thank M. Fisk, T. B. Moorman, R. Mullen, P. Radehaus, and K. M. Scow for helpful comments on earlier drafts of this chapter.

REFERENCES

Alexander, M., and Scow, K. M. (1989), in *Reactions and Movement of Organic Chemicals in Soils*, Soil Science Society of America, Spec. Publ. No. 22, pp. 243–269.

Anderson, D. J., Day, M. J., Russel, N. J., and White, G. F. (1990), *Appl. Environ. Microbiol.*, **56**, 758–763.

Anderson, J. P. E., and Domsch, K. H. (1978), *Soil Biol. Biochem.*, **10**, 215–221.

Andrews, J. F. (1968), *Biotechnol. Bioeng.*, **10**, 707–723.

Ayala, F. J., Gilpin, M. E., and Ehrenfeld, J. G. (1973), *Theor. Pop. Biol.*, **4**, 331–356.

Azam, F., and Hodson, R. E. (1981), *Mar. Ecol. Prog. Ser*, **6**, 213–222.

Bader, F. B. (1978), *Biotechnol. Bioeng.*, **20**, 183–202.

Bader, F. B. (1982), in *Microbial Population Dynamics* (M. J. Bazin, Ed.), CRC Press, Boca Raton, FL, pp. 1–32.

Bard, Y. (1974), *Nonlinear Parameter Estimation*, Academic Press, New York.

Blackman, F. F. (1905), *Ann. Bot.*, **19**, 281–295.

Boethling, R. S. (1985), in *Proceedings of the Workshop: Biodegradation Kinetics* (A. W. Bourquin, P. H. Pritchard, W. W. Walker, and R. Parrish, Eds.), EPA/600/9-85/018, U.S. Environmental Protection Agency, Washington, DC, pp. 40–48.

Boon, B., and Laudelout, H. (1962), *Biochem. J.*, **85**, 440–447.

Brunner, W., and Focht, D. D. (1984), *Appl. Environ. Microbiol*, **47**, 167–172.

Bull, A. T., Bushell, M. E., Mason, T. G., and Slater, J. H. (1975), *Soc. Gen. Microbiol, Proc.*, **3**, 62–63.

Button, D. K. (1985), *Microbiol. Rev.*, **49**, 270–297.

Carlson, T. (1913), *Biochem. Z.*, **57**, 313–334,

Christiansen, F. B., and Fenchel, T. M. (1977), *Theories of Populations in Biological Communities*, Springer-Verlag, New York.

Couteaudier, Y., and Steinberg, C. (1990), *FEMS Microbiol. Ecol.*, **74**, 253–260.

Dabes, J. N., Finn, R. K., and Wilke, C. R. (1973), *Biotechnol. Bioeng.*, **15**, 1159–1177.

D'Adamo, P. D., Rozich, A. F., and Gaudy, A. F. Jr. (1984), *Biotechnol. Bioeng.*, **25**, 397–402.

Dennis, J. E. Jr., and Schnabel, R. B. (1983), *Numerical Methods for Unconstrained Optimization and Nonlinear Equations*, Prentice-Hall, Englewood Cliffs, NJ.

Edwards, V. H. (1970), *Biotechnol. Bioeng.*, **12**, 679–712.

REFERENCES

Focht, D. D., and Shelton, D. (1987), *Appl. Environ. Microbiol.*, **53**, 1846–1849.

Gause, G. F. (1934), *The Struggle for Existence*, Williams & Wilkins, Baltimore (reprinted 1964, Hafner Publishers, New York).

Gilpin, M. E., and Ayala, F. J. (1973), *Proc. Natl. Acad. Sci. USA*, **70**, 3590–3593.

Haldane, J. B. S. (1930), *Enzymes*, Longmans, Green and Co., London (reprinted 1965, MIT Press, Cambridge MA).

Hamaker, J. W. (1972), in *Organic Chemicals in the Soil Environment* (C. A. I. Goring and J. W. Hamaker, Eds.), Marcel Dekker, New York, pp. 253–340.

Harder, W. and Dijkhuizen, L. (1982), *Philos. Trans. R. Soc. Lond. Ser. B*, **297**, 459–479.

Hess, T. F., Schmidt, S. K., Silverstein, J., and Howe, B. (1990), *Appl. Environ. Microbiol.*, **56**, 1551–1558.

Hunt, H. W., Coleman, D. C., Cole, C. V., Ingham, R. E., Elliott, E. T., and Woods, L. E. (1984), in *Current Perspectives in Microbial Ecology* (M. J. Klug and C. A. Reddy, Eds.) American Society for Microbiology, Washington, DC, pp. 346–352.

Hutchinson, G. E. (1978), *An Introduction to Population Ecology*, Yale University Press, New Haven.

Hwang, H.-M., Hodson, R. E., and Lewis, D. L. (1989), *Environ Toxicol. Chem.*, **8**, 65–74.

Jenkinson, D. S. and Ladd, J. N. (1981), in *Soil Biochemistry* (E. A. Paul and J. N. Ladd, Eds.), Marcel Dekker, New York, pp. 451–471.

Jones, S. H., and Alexander, M. (1986), *Appl. Environ. Microbiol.*, **51**, 891–897.

Kingsland, S. E. (1982), *Quart. Rev. Biol.*, **57**, 29–52.

Kingsland, S. E. (1985), *Modeling Nature*, University of Chicago Press, Chicago.

Klecka, G. M., and Maier, W. J. (1985), *Appl. Environ. Microbiol.*, **49**, 46–53.

Koch, A. L. (1971), *Adv. Microb. Physiol.*, **6**, 147–217.

Koch, A. L. (1982), *J. Theor. Biol.*, **98**, 401–417.

Koch, A. L. (1985), in *Bacteria in Their Natural Environments* (M. Fletcher and G. D. Floodgate, Eds.), Academic Press, Orlando, pp. 1–42.

Kompala, D., Ramkrishna, D., and Tsao, G. T. (1984), *Biotechnol. Bioeng.*, **26**, 1272–1281.

LaPat-Polasko, L. T., McCarty, P. L., and Zehnder, A. J. B. (1984), *Appl. Environ. Microbiol.*, **47**, 825–830.

Law, A. T., and Button, D. K. (1977), *J. Bacteriol.*, **129**, 115–123.

Lee, A. L., Ataai, M. M., and Shuler, M. L. (1984), *Biotechnol. Bioeng.*, **26**, 1398–1401.

Levin, S. A. (1982), *Am. Zool.*, **21**, 865–875.

Lewis, D. L.., and Gattie, D. K. (1991), *Ecol. Modelling*, **55**, 27–46.

Lewis, D. L., Hodson, R. E., and Freeman, L. F. (1985), *Appl. Environ. Microbiol.*, **50**, 553–557.

Malthus, T. B. (1798), *First Essay on Population* (reprinted 1926, Macmillan, London).

McKendrick, A. G., and Pai, M. K. (1911), *Proc. R. Soc. Edinburgh*, **31**, 649–655.

Mills, A. L., and Bell, P. E. (1986), in *Microbial Autecology: A Method for Environmental Studies* (R. L. Tate, Ed.), Wiley, New York, pp. 27–60.

Monod, J. (1949), *Annu. Rev. Microbiol.*, **3**, 371–394.

Moorman, T. B. (1990), in *Enhanced Biodegradation of Pesticides in the Environment* (K. D. Racke and J. R. Coats, Eds.), ACS Symposium Ser. No. 426, Washington, DC, pp. 167-180.

Motulsky, H. J., and Ransnas, L. A. (1987), *FASEB J.*, **1**, 365-374.

Neidhardt, F. C., Ingraham, J. L., and Schaechter, M. (1990), *Physiology of the Bacterial Cell: A Molecular Approach*, Sinauer Associates, Sunderland, MA.

Nelson, L. M., Yaron, B., and Nye, P. H. (1982), *Soil Biol. Biochem.*, **14**, 223-227.

Papanastasiou, A. C., and Maier, W. J. (1982), *Biotechnol. Bioeng.*, **24**, 2001-2011.

Paris, D. F., and Rogers, J. E. (1986), *Appl. Environ. Microbiol.*, **51**, 221-225.

Paustian, K., and Schnürer, J. (1987), *Soil Biol. Biochem.*, **19**, 613-620.

Pearl, R. (1922), *The Biology of Death*, Lippincott, Philadelphia.

Pirt, S. J. (1975), *Principles of Microbe and Cell Cultivation*, Blackwell Publications, London.

Richter, O., and Söndgerath, D. (1990), *Parameter Estimation in Ecology: The Link Between Data and Models*, VCH Publishers, New York.

Rittmann, B. E., and McCarty, P. L. (1980), *Biotechnol. Bioeng.*, **22**, 2324-2357.

Robertson, T. B. (1923), *The Chemical Basis of Growth and Senescence*, Lippincott, Philadelphia.

Robinson, J. A. (1985), *Adv. Microb. Ecol.*, **8**, 61-114.

Robinson, J. A. (1986), in *Proceedings, 4th International Symposium on Microbial Ecology* (F. Megusar and M. Gantar, Eds.), Slovene Society for Microbiology, Ljubijana, Yugoslavia, pp. 20-29.

Robinson, J. A., and Tiedje, J. M. (1983), *Appl. Environ. Microbiol.*, **45**, 1453-1456.

Salmon, I., and Bazin, M. J. (1988), in *Handbook of Laboratory Model Systems for Microbial Ecosystems*, Vol. 2 (J. W. T. Wimpenny, Ed.), CRC Press, Boca Raton, FL, pp. 235-252.

Schmidt, S. K. (1988), *J. Chem. Ecol.*, **14**, 1561-1571.

Schmidt, S. K., and Alexander, M. (1985), *Appl. Environ. Microbiol.*, **49**, 822-827.

Schmidt, S. K., and Gier, M. J. (1989), *Microb. Ecol.*, **18**, 285-296.

Schmidt, S. K., and Gier, M. J. (1990), *Appl. Environ. Microbiol.*, **56**, 2692-2697.

Schmidt, S. K., Shuler, M. L., and Alexander, M. (1985a), *J. Theor. Biol.*, **114**, 1-8.

Schmidt, S.K., Simkins, S., and Alexander, M. (1985b), *Appl. Environ. Microbiol.*, **50**, 323-331.

Schmidt, S. K., Scow, K. M., and Alexander, M. (1987), *Appl. Environ. Microbiol.*, **53**, 2617-2623.

Scow, K. M., Simkins, S., and Alexander, M. (1986), *Appl. Environ. Microbiol.*, **51**, 1028-1035.

Scow, K. M., Schmidt, S. K., and Alexander, M. (1989), *Soil Biol. Biochem.*, **21**, 703-708.

Scow, K. M., Li, D., Manilal, V. B., and Alexander, M. (1990a), *Mycol. Res.*, **94**, 793-798.

Scow, K. M., Merica, R. R., and Alexander, M. (1990b), *J. Agric. Food Chem.*, **38**, 908-912.

Shehata, T. E., and Marr, A. G. (1971), *J. Bacteriol.*, **107**, 210–216.
Simkins, S., and Alexander, M. (1984), *Appl. Environ. Microbiol.*, **47**, 1299–1306.
Simkins, S., and Alexander, M. (1985), *Appl. Environ. Microbiol.*, **50**, 816–824.
Simkins, S., Mukherjee, R., and Alexander, M. (1986), *Appl. Environ. Microbiol.*, **51**, 1153–1160.
Smith, O. L. (1982), *Soil Microbiology: A Model of Decomposition and Nutrient Cycling*, CRC Press, Boca Raton, FL.
Smith, J. L. (1989), *Biol. Fertil. Soils*, **8**, 7–12.
Smith, J. L., McNeal, B. L., and Cheng, H. H. (1985), *Soil Biol. Biochem.*, **17**, 11–16.
Sokol, W. (1987), *Biotechnol. Bioeng.*, **30**, 921–927.
Soulas, G. (1982), *Soil Biol. Biochem.*, **14**, 107–115.
Soulas, G., Chaussod, R., and Verguet, A. (1984), *Soil Biol. Biochem.*, **16**, 497–501.
Soulas, G., and Lagacherie, B. (1990), *Philos. Trans. R. Soc. Lond. Ser. B*, **329**, 369–373.
Sparling, G. P., Ord, B. G., and Vaughan, D. (1981), *Soil Biol. Biochem.*, **13**, 99–104.
Stroo, H. F., Bristow, K. L., Elliott, L. F., Papendick, R. I., and Campbell, G. S. (1989), *Soil Sci. Soc. Am. J.*, **53**, 91–99.
Suflita, J. M., Robinson, J. A., and Tiedje, J. M. (1983), *Appl. Environ. Microbiol.*, **45**, 1466–1473.
Suflita, J. M., Smolenski, W. J., and Robinson, J. A. (1987), *Appl. Environ. Microbiol*, **53**, 1064–1068.
Tempest, D. W., and Neijssel, O. M. (1984), *Annu. Rev. Microbiol.*, **38**, 459–486.
Van de Werf, H., and Verstraete, W. (1987a), *Soil Biol. Biochem.*, **19**, 253–260.
Van de Werf, H., and Verstraete, W. (1987b), *Soil Biol. Biochem.*, **19**, 261–265.
Verhulst, P.-F. (1838), *Corresp. Math. Phys.*, **10**, 113–121.
Whittaker, R. H. (1975), *Communities and Ecosystems, 2nd ed.*, Macmillan, New York.
Williams, F. M. (1973), in *Modern Methods in the Study of Microbial Ecology* (T. Rosswall, Ed.), Swedish National Science Research Council, Uppsala, pp. 417–426.
Williams, S. T. (1985), in *Bacteria in Their Natural Environments* (M. Fletcher and G. D. Floodgate, Eds.), Academic Press, Orlando, pp. 81–110.

3 A Compartment Model Approach to Bacterial Population Genetics and Biodegradation

W. MICHAEL CHILDRESS and PETER J. H. SHARPE
Center for Biosystems Modeling,
Department of Industrial Engineering,
Texas A&M University

1. Introduction 62
2. Modeling philosophy 64
3. Compartment models 64
   3.1. Genetic configurations 66
   3.2. Genetic change mechanisms 67
      3.2.1. Cell reproduction and death 67
      3.2.2. Transposable elements 68
      3.2.3. Plasmids 69
      3.2.4. Phage 70
      3.2.5. Free DNA 71
   3.3. Environmental factors 71
   3.4. General compartment design 72
   3.5. Population dynamics 72
4. *mer*-Operon system 73
   4.1. *mer*-Operon compartment model 74
   4.2. Preliminary computer simulations 76
      4.2.1. Equilibrium population simulation 76
      4.2.2. Organomercury contamination simulation 77
5. Model enhancements 79
   5.1. Gene expression 79
   5.2. Induction 79
   5.3. Management objectives 79
   5.4. Optimization methods 80
   5.5. Stochastic analysis of compartment dynamics 81
6. Computer simulation 82
7. Model validation 82
8. Summary 84
   Appendix: Rate data for bacterial genetics models 85
   References 86

1. INTRODUCTION

The genetic plasticity of single-celled organisms presents major complications to the application of traditional ecological population concepts and models to microbial systems. Because of their small size, specific individuals cannot ordinarily be tracked over time. Furthermore, individuals may change their complement of genetic material, and thus their behavior and environmental responses, by any of a number of molecular and physiological mechanisms (e.g., conjugation, transduction, and transformation). Understanding the ecology and dynamics of microbial populations and communities therefore requires understanding of how the genetic configurations of individuals and populations change over time and in response to environmental conditions. Such understanding also has potentially great importance in environmental pollution management. The importance of considering bacterial genetics in biodegradation/bioremediation has been addressed by Olson and Goldstein (1988) and Rittmann et al. (1990). Olson and Goldstein refer to this area of inquiry as genetic ecology.

The genetic ecology approach has several advantages over more typical biotechnology approaches. First, it does not involve introducing "genetically engineered microbes" (GEMs) into the environment. Thus there is no need to be wary of the exchange of exotic genetic material between the introduced GEMs and the natural microbial community. The fate of GEMs in the environment is still difficult to predict and their release has met with considerable public opposition. Such problems can be avoided by the *in situ* manipulation of naturally occurring microbes. Second, indigenous organisms are likely to be better adapted to the existing environment than introduced organisms and therefore more suited to successfully perform biodegradation.

Simulation models for bacterial bioremediation should explicitly represent the genetic diversity among individuals for a variety of reasons:

1. Genetic differences among individuals indicate differences in capacity to do bioremediation "work" and to tolerate adverse conditions such as the presence of contaminants.
2. The distribution of the bioremediation genetic units (genes, alleles, operon, or other units) among individuals in the bacterial population is important. Sequestering multiple copies of the units in a few individuals reduces the amount of work that can be performed because there is not likely to be a proportional increase in gene expression and hence biodegradation capacity per individual with increase in copy number. Such an uneven distribution leaves most of the population vulnerable to toxic or inhibitory effects of the contaminant of interest.
3. Certain management objectives may dictate particular distributions of bioremediation genetic units (henceforth called genes for simplicity) among the individuals. For example, "seeding" a natural population with a few individuals with high copy numbers to spread the genes through the population

requires knowledge of the mechanisms and rates of transfer of the genes. Similarly, manipulation of the genetics of the entire population by altering environmental conditions requires knowledge of the initial distribution of genes in the population and the ideal target distribution to meet management objectives.

4. There are a number of genetic mechanisms by which individuals change their genetic configuration. Manipulating the genetics of the population to enhance its capacity for bioremediation work requires understanding these mechanisms and means for altering the genetic states of individuals in the population.

5. Genes of interest can be located at several sites in the bacterium: chromosome, plasmids, transposons, phage, and free DNA. There are distinct genetic change mechanisms for each site, so manipulation of genetics of individuals and thereby populations requires knowledge of locations of the genes and understanding of mechanisms appropriate to each.

6. Mechanistic effects of most environmental influences (e.g., pollutants, temperature, pH, nutrients) operate at the level of the individual. For example, cell division and conjugation rates of individuals should generally increase with temperature. These effects are in turn mediated by the genetic configuration of the individual.

Biodegradation models should also be useful from the standpoint of general understanding of the ecology and population dynamics of bacterial communities. Condit et al. (1988) and Condit and Levin (1990) utilize a compartment modeling approach to explicitly represent the different possible genetic configurations which individuals in the population can assume. Although their approach was primarily concerned with plasmids and transposons, this approach can be generalized to include all possible known mechanisms. A comprehensive conceptual model can then be simplified and calibrated to suit the observed mechanisms for specific genetic units of interest in particular populations and environments.

General or specific models still depend on understanding the dynamics of each relevant genetic change mechanism. Rittmann et al. (1990) and Smets et al. (1990) examine in detail the kinetics of plasmid dynamics in bacterial populations. Condit et al. (1988) and Condit and Levin (1990) address plasmids and transposable genetic elements, relevant genetic change mechanisms, and criteria for their perseverance in bacterial populations under different selection pressures. Little existing work considers all the factors needed for analysis of genetic change kinetics, and none considers the whole range of possible genetic change mechanisms in a population/community.

In this chapter, we describe a general systems modeling approach for simulating genetic configurations ("states") of bacteria. This approach explicitly represents genetic variation among individuals in the population to provide a mechanistic basis for assessing population responses to environmental conditions. We present as an example a prototype model of the *mer*-operon system, which genetically controls for volatilization of potentially toxic organomercuric compounds by pseudomonad bacteria. We frame much of our presentation in terms of bioremediation scenarios, not only because of their practical importance, but because specific remediation enhancement practices are an area of active research (Thomas and Ward, 1990). Nonetheless, the modeling approach we present here has relevance to microbial genetic ecology in general.

2. MODELING PHILOSOPHY

In addressing the genetic ecology of bacterial populations, we bring a certain modeling philosophy to bear. We consider four general levels of utility of biological models.

1. *Organization of Concepts.* All relevant definitions, concepts, and theories are assembled into a conceptual whole. Qualitative interactions and relationships among components are emphasized. Areas lacking conceptual development are identified.
2. *Organization of Data.* All relevant data (i.e., quantities and rates) are placed within the structure of the conceptual model. This accommodates identification of data relevant to the system which are lacking or inadequate, and can guide future experimental research in obtaining requisite data.
3. *Testing of Concepts and Data.* Model simulations using the available data are performed to assess whether the results match observations of the real system. Where the results are inadequate, necessary revisions of concepts and additional data requirements are identified.
4. *Prediction.* What is most commonly thought of as the most important use of models, prediction, is usually the last stage in the development of models and is often never fully achieved. Prediction is the use of the model to predict future system behavior, often in a management or assessment context.

We present our modeling approach in terms of organization of concepts because many of the relevant data are lacking, especially those concerning dynamics of population genetics. Prediction is not feasible, and testing of concepts and data can only be preliminary. We also present computer simulations of a conceptual model for the *mer*-operon system in pseudomonad bacteria.

3. COMPARTMENT MODELS

Our interest is in tracking the genetic configuration of individuals in a population or community with respect to a particular set of genetic units. The genes could be particular alleles that occur at a fixed location on the chromosome, or, more commonly in bioremediation contexts, an unusual set of genes introduced into the bacterium by means of plasmids, viruses, or uptake of free DNA. In many cases, the genes occur in the system as a single transcriptional unit, that is, an operon.

Because the genetics of individual bacteria can change by means of a number of molecular and cellular mechanisms, it is of interest to simulate changes in numbers of individuals in each genetic configuration. Our modeling objective is to keep track of and predict the proportions of individuals in the population in the various possible genetic configurations, and to optimize these proportions for bioremediation work. In bioremediation applications this would allow us to assess (1) the sensitivity of the population to the contaminant, (2) the genetic configuration of the population with and without the selective pressure of contaminants degraded by the genes of interest, and (3) the ability of the population to do bioremediation "work."

We have chosen compartment modeling as most appropriate for this objective. Each compartment represents the number of individuals in a particular genetic configuration. For example, one compartment could keep track of the number of individuals in the population that have no copies of the genes of interest. Other compartments might include individuals with one copy, two copies, "many" copies (e.g., more than 10), and so on. Separate compartments can also be designated for individuals with copies of the genes at different sites, such as chromosomal and plasmid sites.

The number of individuals in any particular compartment can change by several mechanisms. First, the genetic configuration of a bacterium can change, and the bacterium is "transferred" to another compartment in the population model. Second, bacteria can divide and produce two daughter cells with genetic configurations identical to the parent, effectively increasing the cell number in the compartment by one for each division. Finally, bacteria can die from one of several causes, resulting in loss of individuals from the compartment.

Bacteria can change genetic configurations and thereby their compartment classification by a number of means (Table 3.1). The population compartmental model

TABLE 3.1. Bacterial Genetic Change Mechanisms, Associated DNA Sites, and Net Effect on Number of Gene Copies in Each Bacterium

Mechanism	DNA Sites	Gene Copy Number
Excision	Chromosome	Decreases
Gene amplification	Chromosome	Increases
Conservative transposition	Chromosome, plasmid	No change
Replicative transposition	Chromosome, plasmid	Increases
Conjugation	Plasmid	Donor: no change; receiver: increases
Segregation	Plasmid	Decreases
Transduction	Viral nucleic acid	Increases
Transfection	Viral nucleic acid	Increases
Transformation	Free DNA	Increases
Export	Free DNA	Decreases

simulates these changes as instantaneous transitions of individuals from one compartment (configuration) to another. The genetic mechanisms of these configurational changes can be kinetically described, so that transitions between compartments for a particular mechanism can be mathematically described as rates per individual in the initial compartment per unit time. Because several different mechanisms could account for the transition of an individual to another classification, the total number of transitions from a compartment is the sum of transition rates for the different mechanisms. At the same time, other genetic mechanisms could result in reverse genetic transitions, that is, transfers into the compartment. The "net flow" between any two compartments is therefore the difference between the sum of rates going in one direction and the sum of rates going in the reverse direction.

3.1. Genetic Configurations

Different genetic configurations of individual bacteria relate primarily to the presence or absence of particular genetic units of interest. Configurations are distinguished by two criteria: (1) the number of copies of the genes and (2) the location of the copies in the bacterium. The number of copies is important to determine the genetic potential of the individual with respect to biodegradation capacity and sensitivity to concentrations of contaminants. The location of the copies is important to possible mechanisms of genetic change and to possible differences in transcription rates at different locations. For example, only copies on plasmids can be conjugatively transferred to other bacteria, and only copies on plasmids are removed by means of vegetative segregation.

Different genetic change mechanisms are to be considered in developing models for particular bacterial systems. If the genes of interest are known to be incorporated in a transposable operon, then transposon mechanisms must be included in the model. Similarly, genes thought to be transported via phage will require implementation of phage transfer mechanisms. The mechanisms appropriate to the particular system will usually be a selected subset of the comprehensive list incorporated in the general model.

Construction of a model for any particular real system first requires specification of possible genetic configurations of individuals in the population. This involves two steps: (1) identification of possible sites for the genes (chromosome, plasmid, phage, and free DNA are considered here) and (2) identification of reasonable maximum numbers of genes at each site. Next, possible mechanisms that could result in transitions from one configuration to another are identified and listed for each possible configuration. For example, the configuration "no copies" could be changed to "one copy on a plasmid" by means of a single conjugation event; a reverse transition would result from a single segregation event. Finally, a map of configuration compartments is drawn with arrows between boxes representing possible transition mechanism pathways. It is to be expected that there will be parallel and reverse arrows between boxes, representing parallel and reverse mechanisms. The overall population compartment map then represents a complete conceptual

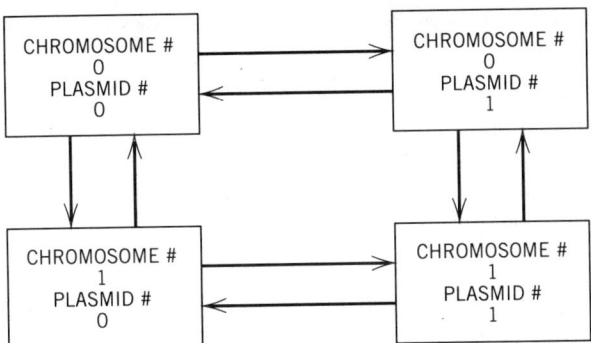

FIGURE 3.1. Simple compartment model of bacterial genetic states. Numbers in each state compartment refer to number of copies of genetic units on the chromosome and plasmid, respectively, of each bacterium. Arrows depict possible genetic change transition paths between compartments.

model of population genetic configurations. An example map is given in Fig. 3.1 for a simple system with one or no gene copies on the chromosome or plasmid.

Subsequent development of the model involves first assigning a quantitative transition rate to each transition arrow in the map; this corresponds to the organization of data level of model development. In order to conduct simulations, initial population counts are assigned to each compartment, and transitions among compartments are calculated and conducted at small time steps.

3.2. Genetic Change Mechanisms

Mechanisms relevant to natural bacterial systems (Table 3.1) are reviewed by Trevors et al. (1987). Not all these mechanisms are appropriate in any given system. The fewer the appropriate configurations and transition mechanisms in a particular system, the simpler the corresponding model will be. In order to conduct simulations, quantitative values for each transition rate must be either experimentally determined or simply estimated using background data or experienced judgment.

We now consider conceptual mathematical representations for the rates of genetic change mechanisms. The mathematical representation for most genetic mechanisms expresses the product of a rate of occurrence per individual per unit time (RATE) and the number of individuals in the compartment (NUMBER-INDIVIDUALS). The rate is equivalently the number of events per unit time in a population or the probability of occurrence for a single individual.

3.2.1. Cell Reproduction and Death.

Cell Division. A bacterium divides, producing two identical daughter cells. In effect, this increases the number of individuals in the compartment by one each time a cell divides.

```
REPRODUCTION = REPROD-RATE * NUMBER-INDIVIDUALS
```

The reproduction increase for the compartment (REPRODUCTION) is the reproduction rate (REPROD-RATE) times the number of individuals in the compartment (NUMBER-INDIVIDUALS). REPROD-RATE may vary among compartments because of the energetic costs of carrying extra genetic material such as large plasmids or multiple copies of plasmids.

Cell Death. A bacterium dies from any cause, reducing the number of individuals in its compartment by one.

```
CELL-DEATH = DEATH-RATE * NUMBER-INDIVIDUALS
```

Death may result from contaminant concentrations whose toxicity (i.e., lethality) may vary among different genetic compartments; therefore DEATH-RATE may be variable among compartments.

3.2.2. Transposable Elements

Conservative Transposition. A transposable element excises itself from its DNA site and inserts itself elsewhere. Only if the element inserts itself at a different DNA site (e.g., moves from a plasmid to the chromosome) is there a change in genetic configuration. This mechanism will usually apply only in systems in which both chromosome and plasmid sites are present. Movement from one site on the chromosome to some other site on the chromosome, or from one site on a plasmid to some other site on a plasmid, would not represent a change in genetic configuration.

```
TRANSPOSITION =    TRANSPOSE-RATE
                 * NUMBER-INDIVIDUALS
                 * SITE-CHANGE-PROB
```

SITE-CHANGE-PROB is the likelihood that the transposon will reinsert itself at a new DNA site rather than back onto its original site. An example algorithm for SITE-CHANGE-PROB is the ratio of the nucleic acid count for the new site to that of the original site, that is, a ratio of target sites.

Replicative Transposition. The transposable element, instead of excising itself from its initial DNA site, replicates and the copy inserts itself elsewhere in the initial site or at some other DNA site. Two expressions are required: one for a one-copy addition to the same DNA site (e.g., the chromosome), and one for a one-copy addition (change) to another DNA site (e.g., a plasmid).

```
REPLICATIVE-TRANSPOSITION-ADDITION  =   REPLIC-TRANSPOSE-
                                        RATE
                                      * NUMBER-INDIVIDUALS
                                      * SAME-SITE-PROB

REPLICATIVE-TRANSPOSITION-CHANGE    =   REPLIC-TRANSPOSE-RATE
                                      * NUMBER-INDIVIDUALS
                                      * SITE-CHANGE-PROB
```

Excision. The transposable element excises itself from its initial site but does not insert itself elsewhere, becoming "lost" in the cell and eventually degraded.

```
EXCISION = EXCISION-RATE * NUMBER-INDIVIDUALS
```

3.2.3. Plasmids

Conjugation. A donor bacterium transfers a copy of the plasmid containing the genes of interest to another bacterium, causing the recipient cell to change genetic configuration, but not the donor. Various factors affecting transfer rates are considered by Freter (1984), and kinetics models have been developed by Levin and co-workers (Levin et al., 1979; Levin and Rice, 1980). In many systems, only a single copy of the plasmid is tolerated in each bacterium, so conjugation to a bacterium only occurs if it contains no copy of the plasmid. The dynamics of this mechanism depend on absolute abundance of both donors and potential recipients, implemented as DENSITY-FACTOR. NUMBER-INDIVIDUALS refers to number of potential recipients.

```
CONJUGATION =   CONJUG-RATE
              * [ NUMBER-INDIVIDUALS
              * NUMBER-DONORS-TOTAL ]
              / DENSITY-FACTOR
```

Plasmid Replication. Ordinarily, plasmids replicate in synchrony with the chromosome prior to cell division. Asynchronous replication results in an increase in the number of plasmids per bacterium and is implemented as PLASMID-REPLICATION:

```
PLASMID-REPLICATION = PLASMID-REPROD-RATE
                    * NUMBER-INDIVIDUALS
```

Vegetative Segregation. This is spontaneous loss of a plasmid, whether by degradation in the cell or transport out of the cell. The dynamics of segregation have been mathematically addressed by Lenski and Bouma (1987). We use a simple expression for segregation:

```
VEGETATIVE-SEGREGATION = VEG-SEGREGAT-RATE
                       * NUMBER-INDIVIDUALS
```

Segregation. This is a special case of cell division in which the plasmid did not replicate first, so that one daughter cell will not have a plasmid.

```
SEGREGATION = SEGREGAT-RATE * NUMBER-INDIVIDUALS
```

This does not result in a loss from the plasmid-carrying compartment, but instead a spontaneous gain to the corresponding no-plasmid compartment.

3.2.4. Phage

Transduction. This is infection of a bacterium by a phage (bacterial virus) carrying the genes of interest. Phage-borne genetic material is usually limited in size because of packaging limitations; that is, the transferred piece of genetic material must not be too large to fit inside the virus capsule. In many systems the material carried is randomly selected from the host DNA. The relative importance of this mechanism in dissemination of bioremediation genes in natural systems is still unclear (Trevors et al., 1987). A key factor is density of free phage in the environment (PHAGE-DENSITY). Where needed data are available, an epidemiological model that explicitly considers number of infectious bacteria (i.e., bacteria infected with the virus, and capable of lysing and releasing new virus) would be appropriate.

```
TRANSDUCTION =    TRANSDUC-RATE
               * NUMBER-INDIVIDUALS
               * PHAGE-DENSITY
```

Phage Replication. This is the reproduction of phage DNA or RNA in the bacterium, that is, the increase in number of gene copies at the phage site. The reproduction rate may be an asymptotic function, reaching some specified maximum density in the cell; therefore the replication rate may vary among compartments classified for different numbers of phage present.

```
PHAGE-REPLICATION = PHAGE-REPLIC-RATE * NUMBER-INDIVIDUALS
```

Transfection. This is the incorporation of phage DNA into the chromosome. The rate for any compartment will depend on the number of phage units present in the bacterium. The effect is to change the site of the genes of interest, but not the total number of copies in the cell.

```
TRANSFECTION = TRANSFEC-RATE * NUMBER-INDIVIDUALS
```

Phage Activation. This is a grab bag of mechanisms that correspond to initiation of other phage mechanisms. Here the phage is assumed to exist in the bacterium in a quiescent state, that is, exhibiting a low reproduction rate and acting much like normal cellular DNA material with respect to transcription and replication. Activation implies that the phage begins an infectious stage of rapid replication and eventual cell lysis. Mathematical representation is not presented here, because a number of submechanisms are involved that need special representation for the particular system of interest.

3.2.5. Free DNA

Transformation. Free DNA adrift in the environment is picked up and incorporated into the chromosome or other DNA site. The dynamics of this mechanism are largely unknown; certainly complex stochastic processes are involved in what DNA segments persist in the environment, how and whether they would be transported across the cell membrane, and whether they would be incorporated into the cellular DNA.

Export. This is the spontaneous loss of genetic material from the cell, except for excision of transposons and vegetative segregation of plasmids. This is another heterogeneous group of possible processes. The rate can only be empirically or theoretically estimated.

```
EXPORT = EXPORT-RATE * NUMBER-INDIVIDUALS
```

3.3. Environmental Factors

Definitions and mathematical representations of the various genetic mechanisms above do not include effects of various environmental factors. Many such factors will influence the various genetic change rates, especially cell division and cell death rates. For example, Bale et al. (1988) demonstrated significant temperature effects on conjugation rates. Nutrient additions are a standard bioremediation practice to encourage bacterial growth and bioremediation activity for natural bacterial communities (Thomas and Ward, 1990). The rates given above for our prototype model are fixed, but variable rates can easily be incorporated for any environmental factor of interest. An important example is temperature effects on cell division rate. Temperature and other physicochemical effects can easily be incorporated by means of Q_{10} factors in the mathematical representations. The cell division rate would then be

```
REPROD-RATE = BASE-REPROD-RATE * TEMP ^ Q10 REPROD
```

Algebraic representations for rate modifications can be derived for various other factors.

In bioremediation situations, the concentration of a toxic contaminant greatly affects bacterial population dynamics.

- Cells with degradation genes are not adversely affected, but noncarriers presumably have greater death rates. Apparently, the concentration differential in sensitivity to organomercuric compounds for carriers versus noncarriers of the *mer*-operon is a factor of 50–100 (Summers, 1984).

- Cells with large gene copy numbers can presumably tolerate greater concentrations of the contaminant than those with few or no copies. However, for the *mer*-operon system, multiple copies apparently do not greatly increase volatilization rates because the limiting step is transport of organomercuric compounds into the cell, not the volatilization step (Summers, 1984).
- The presence of only a few carriers may be sufficient to degrade the contaminant at low concentrations so that all bacteria are relatively unaffected.

The genetic configuration of the individual may itself affect genetic change rates.

- The presence of a plasmid in a bacterium incurs an energetic cost for maintenance and replication (Lenski and Bouma, 1987; Bouma and Lenski, 1988; Nguyen et al., 1989). This means that the cell division rate for such individuals is somewhat reduced.
- Likewise, multiple copies of the genes of interest on the chromosome should incur an energetic cost.
- Conversely, presence of the genes should reduce the sensitivity of the bacterium to presence of a contaminant, and thereby increase its fitness relative to noncarriers.

These considerations can readily be implemented in adjustments to cell division and death rates.

3.4. General Compartment Design

A typical compartment module including the transfer paths in and out is illustrated in Fig. 3.2. All compartments have transition flows both in and out, and all compartments will have cell division and cell death as flows. The presence of other mechanism flows depends on the genetic configuration definition of each compartment. Different compartments are linked by flows according to feasible genetic change mechanisms by which the individuals could conceivably be transformed from the configuration for one compartment to the other.

3.5. Population Dynamics

To model changes in genetic configuration of individuals in the population, we must first specify an initial population configuration, that is, how many individual bacteria there are in each possible genetic configuration. Typical simulations would involve seeding a population with individuals carrying the genes of interest, subjecting a population at genetic equilibrium to environmental stress (e.g., adding a contaminant), or ongoing management manipulations of the population for specified objectives. Some of these can begin from essentially any initial distribution of the population across compartments, but many will require realistic estimates of numbers in each compartment.

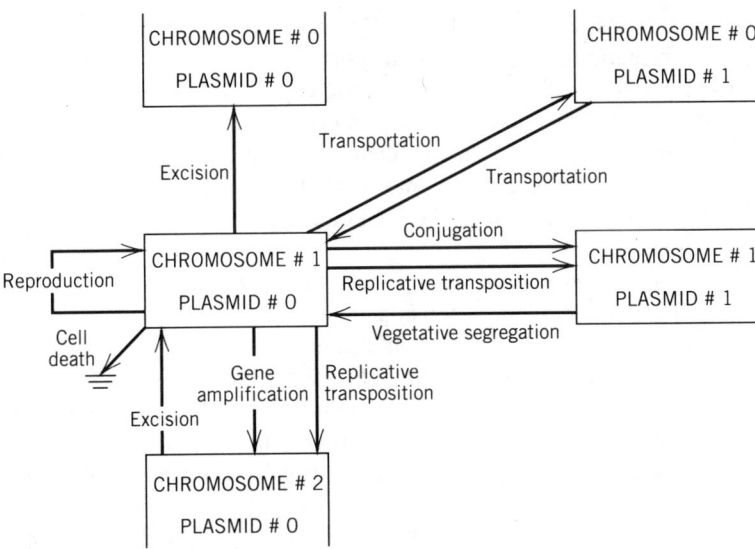

FIGURE 3.2. A typical compartment in the bacterial population genetics model. Numbers in each compartment refer to number of copies of the genetic units on the chromosome and plasmid, respectively, of each bacterium. Possible genetic change mechanisms and corresponding transition paths to and from other compartments are depicted as arrows.

Determining numbers of individuals in each possible genetic configuration for a sample from a bacterial population is often difficult because of the variety of configurations typically possible. A number of assays must be used to determine number and location of gene sets within individuals. Listing these methodologies is beyond the scope of this chapter but is an important consideration in specifying possible model configurations. It may not be reasonable to designate separate compartments for genetic configurations that cannot be resolved in population samples, unless the configurations can be deduced from empirical observations of genetic change dynamics.

Changing environmental conditions should have profound effects on the population configuration. Changes in contaminant concentrations will increase mortality rates in sensitive individuals, allowing individuals with degrading genes to flourish with respect to noncarriers. Temperature changes result in altered rates for cell division and possibly in biodegradation rate and pollutant availability. Other environmental factors that can be incorporated into the model as desired include pH, nutrients, and dissolved gases.

4. *mer*-OPERON SYSTEM

The *mer*-operon system in pseudomonad bacteria encodes several genes involved in the cellular transport and volatilization of organomercuric compounds in soil and aquatic systems (Summers, 1984). Organomercuric compounds are particularly toxic to bacteria because the organic component of the compounds facilitates their transport across cell membranes. In bacterial volatilization, Hg(II) in compounds such as methylmercury and phenylmercuric acetate is converted to metallic mercury vapor by splitting of the covalent carbon–mercury bond and enzymatic reduction of Hg(II). The relatively nontoxic Hg(0) then diffuses rapidly away from the cell. The bacterial operon for volatilization is typically carried as a transposable element on a large plasmid.

A model for the *mer*-operon system must consider a large number of possible transitions among a number of different genetic configurations. The operon can be located on either the chromosome or the plasmid and can occur in multiple copies at either site because of its capacity to be replicatively transposed among sites. One simplification is that the plasmid is usually large, so that although several copies of the operon could occur on the plasmid, only one plasmid will be present per individual bacterium, and this single copy usually replicates in synchrony with cell division.

We present here a simulation model of a pseudomonad bacterial population using the limited data concerning the *mer*-operon system that are available in the literature. Data from other similar systems are used where necessary to implement simulation. We conduct two simple numerical simulations to demonstrate that the concepts incorporated into the model seem to function properly, and to help identify key data needs in manipulating the system.

4.1. *mer*-Operon Compartment Model

To simplify the notation for different genetic configurations among individuals, we have adapted a two-character system for each configuration. The first character refers to the number of operon copies on the chromosome, and the second to the number on the plasmid. For example, "11" would indicate one operon copy on the chromosome and another on the plasmid. This system does not necessarily indicate whether there is a plasmid in the bacterium, since "10" could indicate that a plasmid is present with no copy of the operon; however, we assume this to be very rare. Furthermore, since conjugative inhibition is observed once the plasmid is present in the bacterium, we assume that conjugation will not increase the number of operon copies on the plasmid. Instead, only replicative transposition can result in multiple operon copies on the plasmid.

Because so many copies can be present in a bacterium (over 300 copies per individual have been experimentally induced in the laboratory by Olson and co-workers at the University of California at Irvine; Betty Olson, personal communication), it is not feasible to have separate compartments for each possible copy number. Instead, we use a semiquantitative system for numbering operon copies:

- 0   Zero copies at the site
- 1   One copy
- F   Few copies (two to six)
- M   Many copies (more than six)

We assume that the plasmid, being limited in size, can accommodate no more than F copies, but that the chromosome can contain up to M copies. For both sites considered together, there are 12 different configurations, and therefore 12 compartments in the model (Fig. 3.3).

Because the operon is located on a transposable element and can occur on the chromosome or a conjugative plasmid, a number of genetic transition mechanisms must be represented in the model. These include mechanisms listed in Section 3.2 for cell reproduction and death, transposable elements, and plasmids. Some of these mechanisms are operative only under certain conditions. For example, conjugal transfer of an operon into a cell only occurs for bacteria without a plasmid operon, that is, those compartments with notation "x0," where x means any value for the chromosome operon count. Conversely, segregation only occurs for compartments that have at least one plasmid copy, that is, "x1" or "xF." Note that we assume that the F count refers to about four copies, so that the segregation transition from F to 1 requires loss of three copies. In this case the transition rate for segregation of a single copy SEGREG is raised to the third power to indicate the segregation loss rate from compartment "xF" to "x1." Similar dynamics assumptions are made for the other mechanisms for compartments with F or M operon counts.

Possible transition pathways among the 12 compartments in the *mer*-operon model are depicted in Fig. 3.3. Cell division and cell death are not depicted for simplicity, since these occur for all compartments. We assume that only one mechanism can operate over a unit time, so only one genetic change is allowed per transition, although it is possible that changes could proceed with rapidity in natural and laboratory systems.

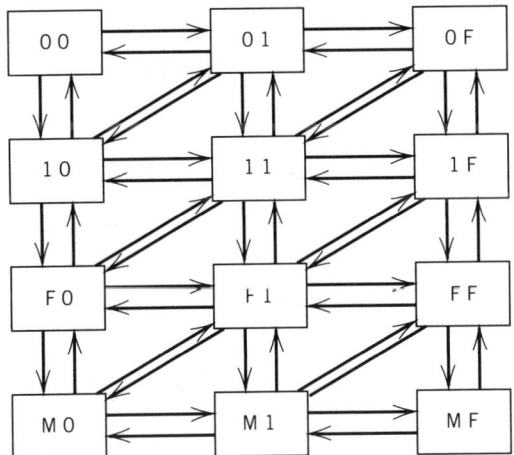

FIGURE 3.3. Compartments and genetic change transitions for the bacterial population genetics model for the *mer*-operon system. Compartment designations indicate number of copies of the operon on the chromosome (first digit) and on the plasmid (second digit) of each bacterium. "F" refers to "few" copies (two to six) and "M" refers to "many" copies (more than six).

4.2. Preliminary Computer Simulations

The conceptual population compartment model has been implemented using Stella (TM of High Performance Systems, Inc.), a simulation-modeling package for Macintosh computers (TM of Apple Computers, Inc.).

4.2.1. Equilibrium Population Simulation. A scenario of interest to environmental scientists is "seeding" a few individuals carrying the *mer*-operon on plasmids into a natural population of pseudomonad bacteria that lack the *mer*-operon. We pretend that a sample of bacteria from the original population was taken, cultured with conjugative donor bacteria carrying the *mer*-operon, and returned to the natural system. The objective of this procedure would be to provide the native population with genes to volatilize organomercuric compounds in the local environment. This approach is important to environmental concerns in that instead of using exotic species, the native population is used for bioremediation. The simulation results show the dynamics of genetic configurations of the natural population after seeding and without the presence of organomercuric compounds in the environment.

We implemented mathematical transition functions as described in Section 3.2 earlier for the 12 compartment *mer*-operon model. Numerical values for various rate constants are presented in the appendix; most rates are in units of number of transitions per individual bacterium per hour. For this simulation, we specify that the initial population was 10 billion bacteria, all in genetic configuration 00 (no *mer*-operon on the chromosome or plasmid), and that the population was seeded with 1000 individuals with one *mer*-operon on the plasmid (configuration 01). The simulation was conducted for 3000 time steps (one step equals one hour) to allow compartment numbers to approach equilibrium.

Initial and equilibrium numbers of bacteria in each compartment are compared in Table 3.2. The total number of individuals carrying the *mer*-operon increased from 1000 to almost 70 million, with the vast majority of these in the same 01 configuration as the seed individuals. These results indicate that limited numbers of seed individuals can propagate the *mer*-operon to a significant although proportionally small number of individuals in the population. This implies that seeding might be a feasible means for maintaining the presence of the *mer*-operon in the system.

The results of this simulation should not be taken literally because of the many limitations of the current model. First, the model is largely conceptual and does not take into consideration possible interactions among the different genetic change mechanisms. Second, most of the transition rate data are only educated guesses and are not based on experimental data specific to the *mer*-operon system. Third, the projected number of *mer*-operon carriers in the population remains a small proportion of the population, so even substantial changes in number of carriers might have little significance to the population as a whole. The simulation results nonetheless agree with our conceptual understanding of the system given the initial population and environmental conditions.

TABLE 3.2. Number of Bacteria in Genetic State Compartments in a Stella Simulation Model After Seeding the Natural Population with 1000 Bacteria Carrying the *mer*-Operon

Genetic State	Log Initial Number	Log Equilibrium Number	Equilibrium Number
00	10.00	10.00	10,000,000,000
01	3.00	7.84	69,180,000
0F	−inf	3.70	5,012
10	−inf	5.10	125,900
11	−inf	5.00	100,000
1F	−inf	1.71	51
F0	−inf	2.40	251
F1	−inf	2.28	191
FF	−inf	−0.350	<1
M0	−inf	−25.32	<1
M1	−inf	−24.79	<1
MF	−inf	−25.61	<1

Total simulation time is 3000 time steps. Log 0 is indicated by "−inf" for minus infinity.

4.2.2. Organomercury Contamination Simulation. Another scenario of environmental interest is addition of organomercuric compounds to the equilibrium population simulated above. We assume that the organomercuric compounds increase the death rate of individual bacteria without the *mer*-operon, and that increasing concentrations of the compounds would increase this effect. On the other hand, greater copy numbers of the *mer*-operon should reduce the sensitivity of bacteria to organomercuric poisoning. The expected behavior of the population is to reduce the total numbers and proportion of noncarrier genetic configurations in the population over time and with increased concentration of organomercuric compounds.

The effect of organomercuric compounds on death rate is implemented in the *mer*-operon model as a factor in the death rate computation:

```
DEATH_x = BASE_DEATH-RATE * MERC_ADJ_x
```

where DEATH_x is the death rate for individuals with x copies of the *mer*-operon and MERC_ADJ_x is the adjustment factor for individuals with x copies (Table 3.3). As the mercury concentration increases, MERC_ADJ_x increases from 1.0 to a maximum of 10.0, effectively increasing the death rate by a factor of 10. However, the rate of this increase is slower for individuals with greater numbers of *mer*-operon copies. The mercury concentration scale used in these simulations is an abstract scale from 0 to 1. This scale and the death rate adjustment are a strictly hypothetical implementation and are not based on experimental data.

Results of 48-hour simulations of the bacterial population with different concentrations of organomercuric compounds are presented in Table 3.4. Total pop-

TABLE 3.3. Hypothetical Mercury Adjustment Factors, MERC-ADJ-x, Used to Adjust Mortality Rates for Different Concentrations of Organomercuric Compounds (OMC) and Different Copy Numbers of the *mer*-Operon per Individual Bacterium

Relative OMC Concentration	*mer*-Operon Copy Number				
	0	1	2	F	M
0.0	1.00	1.00	1.00	1.00	1.00
0.1	1.50	1.00	1.00	1.00	1.00
0.2	3.48	1.23	1.00	1.00	1.00
0.3	8.24	1.63	1.00	1.00	1.00
0.4	9.51	2.44	1.18	1.00	1.00
0.5	10.00	3.79	2.35	1.00	1.00
0.6	10.00	5.37	3.70	1.18	1.00
0.7	10.00	8.06	5.37	2.35	1.00
0.8	10.00	9.46	7.25	3.70	1.00
0.9	10.00	10.00	9.50	7.25	3.70
1.0	10.00	10.00	10.00	10.00	10.00

TABLE 3.4. Logarithms (Base 10) of Numbers of Bacteria in Genetic State Compartments After 48-hour Model Simulations Under Various Organomercuric Compound (OMC) Concentrations

Genetic State	OMC Concentration					
	0.0	0.1	0.3	0.5	0.8	1.0
00	10.00	8.38	2.36	−2.99	−11.23	0.097
01	7.84	8.02	6.23	0.823	−12.06	−2.06
0F	3.71	3.89	4.54	5.40	−2.30	−6.20
10	5.11	5.29	3.57	−1.40	−13.06	−4.80
11	5.01	5.19	5.84	1.69	−9.71	−4.90
1F	1.72	1.91	2.57	3.42	−4.28	−8.18
F0	2.41	2.60	3.25	4.10	−3.60	−7.49
F1	2.29	2.48	3.14	3.98	−3.72	−7.61
FF	−0.330	−0.146	0.513	1.37	4.15	−10.23
M0	−25.30	−25.11	−24.45	−23.60	−20.81	−35.20
M1	−24.77	−24.50	−23.93	−23.07	−20.29	−34.67
MF	−25.58	−25.40	−24.74	−23.88	−21.08	−35.48
Totals	10.00	8.54	6.39	5.44	4.15	0.10

ulation numbers and numbers in configuration 00 (no *mer*-operon) decrease with each increase in concentration. All compartments with individuals carrying at least one copy of the *mer*-operon increase from concentrations 0.0 to 0.1; numbers of individuals with multiple (F or greater) copies persist in the population up to concentration of 0.5, but decrease thereafter. These results conform to the expected behavior of the natural population, based on our current understanding of the *mer*-operon system.

5. MODEL ENHANCEMENTS

The compartment model described above considers only the genetics of the population and environmental factors that directly affect gene copy numbers per individual and in the total population. However, other factors such as the level of gene expression and environmental manipulations are also important considerations in the total ecology of bacterial populations and certainly in bioremediation scenarios. These factors can be added to the genetic compartment model as enhancements to meet different modeling objectives. Some of these are discussed below.

5.1. Gene Expression

Gene expression refers to the phenotypic expression of the genes present. In the context of the model presented here, we refer to the amount of physiological work (i.e., biodegradation) that a bacterium will perform per copy of the gene present. It is not likely that bioremediation capacity will increase in exact proportion to the number of gene copies because of energetic costs to the individual in maintaining and expressing the genes. This means that not only the total number of gene copies in the population but the distribution of copies is important in estimating capacity to do bioremediation work. We postulate asymptotic expression with number of copies, so that above a certain number of copies per individual the bacterium will not have increased bioremediation capacity. An appropriate function could be empirically determined and implemented in the model to more accurately predict true bioremediation capacity of the population given its observed distribution of gene copies among individuals.

5.2. Induction

Presence of gene copies does not necessarily mean that the genetic capacity for biodegradation will actually be expressed. In many prokaryotic systems, the genes of interest must be induced into action by environmental factors such as pollutant concentration, population density, or physicochemical factors. Presumably, in bioremediation systems, certain minimum concentrations of specific chemical species are required before transcription (i.e., the process in which messenger RNA is produced to provide a template for protein synthesis) of the genes is initiated, and transcription rates may also be functionally related to contaminant concentrations. Weiss et al. (1977) reported an increase in volatilization activity for the *mer*-operon system of 100 times after induction with mercuric chloride. This requires adding to the model a threshold function that uses contaminant concentration to predict induction and modify gene expression.

5.3. Management Objectives

The population compartment model has the potential for making predictions about the capacity of the bacterial population to bioremediate a contaminated environment. However, there are other important considerations in bioremediation efforts besides simple degradation rate. For example, chronic contamination sources require continuous and ongoing bioremediation, which requires that the degrading capacity of the bacterial population be maintained over time. Laboratory production of seed bacteria would emphasize genetic configurations that have high copy numbers of genes placed on plasmids or phage DNA for ready transfer to other bacteria. The management objectives therefore dictate subtle variations in the desired or target genetic configuration of the bacterial population or community.

In the context of bioremediation, three general management objectives have been identified.

1. Enhancing or maximizing biodegradation capacity. A typical situation would be in surface soils after a contaminant spill. Here the one-time clean-up of the site would simply require maximum degradation capacity of the native bacterial community to remove the contaminant.

2. Increasing or maximizing gene copy number per individual. An example would be industrial production of seed bacteria to be placed in a contaminated site. The remediation capacity of the bacterial community would be enhanced by culturing a sample of the site material to increase copy number per individual. The culture would then be returned to the site as a seed to enhance native remediation capacity. Not only is the copy number per individual a criterion for culturing, but having copies on transfer sites such as transposons, plasmids, and/or phage DNA would be important in specifying the target or optimal genetic configuration for the culture.

3. Enhancing persistence of remediation genes in the population. Carrying extra-chromosomal remediation genes would ordinarily be energetically expensive to natural bacterial populations, so that in the absence of the contaminant, the genes would be expected to disappear from the population over time. In remediation situations where the contamination is chronic or periodic, a management objective would be to maintain the remedial capacity of the native bacterial community over time. This means that genetic configurations that facilitate the persistence of the genes in the population under conditions of low or infrequent contaminant concentrations would be valuable.

5.4. Optimization Methods

Standard engineering optimization methodologies can be applied to the model at any level to serve two functions:

1. Identifying optimal configurations to meet management objectives. For example, what population genetic configuration yields maximum biodegradation capacity? Alternatively, what configuration is most persistent in the absence of the contaminant?
2. Evaluating different means for reaching the optimal configuration. For ex-

ample, should the population be seeded with gene carriers, or will simple environmental manipulations (e.g., adding nutrients to enhance bacterial growth and replication) suffice to increase gene frequencies? Should contaminant release, if controllable, be slow and chronic or in pulses to maximize remediation activity?

A number of methodologies can be applied in addressing such questions. For example, each genetic compartment in the model can be assigned a function to calculate a utility value with respect to the specified management objectives. The utility of each compartment is the number of individuals in the compartment times the utility value per individual, and the utility of the population as a whole is the sum of utilities for all compartments. The management objective would then best be met by maximizing the population utility value by various environmental manipulations or genetic enhancements. Management alternatives could similarly be evaluated by comparing changes in utility value. This approach is similar to linear programming, where the objective function indicates the optimal configuration given various utility functions and constraints.

5.5. Stochastic Analysis of Compartment Dynamics

Simulations of compartment models are not limited to deterministic numerical methods such as those implemented in most systems model packages like Stella. Considerable recent research has centered on stochastic analysis of compartment dynamics (Godfrey, 1983). For simple models (up to, say, four compartments) and simple processes (e.g., exponential birth and death rates), mean and variance in compartment numbers and transitions can be analytically calculated as functions of time and initial compartment numbers (Matis et al., 1991). The procedures involved are based on treating transitions in the compartment as either discrete or continuous-time Markov processes (Chiang, 1980) and deriving solutions using matrix or partial-differential equation calculations, respectively.

Although such analyses are not tractable for large compartment models such as the *mer*-operon system presented here, they are still useful in model development in two ways. First, submodels of a few compartments can adequately be analyzed separately. This allows focus on the dynamics of particular genetic change mechanisms (e.g., conjugation) between adjacent compartments in the model. Different solutions for different transition processes (e.g., exponential, Poisson, or binomial) can be derived analytically to determine which process best fits the observed dynamics for the mechanism. Second, all mechanisms linking any two compartments can be analyzed simultaneously to describe all possible interactions. Most importantly, such analysis yields both means and variances in compartment numbers and transitions between the two compartments. Third, if analytical solutions can be found for submodels, then experimental data for particular mechanisms can be used to parameterize these solutions for statistical projections.

Deriving statistical means and variances for compartments and transitions is a tremendous advance over numerical simulations, which can only give deterministic results. However, stochastic analysis requires good understanding of genetic change mechanisms to select appropriate analytical process models, and extensive experimental data to parameterize the resulting analytical solutions. At present, few data are available and dynamics of genetic changes are poorly understood. Considerable research is required before stochastic methods can be applied, but it is encouraging that powerful analytical techniques can be brought to bear in the future.

6. COMPUTER SIMULATION

Compartment models can readily be implemented in most computer systems for simulation using existing simulation packages or specially written program code. Initial development of the *mer*-operon model presented here was conducted using Stella on a Macintosh II computer. Such off-the-shelf packages typically provide numerical solutions for the differential rate equations for all compartments simultaneously, and the results are plotted in simple graphs.

Even though the modeling approach presented here describes only a prototype in the course of development, certain analyses of the model can be conducted to investigate the validity and dynamics of the current model version. In particular, sensitivity analysis can be used to determine the sensitivity of prototype simulation results to variations in initial compartment values and genetic transition rates. Such analysis serves two important functions. First, by conducting simulations with a wide range of values for those variables for which firm estimates are not available, simulation results may help in selecting reasonable working estimates for subsequent modeling efforts. Second, the validity of the model itself can be tested across various values to determine under which conditions the model appears to perform adequately.

A successful predictive model for bacterial population genetics in an environmental remediation situation would probably be a complex system requiring a large number of inputs about the bacterial population, the genetics of the bacterial taxon, the environmental structural habitat, and various physicochemical conditions. Such a large model would require extensive simulation testing, experimental research into manipulation of bacterial populations and genetics, and on-site validation. However, the potential utility of predicting and assessing effects of various expensive management alternatives should make model development and testing cost-effective.

7. MODEL VALIDATION

A key step in model development is validation. Validation has been variously defined in the modeling literature, so it is important to specify our intent here. We hold that validation means demonstrating the success in meeting modeling objectives. As noted earlier, our objective at this time is to develop a conceptual and organizational level model of the population genetic ecology of bacterial populations. Caswell (1976) distinguishes two categories of modeling objectives that require model validation: (1) models designed to predict behavior of a system and (2) models designed to gain insight into system behavior. Clearly, our present intent is the latter, to compile concepts and data into a model in an attempt to cohesively articulate our current understanding of bacterial genetic ecology. Nonetheless, one of the driving forces in development of such models is in bioremediation applications, wherein we wish to predict the results of alternative management practices.

Caswell (1976) further suggests that the goal of validation should be different for each category. He applies the term "corroboration" for validation of insightful or conceptual models. This category would be like that of the classical scientific method: our theory or concepts cannot be proved true, but can only be invalidated and supplanted by other theories or concepts. He maintains use of "validation" for the first category above, because in making predictions, the model objective is explicitly to produce reasonably correct or "valid" predictions. We use "validation" here for our conceptual model simply because of its prevalence as a model development term, although we keep in mind that at the current level of model development, we do not yet attempt prediction, only ongoing demonstration of the validity and generality of our concepts.

Validation of our model therefore involves demonstrating that observed or conceivable phenomena can be conceptually accounted for given the concepts and mechanisms incorporated into the model. For the most part, this involves incorporating as much generality as possible, and including all known and appropriate genetic mechanisms. It will hopefully be possible to simplify the general model to suit any specific system of interest, and in so doing to reduce its complexity to manageable levels for application. For example, we would ask: What genetic configurations are known for a particular system of interest, what mechanisms of genetic change are involved, and how are these mechanisms linked together? Essentially, we would want to be able to construct specific conceptual models from the more general conceptual model presented here. Our success in doing so would be validation of the general model.

Development of predictive models in the future will require quantitative validation of model predictions by laboratory, microcosm, and field studies, and case histories of actual remediation efforts. Initial development of the models will require adaptation of literature data on particular mechanisms taken one at a time. The predictive model will require linkage with physical models to consider location of bacteria with respect to environmental conditions and contaminant concentrations (Baek et al., 1989). Interactions among mechanisms (e.g., whether competitive or mutualistic) should be addressed quantitatively (Hickman and Novak, 1989; Hwang et al., 1989a,b). Finally, real-world experiments and efforts will provide validation of models of entire systems, including all relevant genetic configurations and change mechanisms. Again, criteria for validation in this context involve meeting modeling objectives. For example, a valid model need not necessarily predict to within so many percentage points exactly how much bioremediation of a contaminant is achieved in a particular trial, but perhaps only which management alternatives will achieve the most remediation, and whether, in fact, any alternatives will meet desired remediation levels.

8. SUMMARY

Understanding the ecology and dynamics of bacterial populations requires understanding how genetic configurations of individuals as well as populations change over time and how this occurs in response to environmental conditions. We present a conceptual compartment model of the population genetics of pseudomonad bacteria involved in volatilization of organomercuric compounds. Each compartment represents the number of individual bacteria with similar genetic configurations with respect to the genetic units (the *mer*-operon) that control volatilization. This operon is a transposable element that is ordinarily transferred among bacteria by means of a conjugative plasmid. The various possible mechanisms for genetic change with respect to the operon are implemented in the compartment model as transitions or flows of individuals between compartments. Transition rates are based on simple kinetics equations. The compartment model simulates changes in numbers of individuals in the different genetic configurations resulting from the dynamics of those mechanisms that confer genetic change.

We present results of two simulations using the prototype compartment model, for scenarios with and without the environmental presence of organomercuric contaminants. In the first simulation, seeding of the population with a few individuals carrying the operon resulted in spread of the operon among a small proportion of the population. The second simulation demonstrated that with the selective pressure of the contaminants, bacteria able to volatilize the contaminant increased drastically in proportion to dominate the population. We consider a number of possible enhancements for model development, including gene expression and induction, implementing management objectives, optimization techniques, stochastic methods, and alternatives for model simulation and validation.

ACKNOWLEDGMENTS

We thank Edward J. Rykiel, Jr., Paul A. Rochelle, and Betty H. Olson for reviewing an earlier version of this chapter. Support during model development was provided in part by subcontract through the Program in Social Ecology, University of California at Irvine, for Electric Power Research Institute, Palo Alto, California.

APPENDIX: RATE DATA FOR BACTERIAL GENETICS MODELS

The following narrative outline describes various genetic transition rate constants used in *mer*-operon system bacterial population genetics models. All rates are per individual bacterium (per "ind") per hour except where noted otherwise. Literature sources are indicated in parentheses.

Reproduction/Mortality Rates

Cell division rate: REPROD-RATE
  0.10/ind/h
Notes: Estimated range is 0.02 to 0.50 *in situ*; laboratory maximum is about 3.0.

Cell death rate: DEATH-RATE
  0.10/ind/h
Notes: We assume an equilibrium population in which the birth and death rates are equal.

Segregation rate: SEGREGAT-RATE
  0/ind/h
Notes: We assume no reproductive segregation; that is, the plasmids always replicate in synchrony with the chromosome and each daughter cell receives exactly equal numbers of plasmid copies.

Conjugation Rates

Conjugation rate: CONJUG-RATE
  8.0×10^{-4}/recipient/donor/h
Notes: This rate is a nonlinear function dependent on both numbers of donor and potential recipient bacteria in the population. The estimate is for *in situ* conditions. Other estimates from the literature include:

For donor and recipient ratio of 1:1
  4.9×10^{-6}/5 h
  16×10^{-6}/24 h
  (Gauthier et al., 1985)
  6.9×10^{-9} to 1.5×10^{-2}/recipient/24 h
  (Bale et al., 1988)

Vegetative segregation rate: VEG-SEGREGAT-RATE
  1×10^{-4}/h
  (estimate from Condit, 1988)

Transposition Rates

Transposition rate: TRANSPOSE-RATE
$7.1 \times 10^{-7}/h$
(estimate from Condit, 1988)
Notes: Another estimate is a maximum of 1×10^{-3} per cell division.

Replicative transposition rate: REPLIC-TRANSPOSE-RATE
$7.1 \times 10^{-7}/h$
(estimate from Condit, 1988)
Notes: Condit (1988) assumes replicative transposition rate is equal to the simple transposition rate.

Excision rate: EXCISION-RATE
$1 \times 10^{-4}/h$
(Condit et al., 1988)

Pseudomonad chromosome mass: CHROMOSOME-MASS
2.3×10^9 daltons
Pseudomonad *mer*-plasmid mass: PLASMID-MASS
9.0×10^7 daltons
Notes: The ratio of these masses indicates the probability that the transposon will reinsert itself on the chromosome; (1 − ratio) is also the probability for reinsertion on the plasmid.

REFERENCES

Baek, N. H., Clesceri, L. S., and Clesceri, N. L. (1989), *J. Environ. Eng.*, **115,** 150–172.
Bale, M. J., Fry, J. C., and Day, B. R. (1988), *Appl. Environ. Microbiol.*, **54,** 972–978.
Bouma, J. E., and Lenski, R. E. (1988), *Nature*, **335,** 351–352.
Caswell, H. (1976), in *Systems Analysis and Simulation in Ecology* (B. C. Patten, Ed.), Vol. IV, Academic Press, New York, pp. 313–325.
Chiang, C. L. (1980), *An Introduction to Stochastic Process and Their Applications*, Krieger, Huntington, NY.
Condit, R., Stewart, F. M., and Levin, B. R. (1988), *Am. Nat.*, **132,** 129–147.
Condit, R., and Levin, B. R. (1990), *Am. Nat.*, **135,** 573–596.
Condit, R. (1992), *Am. Nat.*, in press.
Freter, R. (1984), in *Current Perspectives in Microbial Ecology* (M. J. Klug and C. A. Reddy, Eds.), American Society for Microbiology, Washington, DC, pp. 105–114.
Gauthier, M. J., Cauvin, F., and Breittmayer, J.-P. (1985), *Appl. Environ. Microbiol.*, **50,** 38–40.

REFERENCES

Godfrey, K. (1983), *Compartment Models and Their Applications*, Academic Press, London.

Hickman, G. T., and Novak, J. T. (1989), *Environ. Sci. Technol.*, **23**, 525–532.

Hwang, H.-M., Hodson, R. E., and Lewis, D. L. (1989a), *Environ. Tox. Chem.*, **8**, 65–74.

Hwang, H.-M., Hodson, R. E., and Lewis, D. L. (1989b), *Environ. Tox. Chem.*, **8**, 209–214.

Larson, R. J. (1984), in *Current Perspectives in Microbial Ecology* (M. J. Klug and C. A. Reddy, Eds.), American Society for Microbiology, Washington, DC, pp. 677–686.

Lenski, R. E., and Bouma, J. E. (1987), *J. Bacteriol.*, **169**, 5314–5316.

Levin, B. R., and Rice, V. A. (1980), *Genet. Res.*, **35**, 241–259.

Levin, B. R., Stewart, F. M., and Rice, V. A. (1979), *Plasmid*, **2**, 247–260.

Matis, J. H., Miller, T. H., and Allen, D. M. (1991), in *Ecotoxicology of Metals: Current Concepts and Application* (M. C. Newman, Ed.), Lewis Publishers, Detroit, MI.

Nguyen, T. N. M., Phan, Q. G., Duong, L. P., Bertrand, K. P., and Lenski, R. E. (1989), *Mol. Biol. Evol.*, **6**, 213–225.

Olson, B. H., and Goldstein, R. A. (1988), *Environ. Sci. Technol.*, **22**, 370–372.

Rittmann, B. E., Smets, B. B., and Stahl, D. A. (1990), *Environ. Sci. Technol.*, **24**, 23–29.

Smets, B. F., Rittmann, B. E., and Stahl, D. A. (1990), *Environ. Sci. Technol.*, **24**, 162–169.

Summers, A. O. (1984), in *Current Perspectives in Microbial Ecology* (M. J. Klug and C. A. Reddy, Eds.), American Society for Microbiology, Washington, DC, pp. 94–104.

Thomas, J. M., and Ward, C. H. (1990), *Environ. Sci. Technol.*, **23**, 760–766.

Trevors, J. T., Barkay, T., and Bourquin, A. W. (1987), *Can. J. Microbiol.*, **33**, 191–198.

Weiss, A. A., Murphy, S., and Silver, S. (1977), *J. Bacteriol.*, **132**, 197–208.

4 Modeling the Growth of Cellulase-Producing Organisms

DAVIS W. HUBBARD and TOMAS B. CO
Department of Chemical Engineering,
Michigan Technological University

1. Introduction 89
   1.1. Cellulose and cellulase 90
   1.2. Types of growth models 90
      1.2.1. Unstructured models 91
      1.2.2. Structured models 91
2. Microbial growth 91
3. Modeling microbial growth 92
   3.1. The lag phase 92
   3.2. The exponential growth phase 93
   3.3. The stationary phase 96
   3.4. The death phase 96
   3.5. Growth of filamentous organisms 98
4. Models developed from modified logistic equations 99
5. Model verification 103
   5.1. *Trichoderma viride (Trichoderma reesei)* 103
   5.2. *Acidothermus cellulolyticus* 107
   5.3. *Thermomonospora fusca* 108
6. Conclusions 110
   Nomenclature used in equations 110
   Appendix: The modulating function method 111
   References 113

1. INTRODUCTION

There is great interest in microorganisms that use cellulose as a source of carbon and energy. Waste cellulose is an abundant source of nutrients, and a fermentation process involving such organisms can be operated to produce protein, glucose, or excess enzymes. Mitra and Wilke (1975) used spruce wood cellulose to produce enzymes. Wilke and Yang (1975) and Grethlein (1978) studied the hydrolysis of newsprint. Andren et al. (1975) used hydropulped government documents to produce sugars. Callihan et al. (1975) used bagasse, Crawford et al. (1973) used paper pulping fines, Peitersen (1975) used barley straw, and Eriksson and Larsson (1975) used pulp mill waste fibers to produce proteins. Vallander and Eriksson (1990) review the suitability of a variety of different lignocellulosic materials for producing ethanol and enzymes. To develop useful large-scale fermentation processes for using waste cellulose, mathematical models for the growth of the microorganisms are needed. These models are used in developing process control strategies and in process optimization. The purpose of this work is to review techniques for modeling microbial growth and to show how the different techniques can be combined and used to model the growth of microorganisms that use cellulose.

1.1. Cellulose and Cellulase

Cellulose is a polymer of glucose in which the glucose units are joined by a β-1,4-glucosidic linkage. The repeating unit in the cellulose polymer is the disaccharide cellobiose. Microorganisms depolymerize cellulose to glucose by enzyme action. They produce a mixture of enzymes known collectively as cellulase, which contains the following components:

- Exocellulase (exo-β-1,4-cellobiohydrolase; BH or C_1), which converts crystalline cellulose to amorphous cellulose
- Endocellulase (endo-β-1,4-glucosidase; C_x), which converts amorphous cellulose to cellobiose
- Cellobiase, which converts cellobiose to glucose

In order for cellulose to be completely depolymerized, all three enzymes must be present. The enzymes all hydrolyze the β-1,4-glucosidic linkage, but each acts on a different size and form of polymer substrate. Howell and Stuck (1975), Howell (1978), and Okazaki and Moo-Young (1978) have studied various aspects of enzyme action on cellulose. This mixture of enzymes, which must be present in different proportions depending on the degree of crystallinity of the cellulose used, makes modeling the growth less straightforward than if only a single enzyme were involved. Cellulase-producing microorganisms have other interesting characteristics. Many are thermophilic and grow best at temperatures up to 55°C. Some grow best in acidic media.

1.2. Types of Growth Models

Models to describe growth can be established with different degrees of biological information included.

1.2.1. Unstructured Models. These models are developed from experimental data for microbe growth. Data are presented in terms of cells per unit volume of medium or biomass per unit volume of medium. Measurements may be made by an optical absorption or light-scattering method or by collecting samples and separating, drying, and weighing the cells. Unstructured models contain very little biological information besides experimental data about cell concentration and medium composition measured for a particular organism under particular growth conditions. All the biomass present is considered to be a single entity. Data useful for developing unstructured models is the information most readily available.

1.2.2. Structured Models. Cell growth involves many components of the cell structure, each of which may be associated with different growth capabilities. Structured models contain separate information about different components of the biomass and how each component contributes to growth. Bailey and Ollis (1986a) review some structured models that contain two "cells" or "components." Such two-cell compartment models and metabolic models have a component associated with nutrient consumption and metabolism and a component associated with cell synthesis and cell division. Often, insufficient data are available to enable verifying structured model parameters.

Data about cell growth are most frequently available for batch culture experiments in which the fermentation vessel is loaded with the nutrient mixture or broth and inoculated with living cells. The temperature, pH, and other growth conditions are maintained at constant values, and cell growth takes place until one or more nutrients are depleted.

2. MICROBIAL GROWTH

Four phases are observed for microbial growth in a batch culture.

- *Lag Phase.* Cell division is suppressed, because the organisms are getting ready for sustained growth by synthesizing the enzymes needed to use the energy sources available. Part of the lag may occur because nutrients and other materials cannot be transported instantaneously across the cell membrane.
- *Exponential Growth Phase.* All mechanisms required for sustained cell division are functioning at normal rates.
- *Stationary Phase.* The maximum population density is reached, and cell division is greatly reduced.
- *Death Phase.* Organisms die because of lack of critical nutrients or because of toxins in the medium.

92   MODELING THE GROWTH OF CELLULASE-PRODUCING ORGANISMS

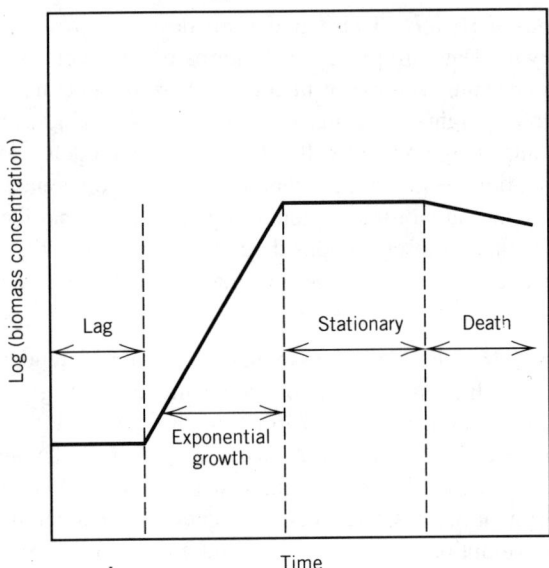

FIGURE 4.1. Schematic representation of microorganism growth.

A sketch illustrating these phases schematically is shown in Fig. 4.1. These phases are observed for most microorganisms including those that produce cellulase.

For batch fermenter design, three points are important.

1. Minimize the lag phase.
2. Maximize the rate of exponential growth.
3. Harvest products or cells at the beginning of the stationary phase.

Because of these points, data for all growth phases are not usually available for industrial fermentation processes. It is not economical to allow growth to continue past the end of the exponential growth phase. Data for the death phase may be available from sterilization studies. Bailey and Ollis (1986b) give a summary of batch growth models.

3. MODELING MICROBIAL GROWTH

3.1. The Lag Phase

A lag phase may follow a rapid change in medium composition. Assimilation of different nutrients may require synthesizing new enzymes. Changes in nutrient concentration may require synthesizing different enzyme concentrations. Vitamins, cofactors, and ions needed for growth may have some finite permeability through the cell membrane. Vitamins, cofactors, and ions already present in the cells may diffuse out when cells are transferred to a new medium reducing the concentration to a value below that required for sustained growth. Multiple lag phases—diauxic growth—may occur if more than one limiting nutrient is present. The nutrients may be depleted sequentially requiring adjustments in the cell metabolism when each is depleted. Pamment et al. (1978) discuss a structured model for the lag phases and present some data illustrating diauxic growth. Their experiments with *S. cerevisiae* show that glycocytic enzymes are prevalent at the beginning of growth when the organisms are using glucose for their growth. Ethanol is produced during this growth. When the concentration of ethanol is large enough to inhibit growth by the glycolytic mechanism, new enzymes associated with respiratory metabolism are synthesized so the organisms can continue to grow using ethanol. The data to verify this combination of mechanisms were obtained for growth limited by glucose with an excess of oxygen present. The model used for growth is a structured model incorporating two types of cell mass:

1. Mass A associated with substrate uptake and energy production.
2. Mass B associated with reproduction and cell division.

The model also includes changing glycolytic and respiratory enzyme activities and the effect of these changes on the growth process. Enzymes are postulated to be produced at a rate proportional to the ratio of the existing enzyme concentration to some target enzyme concentration—the enzyme concentration at some steady state. These target enzyme concentrations are calculated from the rate constants for the metabolic reactions postulated and the substrate—ethanol and glucose—concentrations. The model developed contains 12 parameters that must be evaluated from experimental data, and there are also six concentrations for which initial conditions must be established. The model successfully fits the growth data for *S. cerevisiae*, which shows an intermediate lag phase. Bajpai and Ghose (1978) examine yeast growing on a mixture of glucose and cellobiose and develop a model that includes diauxic growth.

3.2. The Exponential Growth Phase

When the cells present have synthesized the correct concentrations of all enzymes necessary for sustained cell division, then growth can begin and continue unchecked as long as nutrients are available. Cell growth and division are complex processes, and mathematical models of these processes are greatly simplified in order that the equations can be solved.

The growth rate for a number of cells or for biomass is postulated to be a function only of the mass of cells present:

$$\frac{dx}{dt} = f(x), \tag{1}$$

where x is cell concentration (cell mass per unit volume).

If $f(x) = \mu x$, then the exponential growth expression is obtained by making a cell mass balance for the entire volume of medium, V:

$$\frac{d(xV)}{dt} = \mu x V \rightarrow \frac{dx}{dt} = \mu x, \quad x(t_{\text{lag}}) = x_0. \tag{2}$$

When integrated with the initial condition indicated and with the specific growth rate μ held constant, the result is $x(t)$:

$$\ln\left(\frac{x}{x_0}\right) = \mu(t - t_{\text{lag}}), \tag{3}$$

or

$$x = x_0 \exp[\mu(t - t_{\text{lag}})]. \tag{4}$$

The concentration of nutrients and other species used for cell metabolism can be modeled in a similar way. For example, let the nutrient consumption rate be proportional to the cell concentration. A nutrient balance for the entire volume of medium in which the growth is taking place is as follows:

$$\frac{d(aV)}{dt} = -k_a x V, \quad a(t_{\text{lag}}) = a_0, \tag{5}$$

or

$$\frac{da}{dt} = -k_a x, \quad a(t_{\text{lag}}) = a_0. \tag{6}$$

By substituting Eq. (4) for $x(t)$ into Eq. (6) and integrating, we get

$$(a - a_0) = \frac{k_a}{\mu} x_0 (1 - \exp[\mu(t - t_{\text{lag}})]). \tag{7}$$

At the beginning of the stationary phase, all the nutrient is consumed and

$$x = x_s = x_0 \exp[\mu(t_s - t_{\text{lag}})]. \tag{8}$$

Then

$$0 - a_0 = \frac{k_a}{\mu} x_0 (1 - \exp[\mu(t - t_{\text{lag}})]) \tag{9}$$

$$-a_0 = \frac{k_a}{\mu} x_0 \left(1 - \frac{x_s}{x_0}\right), \tag{10}$$

or

$$x_s = x_0 + \frac{a_0 \mu}{k_a}. \tag{11}$$

This shows that the maximum cell concentration observed is proportional to the initial cell concentration and initial nutrient concentration. Data reported by Stanier et al. (1970) illustrate this.

Sometimes, inhibiting agents formed during the growth may slow the growth during the exponential growth phase. That is, the specific growth rate is reduced by the presence of the toxin. The mass balance for cells is then given by

$$\frac{dx}{dt} = kx[1 - f_t(c_t)], \tag{12}$$

where c_t is the concentration of the inhibiting agent.

If the inhibiting agent function f_t is proportional to the inhibiting agent concentration, then

$$\frac{dx}{dt} = kx(1 - bc_t). \tag{13}$$

Furthermore, if the inhibiting agent production rate is proportional to the cell concentration, we get

$$\frac{dc_t}{dt} = qx \quad \text{with} \quad c_t(t_{\text{lag}}) = 0 \tag{14}$$

$$c_t = q \int_{t_{\text{lag}}}^{t} x \, dt. \tag{15}$$

Thus

$$\frac{dx}{dt} = k\left(1 - bq \int_{t_{\text{lag}}}^{t} x \, dt\right) = \mu_{\text{eff}} x, \tag{16}$$

where the effective specific growth rate μ_{eff} is a function of time.

The growth stops when $dx/dt = 0$, and this occurs when $c_t = 1/b$. If growth stops in a batch culture because of inhibiting agent formation, diluting the medium with a nonnutrient will allow more growth since this reduces the inhibiting agent concentration. However, if growth stops due to nutrient depletion, dilution with a nonnutrient will not enable further growth. An equivalent criterion for zero growth is

$$1 - bq \int_{t_{lag}}^{t} x\, dt = 0, \tag{17}$$

or

$$\frac{1}{bq} = \int_{t_{lag}}^{t} x\, dt. \tag{18}$$

3.3. The Stationary Phase

The only modeling task here is to predict the duration of the phase. The cell concentration remains at the maximum value attained at the end of the exponential growth period.

3.4. The Death Phase

The death phase is usually not important for commercial processes. A batch process is terminated when the maximum cell concentration is attained. Modeling the death phase has some importance in analyzing sterilization methods.

Let the death rate be proportional to the cell concentration. A mass balance for cells for the entire volume of the medium gives the following equation:

$$\frac{dx}{dt} = -k_d x, \quad x(t_{stat}) = x_s, \tag{19}$$

where t_{stat} is the time when the stationary phase ends and the death phase commences. After integration, we get

$$\ln\left(\frac{x}{x_s}\right) = -k_d(t - t_{stat}), \tag{20}$$

or

$$x = x_s \exp\left[-k_d(t - t_{stat})\right]. \tag{21}$$

The above discussion applies to modeling each growth phase separately. It would be better if the entire growth period could be modeled. This can be done by modifying the function in Eq. (1) to include other biological factors. For example, let the inhibiting agent production rate depend on the cell growth rate as follows:

$$\frac{dc_t}{dt} = \alpha \frac{dx}{dt} \tag{22}$$

with

$$c_t(t_{\text{lag}}) = 0 \tag{23}$$

and

$$x(t_{\text{lag}}) = x_0. \tag{24}$$

Equation (22) can be integrated to give

$$c_t = \alpha(x - x_0). \tag{25}$$

By letting $x \gg x_0$, and substituting c_t from Eq. (25) into Eq. (13), we obtain

$$\frac{1}{x}\frac{dx}{dt} = k(1 - b\alpha x), \tag{26}$$

or

$$\frac{dx}{dt} = kx(1 - \beta x). \tag{27}$$

This is a nonlinear equation that has an inhibition factor proportional to x^2. If this equation is solved with the initial condition $x(0) = x_0$, then

$$x = \frac{x_0 \exp(kt)}{1 - \beta x_0 (1 - \exp(kt))} \tag{28}$$

This is the logistic curve, a sigmoidal curve that simulates a lag phase and that rises rapidly to a maximum value:

$$x_{\max} = \frac{1}{\beta}. \tag{29}$$

The parameters β and k may be evaluated from experimental data using a linear regression technique.

However, the logistic model does not predict a decline in cell concentration after all resources are exhausted. This decline can be included in the mathematical model by adding a memory integral to the differential equation. Suppose there is a component in the medium that has a negative influence on the growth whose concentration is given by

$$\frac{dc}{dt} = K(t)x(t), \qquad c(0) = 0. \tag{30}$$

Then

$$c = \int_0^t K(\tau)x(\tau)\,d\tau. \tag{31}$$

The logistic model can thus be modified by adding this integral term, which gives a Volterra equation:

$$\frac{dx}{dt} = kx(1 - \beta x) - \int_0^t K(\tau)x(\tau)\,d\tau. \tag{32}$$

A population decline is incorporated into the model by letting $K(t) > 0$. However, in this equation, the form of $K(t)$ cannot be determined easily. Frequency response experiments are usually required to obtain the parameters. These are often done easily for chemical reaction processes, but they are not easy to do for cell growth studies. We show later a modification of the logistic equation that is loosely based on the Volterra model, which models all the important phases of cell growth and which contains parameters that are easier to evaluate.

3.5. Growth of Filamentous Organisms

Some organisms grow in filamentous colonies, which have a changing morphology as the growth proceeds. The growth is usually slower than exponential growth. This idea is reviewed in Bailey and Ollis (1986c).

Suppose the rate of increase of the characteristic length of a colony is constant, and that the colony is spherical. Then

$$\frac{dR}{dt} = k_g, \tag{33}$$

where $R(t)$, the colony radius, is selected as the characteristic length, and k_g is the growth parameter for the colony radius. The biomass in a colony is then given by

$$M = \frac{4}{3}\pi R^3 \rho \quad \text{or} \quad R = \left(\frac{3M}{4\pi\rho}\right)^{1/3}. \tag{34}$$

Thus

$$\frac{dM}{dt} = 4\pi\rho R^2 k_g = [(36\pi\rho)^{1/3} k_g] M^{2/3}, \tag{35}$$

or

$$\frac{dM}{dt} = \zeta M^{2/3} \tag{36}$$

with initial biomass given by $M(0) = M_0$. This equation can be integrated to give

$$M = \left(\frac{\zeta}{3} t + M_0^{1/3}\right)^3. \tag{37}$$

This shows that the mass for a spherical colony increases according to t^3. This model is only for the growth of spherical colonies, and there is no information about colony formation.

The models above have very little biological information included. Growth is observed, and a mathematical function is devised to fit the form of $x(t)$. Some reality is introduced by arranging that the model function is related to mass balances for the organisms in the medium in which they are growing and to some observed principles of population dynamics. Using this procedure, any function or differential equation can be analyzed for suitability in modeling the growth of biomass in a batch culture. Computer methods make it quite easy to evaluate the model parameters. Some of the models described above cover a single growth phase and others cover more than one phase. Modeling the entire growth history requires using several parameters whether the modeling is done piecewise or by using a single equation. Using a single equation may give a smoother fit to the growth data from which the model parameters are evaluated. In the section that follows, a model that can be used to describe all growth phases is developed.

4. MODELS DEVELOPED FROM MODIFIED LOGISTIC EQUATIONS

As mentioned earlier, an important model that consists of a lag and a stationary phase is the logistic curve, described by

$$\frac{dx}{dt} = k_b x(1 - \beta x), \tag{38}$$

where k_b and β are constants and $x(0) > 0$. As shown in Fig. 4.2, the stationary phase is achieved when x approaches the value $1/\beta$. However, the logistic model

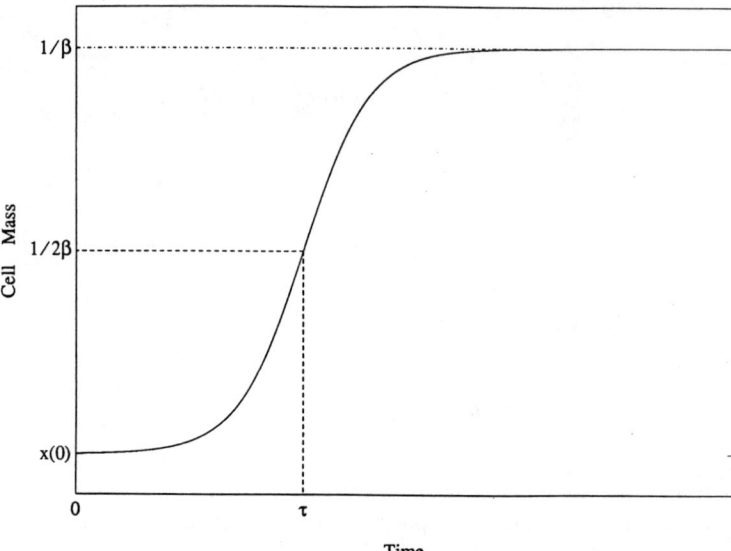

FIGURE 4.2. The logistic equation: $dx/dt = k_b x(1 - \beta x)$. The inflection point occurs at $t = (1/k) \ln ([1 - \beta x_0]/[\beta x_0])$.

fails to describe the death phase. Furthermore, the maximum rate of growth always occurs at $x = 1/2\beta$. Additional modifications are needed to overcome these limitations. One improvement is to introduce different exponential powers into the logistic equation to change the inflection point, that is,

$$\frac{dx}{dt} = k_b x(1 - \beta x)^b. \tag{39}$$

From Eq. (39), the maximum rate of growth occurs at

$$x = \frac{1}{\beta} \frac{1}{1 + b}. \tag{40}$$

The parameter b can be altered to move the inflection point up or down, so that it occurs somewhere between $x = x(0)$ and $x = 1/\beta$, as shown in Fig. 4.3.

Another important modification is to include an additional term to represent the exponential death phase, which in turn depends on two factors: a growth inhibition factor, $C(x, t)$, and a logistic response:

$$\text{death rate} = [C(x, t)] \cdot [x(1 - \beta x)^p]. \tag{41}$$

4. MODELS DEVELOPED FROM MODIFIED LOGISTIC EQUATIONS

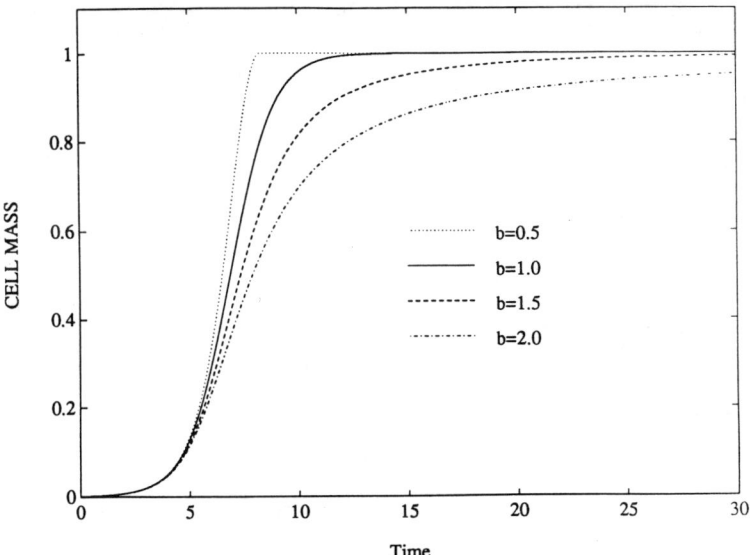

FIGURE 4.3. The logistic equation modified to allow different exponent powers b in the equation $dx/dt = x(1 - x)^b$.

The additional growth inhibition factor, $C(x, t)$, can be modeled to be proportional to the number of cells accumulated; that is,

$$C(x, t) = k_d \int_0^t x(\tau) \, d\tau, \tag{42}$$

where k_d is a proportionality constant.

Lastly, we note that the initial cell concentration does not necessarily start at zero. Also, when using other measurements of cell growth such as cell protein, an equilibrium concentration may be present that is also not necessarily zero. Thus two more modifications are needed to represent this information. The resulting modified logistic equation will need two deviation variables: the first is $(x - x_0)$, which is the deviation with respect to the initial condition, and the second is $(x - x_e)$, which is the deviation with respect to the equilibrium concentration. After combining all these modifications, we obtain

$$\frac{dx}{dt} = k_b(x - x_0)(1 - \beta(x - x_0))^b$$

$$- k_d(x - x_e)(1 - \gamma(x - x_e))^p \int_0^t x(\tau) \, d\tau. \tag{43}$$

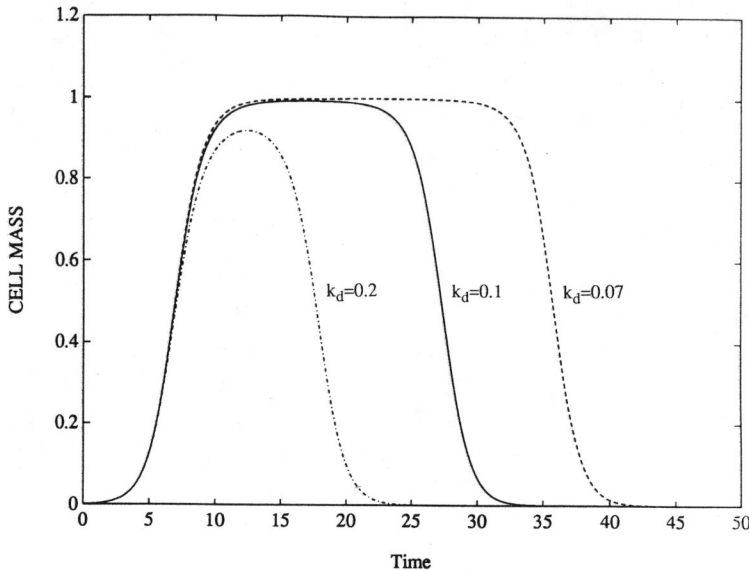

FIGURE 4.4. The modified logistic equation including the death phase: $dx/dt = x(1 - x) - k_d x(x - 1) \int_0^t x(\tau)\, d\tau$.

For the case where the exponential powers are all unity, that is, $b = d = 1$, constants $k_b = \beta = 1$ and terminal conditions $x_0 = x_e = 0$, Fig. 4.4 shows the effects of varying k_d.

In summary, the logistic equation (38) can be modified to described the four main phases of cell growth dynamics: the lag phase, the exponential growth phase, the stationary phase, and the death phase. The important changes include only adding a death rate term and allowing different exponential powers in the original logistic growth rates. The price of increasing model fidelity for cell growth is needed to the identify more parameters. These are:

- k_b Proportionality constant for growth phase
- k_d Proportionality constant for death phase
- β Reciprocal of asymptotic cell population in growth phase
- γ Reciprocal of asymptotic cell population in death phase
- b Exponent power (or order) in growth phase
- p Exponent power (or order) in death phase
- x_0 Initial condition
- x_e Final condition

These parameters can be obtained directly from experimental data using a method for nonlinear model identification known as the modulating function method. This is summarized in the appendix.

5. MODEL VERIFICATION

Several different organisms produce cellulase and metabolize cellulose. Some data are available in the literature that can be used to test the model given in Eq. (43). Biological growth data are sparse compared to data that are often available for modeling a batch chemical reactor. Reasons for this include the fact that fermentation experiments usually take much longer than chemical reaction experiments, and chemical reaction data can usually be obtained by continuous automated monitoring. Assaying biomass during the course of a fermentation is more time consuming than determining chemical composition, and in turbid cellulose-containing broths, automated biomass determination using optical density or spectrophotometric methods cannot be used.

The method of modulating functions has been studied in connection with stirred tank reactors by Co and Ydstie (1990). Some details of the application of this method are given in the appendix. Special methods are developed here for handling sparse data sets available from fermentation experiments. The modulating function method can be used, but the lack of numerous data introduces some extra uncertainty. The method has been applied to three different microorganisms growing on different forms of cellulose. For all the experiments studied, the data do not cover very much of the death phase, so that part of the model is not very certain.

One important consideration for determining model parameters is computational efficacy and efficiency. Even though using different values of the exponents b and p would give a more accurate description of cell growth, evaluation of the exponents involves elaborate nonlinear optimization techniques, and the cost in computation time is increased tremendously. For the modeling results presented here, both b and p have been set equal to unity and simple optimization schemes have been used. Even with this simplification, the method still represents the growth data closely.

5.1. *Trichoderma viride (Trichoderma reesei)*

Peitersen (1975) measured the growth of the fungus *Trichoderma viride* on treated barley straw in a 5-liter batch fermenter. After a lag phase of 0–2 days, growth continued for 2–6 days with up to 70% of the straw consumed. After the maximum biomass concentration (measured by cell protein) was attained, the biomass concentration decreased and approached an equilibrium value, x_e. Figures 4.5, 4.6, 4.7, and 4.8 show the growth data reported by Peitersen along with a line representing the model equation, which is the solution to Eq. (43) with the initial condition that $x = x_0$ at $t = 0$. The model parameters determined for each case are given in Table 4.1.

In Table 4.2, the values of the maximum biomass concentration x_m, calculated using the model parameters, are compared with the experimentally measured value. This gives a test of the accuracy of the modeling method developed.

104 MODELING THE GROWTH OF CELLULASE-PRODUCING ORGANISMS

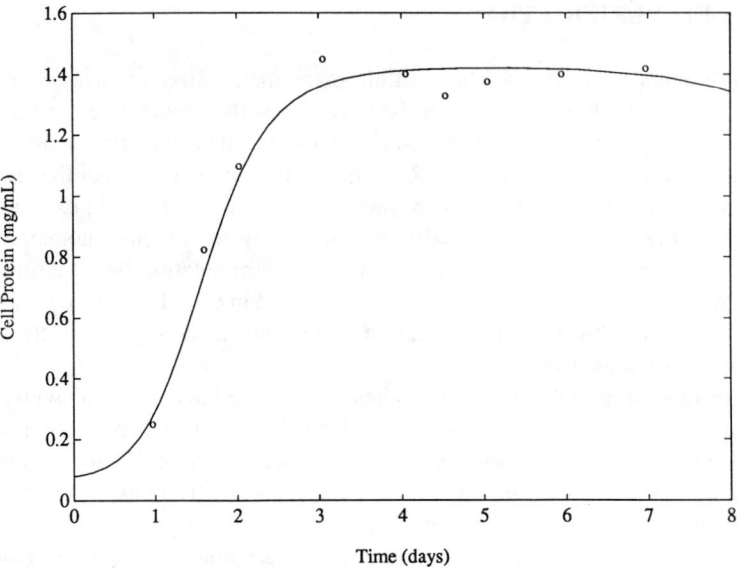

FIGURE 4.5. Growth of *T. viride* on 1% washed straw (agitator speed = 200 min^{-1}). Symbols shown represent actual data points, while the lines represent model functions.

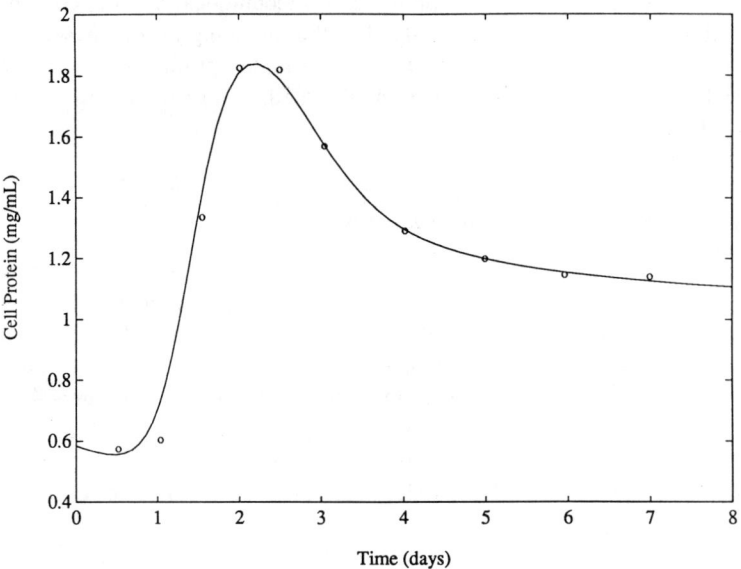

FIGURE 4.6. Growth of *T. viride* on 1% washed straw (agitator speed = 350 min^{-1}). Symbols shown represent actual data points, while the lines represent model functions.

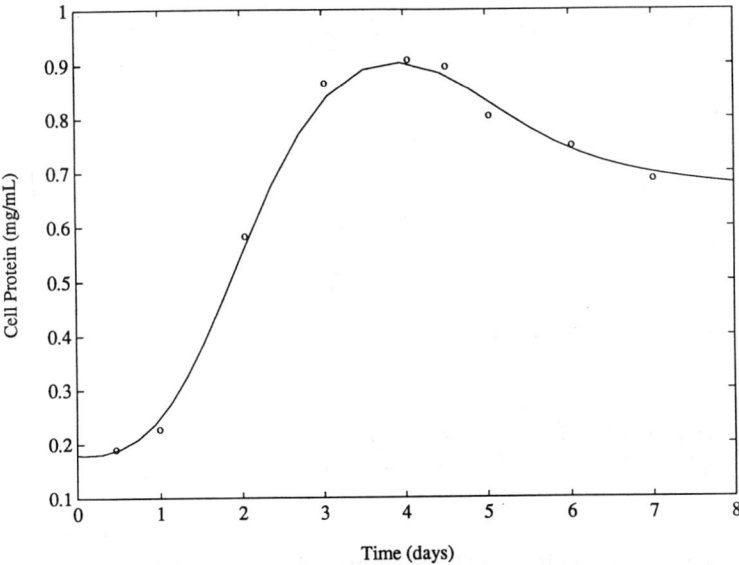

FIGURE 4.7. Growth of *T. viride* on 1% NaOH straw (agitator speed = 300 min^{-1}). Symbols shown represent actual data points, while the lines represent model functions.

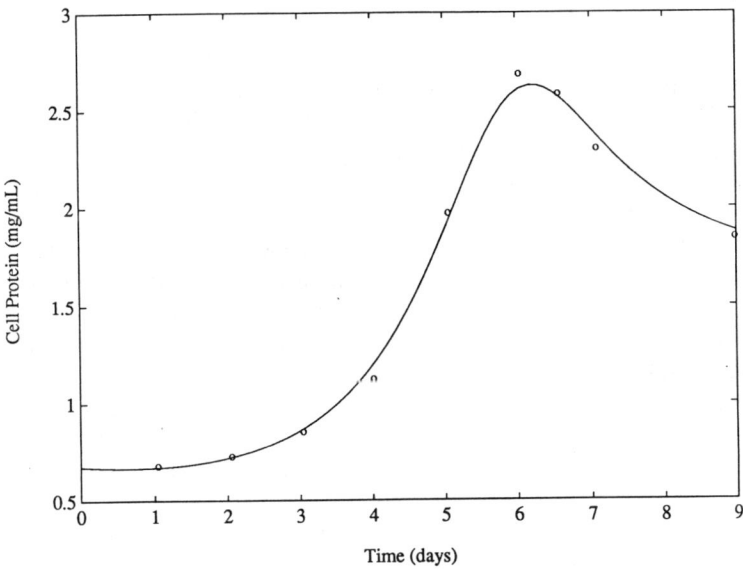

FIGURE 4.8. Growth of *T. viride* on 2% NaOH straw (agitator speed = 300 min^{-1}). Symbols shown represent actual data points, while the lines represent model functions.

TABLE 4.1. Model Parameters for *T. viride* Growing on Barley Straw

Figure number	5	6	7	8
Straw treatment[a]	Washed	Washed	NaOH	NaOH
Straw concentration	1%	1%	1%	2%
Impeller speed (min^{-1})	200	350	300	300
Aeration rate (V/V/M)	0.7	0.7	0.7	0.7
k_b	2.339	4.886	1.385	−0.468
x_0 (mg/mL)	0.0656	0.619	0.191	0.159
β	0.731	0.597	0.909	1.396
k_d	0.420	1.69	1.022	−0.199
x_e (mg/mL)	0.948	0.997	0.595	0.0693
γ	2.048	0.581	1.519	0.723

[a]NaOH means straw treated with 5.7% aqueous NaOH (2-6 kg NaOH/kg straw) and pressed into cakes. Washed means NaOH straw leached with water for 15 min at 115°C and vacuum-dried at 40°C. In either case, the straw was milled to 40 mesh and dried for 24 hours at 105°C.

TABLE 4.2. Maximum Biomass Concentration from *T. viride* Fermentation on Barley Straw

Figure number	5	6	7	8
x_{max} (calc)	1.44	2.11	0.74	4.64
x_{max} (data)	1.48	1.85	0.91	2.15
Deviation (%)	3.7	14.1	18.7	116

The results from Figs. 4.5, 4.6, and 4.7 illustrate the accuracy of the modeling method when applied to sparse data sets. The characteristics of the data—x_{max}, x_0, and x_e—are reproduced with an uncertainty of from ±4% to ±19%. The results from Fig. 4.8 show some of the problems that can occur in using the model. If the death phase begins immediately at the time the maximum biomass concentration is reached, no stationary phase develops, and the maximum in the growth function is sharp. The model is developed by adding two logistic equations, and the two parts may interact strongly if there is no detectable lag phase. The six model parameters are determined in the fitting procedure, and the parameter values are influenced by the way in which the two parts of the model interact. A stringent test of the effect of this interaction is to check whether or not the maximum biomass concentration is reproduced by the model. Table 4.2 shows that for the data given in Fig. 4.8, x_{max} = 2.15 mg/mL, but the model equation predicts that x_{max} = 4.64 mg/mL. This indicates that the model parameter values are uncertain in that particular case. To fit the sharp maximum in the growth data, the fitting procedure produced negative values for k_b and k_d. This is a further indication that the model does not work for this particular data set. The model fails if there is no appreciable stationary phase before the death phase begins, and some modifications of the model will be needed in such instances.

5.2. Acidothermus cellulolyticus

Mohagheghi et al. (1986) reported growth data for aerobic, thermophilic bacteria isolated from acidic hot springs in Yellowstone National Park in Wyoming. Figure 4.9 shows the growth data and the model function that is the solution to Eq. (43) with the following model parameter values: $k_b = 0.0622$, $k_d = -5.21 \times 10^{-5}$, $x_0 = 33.9$, $x_e = -1.065 \times 10^4$, $\beta = 8.03 \times 10^{-3}$, and $\gamma = 9.13 \times 10^{-5}$. The parameters k_b and k_d are negative here. This indicates that the model is not truly applicable here, probably because the biomass concentration begins to decrease just after the maximum value was attained. The parameter x_e has a large negative value, which is a result of the sparseness of the data in the death phase. If there are too few data for biomass concentration in the death phase, then the value of x_e found may not be physically meaningful.

If the modified logistic equation model is not suitable for representing a particular set of growth data, the model for filamentous organisms, Eq. (37), may be used. Standard linear regression analysis techniques can be used to fit a straight line to a set of data if Eq. (37) is rewritten as

$$x^{1/3} = \frac{\zeta}{3}(t - t_{\text{lag}}) + x_0^{1/3}. \tag{44}$$

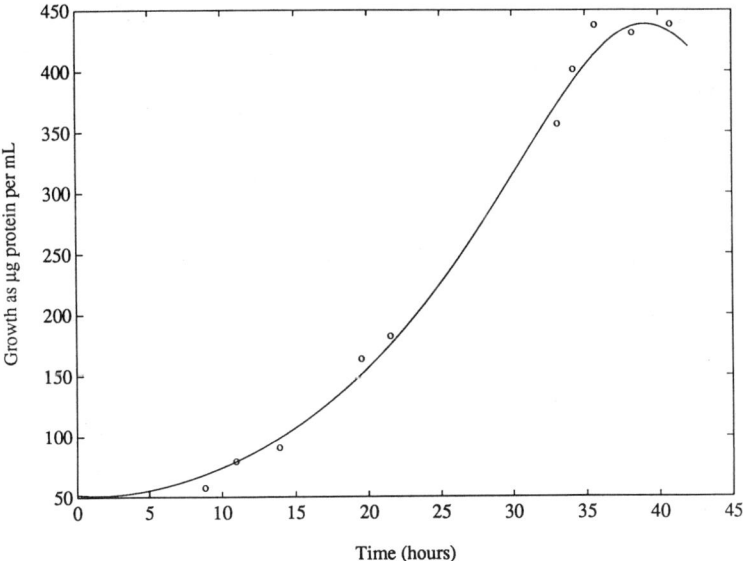

FIGURE 4.9. Growth of *A. cellulolyticus* at 55°C and pH = 5.0: comparison with the logistic model. Symbols shown represent actual data points, while the lines represent model functions.

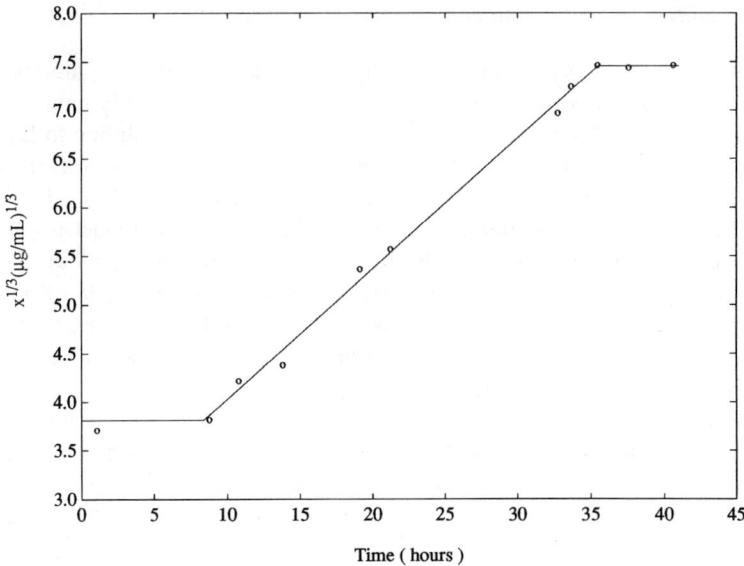

FIGURE 4.10. Growth of *A. cellulolyticus* at 55°C and pH = 5.0: comparison with filament growth model. Symbols shown represent actual data points, while the lines represent model functions.

The duration of the lag phase, t_{lag}, is introduced, because the model only applies during the rapid growth period. This allows Eq. (44) to give a physically meaningful value for the initial biomass concentration x_0. When applied to the data of Mohagheghi et al. (Fig. 4.9), the model parameters in Eq. (44) are t_{lag} = 8.4 h, x_0 = 55.36 μg/mL, and ζ = 0.40 (μg/mL)$^{1/3}$/h.

The data and Eq. (44) are plotted in Fig. 4.10. Plotting the data this way makes it easy to identify the different stationary phases and the period of rapid growth. Only the data for the rapid growth period have been used to evaluate t_{lag}, ζ, and x_0. The model fits the data in this range with a correlation coefficient of 0.998. The cubic dependence on time in this type of model is based on the formation of pellets during the fermentation, but the evaluation of the model parameters is purely mathematical.

5.3. *Thermomonospora fusca*

Crawford et al. (1973) studied the growth of *Thermomonospora fusca* using pulp mill fines as the source of carbon. The growth data for different concentrations of fines from 0.25 to 1.0% on a dry basis are shown in Fig. 4.11 along with solid lines representing the modified logistic model—the solution to Eq. (43). The cell biomass is represented as percent nitrogen determined by the Kjeldahl method, which was indicative of the amino acid content of the product. The cell protein

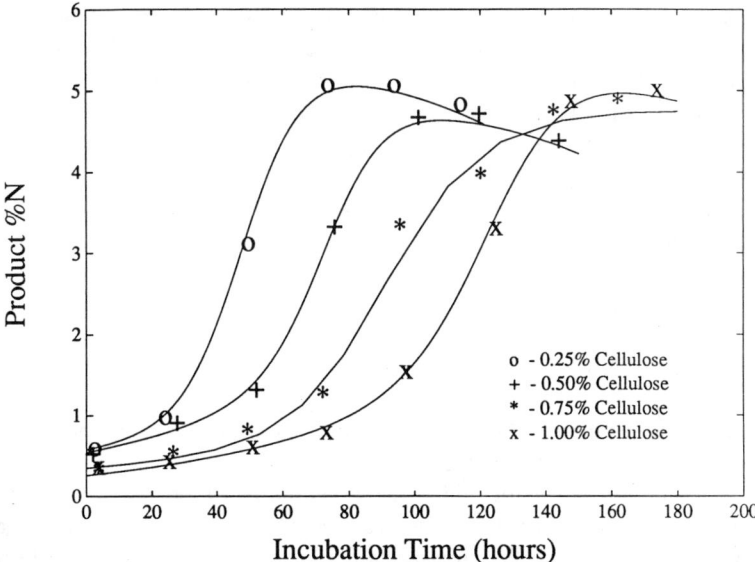

FIGURE 4.11. Growth of *T. fusca* on paper pulp fines: modeling the effect of fines concentration. Symbols shown represent actual data points, while the lines represent model functions.

was $6.25 \times \%$ N. The model parameters determined from these data are given in Table 4.3.

Only the model parameters for the growth phase are reported in Table 4.3. There are not enough data available for the death phase to evaluate the other three parameters, so only the growth half of the modified logistic model has been used. The values of the parameters k_d, x_e, and γ—particularly x_e—are expected to be physically unrealistic because of the few data on which to base the function fitting procedure. The numerical value of the maximum biomass concentration is predicted with an accuracy of from ± 1 to $\pm 11\%$. This is indicated in the last three lines in Table 4.3, where the predicted values of x_{max} are compared with the experimental values reported by Crawford et al. The model accurately represents the

TABLE 4.3. Model Parameters for *T. fusca* Growing on Pulp Mill Fines

Fines (mass %)	0.25	0.50	0.75	1.00
k_b	0.0246	0.0131	0.0926	0.0916
x_0	0.510	0.481	0.314	0.214
β	0.199	0.219	0.220	0.182
x_{max} (calc) $= 1/\beta$	5.02	4.56	4.54	5.49
x_{max} (data)	5.07	4.64	4.78	4.95
Deviation (%)	1.0	1.7	2.9	10.9

lag phase in the growth of *T. fusca*, which depends on the fines concentration. This dependence is attributed to the presence of sulfite in the pulp mill fines. At the growth temperature of 55°C, the sulfite is removed or altered in such a way that the concentration is reduced to a noninhibiting level. Then growth can begin. The batch volume for each fermentation experiment was the same, so higher concentrations of fines would produce higher initial concentrations of sulfite. More time would be required to reduce the higher concentration to the noninhibiting level at which growth could begin, so the lag phase increases with increasing fines concentration.

6. CONCLUSIONS

The modified logistic model developed by combining two logistic functions works well for describing growth data that include lag phases and stationary phases. Exceptions to this are cases for which the maximum in biomass concentration is sharp. If few data are available for the death phase, the standard logistic model or the cubic polynomial model is appropriate. In order to apply the complete model with nonunity values for the exponents b and p in Eq. (43), a more efficient, less time-consuming optimum seeking technique is needed for evaluating the model parameters.

NOMENCLATURE USED IN EQUATIONS

a	Nutrient concentration
a_0	Initial nutrient concentration
b	Exponent for growth phase logistic model or model constant
c	Growth inhibiting component used to develop the Volterra equation
c_t	Concentration of inhibiting substance
$f_t(c_t)$	Function for the effect inhibiting agent on growth rate
k	Model parameter (appears with different subscripts)
k_a	Model parameter for nutrient depletion rate
k_g	Rate of increase for colony radius
k_b	Proportionality constant for growth phase in modified logistic equation
k_d	Proportionality constant for death phase in modified logistic equation
$K(t)$	Memory function in Voterra equation
M	Mass of filamentous colony
M_0	Initial mass of filamentous colony
p	Exponent for death phase in logistic model
q	Model parameter
R	Radius of filamentous colony
t	Time
T	Uniform sampling time used in the experiment

t_{lag}	Time to the end of the lag phase
t_s	Time at the start of the stationary phase
t_{stat}	Time at the end of the stationary phase
u	Matrix of inputs used in modulating function method
V	Volume of medium
x	Biomass concentration
x_e	Biomass concentration after long times
x_0	Initial biomass concentration
x_{max}	Maximum biomass concentration
x_s	Biomass concentration at the start of the stationary phase
y	Time series of output data used in modulating function method

Greek Symbols

β	Model parameter (logistic model)
γ	Model parameter (logistic model)
ζ	Model constant (cubic polynomial model)
μ	Specific growth rate
μ_{eff}	Effective specific growth rate
ρ	Density
τ	Dummy integration variable for time
θ	Model parameter/coefficient for nonlinear models

APPENDIX: THE MODULATING FUNCTION METHOD

The parameters of the modified logistic equation can easily be identified using the modulating function method. The main feature of this method is the ability to use data directly. For this specific identification of cell growth using Eq. (43), the following procedure is needed:

1. Time series data of the dynamic variable $x(t)$ (cell protein, viable cell mass, etc.) are collected using uniform sampling times.
2. The accumulation term $y(t)$ has to be evaluated by integrating $x(t)$ over time,

$$y(kT) = \int_0^{kT} x(\tau) \, d\tau \qquad (45)$$

via quadrature (e.g., trapezoidal rule or Simpson's rule).
3. The matrix of inputs thus becomes

$$u = [x, x^2, y, xy, x^2 y] \qquad (46)$$

and the modulating function method as outlined in Co and Ydstie (1990) can then be used directly to identify a set of parameters, θ, appearing in the equation below:

$$\frac{dx}{dt} = \theta_1 + \theta_2 x + \theta_3 x^2 + \theta_4 y + \theta_5 xy + \theta_6 x^2 y. \tag{47}$$

These parameters can be further combined to obtain the primary model parameters given in Eq. (43),

$$\frac{dx}{dt} = k_b(x - x_0)(1 - \beta(x - x_0)) - k_d y(x - x_e)(1 - \gamma(x - x_e)), \tag{48}$$

using the following relationships:

$$\beta = \frac{|\theta_3|}{\sqrt{\theta_2^2 - 4\theta_1\theta_3}}, \tag{49}$$

$$k_b = -\frac{\theta_3}{\beta}, \tag{50}$$

$$x_0 = -\frac{1}{2\beta}\left(1 - \frac{\beta\theta_2}{\theta_3}\right), \tag{51}$$

$$\gamma = \frac{|\theta_6|}{\sqrt{\theta_5^2 - 4\theta_4\theta_6}}, \tag{52}$$

$$k_d = -\frac{\theta_6}{\gamma}, \tag{53}$$

$$x_e = -\frac{1}{2\gamma}\left(1 - \frac{\gamma\theta_5}{\theta_6}\right). \tag{54}$$

The conditions for the relationship of both parameter sets to be valid are $\theta_2^2 > 4\theta_1\theta_3$, $\theta_3 < 0$, $\theta_5^2 > 4\theta_4\theta_6$, and $\theta_6 < 0$. Otherwise, complex values for k_b, β, γ, and so on will result.

Finally, note that the method requires uniform sampling time. However, due to long and careful sampling involved in cell growth studies, data tend to be sparse and nonuniform. This notwithstanding, the method can still be applied by using piecewise linear interpolation of values between samples. The result is a jagged time path that is later smoothed by the filtering that occurs during the use of the modulating function method.

REFERENCES

Andren, R. K., Mandels, M. H., and Medieros, J. E. (1975), in *Applied Polymer Symposia No. 28* (T. E. Timell, Ed.), Wiley, New York, pp. 205–219.

Bailey, J. E., and Ollis, D. F. (1986a), *Biochemical Engineering Fundamentals*, 2nd ed., McGraw-Hill, New York, pp. 408–421.

Bailey, J. E., and Ollis, D. F. (1986b), *Biochemical Engineering Fundamentals*, 2nd ed., McGraw-Hill, New York, pp. 394–408.

Bailey, J. E., and Ollis, D. F. (1986c), *Biochemical Engineering Fundamentals*, 2nd ed., McGraw-Hill, New York, pp. 405–406.

Bajpai, R. K., and Ghose, T. K. (1978), *Biotechnol. Bioeng.*, **20,** 927–935.

Callihan, C. D., Irwin, G. H., Clemmer, J. E., and Hargrove, O. W. (1975), in *Applied Polymer Symposia No. 28* (T. E. Timell, Ed.), Wiley, New York, pp. 189–196.

Co, T. B., and Ydstie, B. E. (1990), *Computers Chem. Eng.*, **14,** 1051–1066.

Crawford, D. L., McCoy, E., Harkin, J. M., and Jones, P. (1973), *Biotechnol. Bioeng.*, **15,** 833–843.

Eriksson, K.-E., and Larsson, K. (1975), *Biotechnol. Bioeng.*, **17,** 327–348.

Grethlein, H. E. (1978), *Biotechnol. Bioeng.*, **20,** 503–525.

Howell, J. A. (1978), *Biotechnol. Bioeng.*, **20,** 847–863.

Howell, J. A., and Stuck, J. D. (1975), *Biotechnol. Bioeng.*, **17,** 873–893.

Mitra, G., and Wilke, C. R. (1975), *Biotechnol Bioeng.*, **17,** 1–13.

Mohagheghi, A., Grohmann, K., Himmel, M., Leighton, L., and Updegraff, D. M. (1986), *Int. J. Syst. Bacteriol.*, **36,** 435–443.

Okazaki, M. and Moo-Young, M. (1978), *Biotechnol. Bioeng.*, **20,** 637–663.

Pamment, N. B., Hall, R. J., and Barford, J. P. (1978), *Biotechnol. Bioeng.*, **20,** 349–381.

Peitersen, N. (1975), *Biotechnol. Bioeng.*, **17,** 361–374.

Peitersen, N. (1977), *Biotechnol. Bioeng.*, **19,** 337–348.

Stanier, R. Y., Douderoff, M., and Adelberg, E. A. (1970), *The Microbial World*, 3rd ed., Prentice-Hall, Englewood Cliffs, NJ, p. 313.

Vallander, L., and Eriksson, K.-E. L. (1990), in *Advances in Biochemical Engineering/Biotechnology* (A. Fiechter, Ed.), Vol. 42, Springer-Verlag, Berlin, pp. 63–95.

Wilke, C. R., and Yang, R. D. (1975), in *Applied Polymer Symposia No. 28* (T. E. Timell, Ed.), Wiley, New York, pp. 175–188.

5 Kinetic Model of Antibiotic Production by *Streptomyces clavuligerus*

P. K. NAMDEV and M. R. GRAY
Department of Chemical Engineering, University of Alberta

D. W. S. WESTLAKE
Department of Microbiology, University of Alberta

1. Introduction 115
2. Biosynthesis of cephamycin C 116
3. Regulation of antibiotic synthesis 118
   3.1. Regulation of amino acid precursors 119
   3.2. Control of enzyme expression 120
   3.3. Inhibition of biosynthetic enzymes.. 121
4. Effect of dissolved oxygen levels... 123
5. Framework for the kinetic model... 124
6. Kinetics of transcription and inactivation of enzymes........................... 125
7. Regulation of enzyme transcription 132
8. Kinetics of cephamycin C synthesis... 135
9. Kinetics of ACV formation and conversion... 137
10. Predictions of the kinetic model 139
    10.1. Addition of carbon source to resting cells.................................... 139
    10.2. Prediction of cephamycin C formation at 100% DO 141
    10.3. Enhancement of enzyme activity by prolonged transcription 142
    10.4. Optimization of fermentation conditions 143
    10.5. Modification of cellular regulation 143
11. Conclusion 145
    References... 146

1. INTRODUCTION

Although the commercial production of antibiotics by *Streptomyces* spp. accounts for a significant portion of the total amount of antibiotics produced, details of the control and regulation of the antibiotic production pathways are just beginning to appear in the literature. Possibly the best understood pathway is for cephamycin C biosynthesis by *Streptomyces clavuligerus*, as this organism has been widely used for enzymatic and molecular biological investigations of factors controlling this pathway. The characteristics of the enzymes common to this pathway in streptomycetes and fungi, involved in conversion of the tripeptide δ-(L-α-aminoadipyl)-L-cysteine-D-valine (ACV) to desacetylcephalosporin C, are reviewed by Wolfe et al. (1984). The expression of these enzymes in *S. clavuligerus* has been measured in response to variations in medium composition, including phosphate level (Lebrihi et al., 1987; Lubbe et al., 1985a), nitrogen source (Brana et al., 1985; Castro et al., 1985; Vining et al., 1987), carbon source (Cortes et al., 1986; Lebrihi et al., 1988a; Vining et al., 1987), and oxygen levels (Rollins et al., 1989a, 1990, 1991).

Several kinetic models have been proposed for the production of antibiotics by fermentation, for example, penicillin (Bajpai and Reuss, 1980; Nestass and Wang, 1983) and cephalosporin C (Matsumura et al., 1981). The growth of mycelia has been modeled using Monod kinetics (Bajpai and Reuss, 1980), or a structured approach to allow for differentiation of fungi into swollen hyphae and arthrospores (Matsumura et al., 1981). The latter model for production of cephalosporin C included an equation for repression of the pool of biosynthetic enzymes by methionine, but no data for enzyme activities were reported. None of these models linked the kinetics and regulation of enzyme expression to the production of antibiotic.

The objective of this chapter is to use the information made available by the recent studies on the effects of environmental factors to model the biosynthetic pathway for cephamycin C in *S. clavuligerus*, focusing on the rates of enzyme transcription and inactivation, and the regulation of synthesis of those enzymes that are part of the common biosynthetic pathway. Although gaps remain in the details of the location and control of the genes for the biosynthetic pathway, a mathematical model for the biosynthesis of cephamycin C serves several important functions. A model provides a quantitative framework for interpreting the results of the various experimental studies, it serves to test hypotheses regarding the biosynthetic pathway, it identifies gaps in our understanding, and it makes predictions. The latter two results can serve as the basis for future experimental work.

2. BIOSYNTHESIS OF CEPHAMYCIN C

The common enzymatic reactions in streptomyces and fungi involved in the biosynthesis of penicillins and cephalosporins are shown in Fig. 5.1. The enzymes of interest relative to this chapter are ACV synthetase (ACVS), isopenicillin N synthase (IPNS), and desacetoxycephalosporin C synthase (DAOCS), as they have been the subject of many investigations as to stability and expression. The IPNS and DAOCS complex has been cloned. ACVS has a requirement for ATP, and the

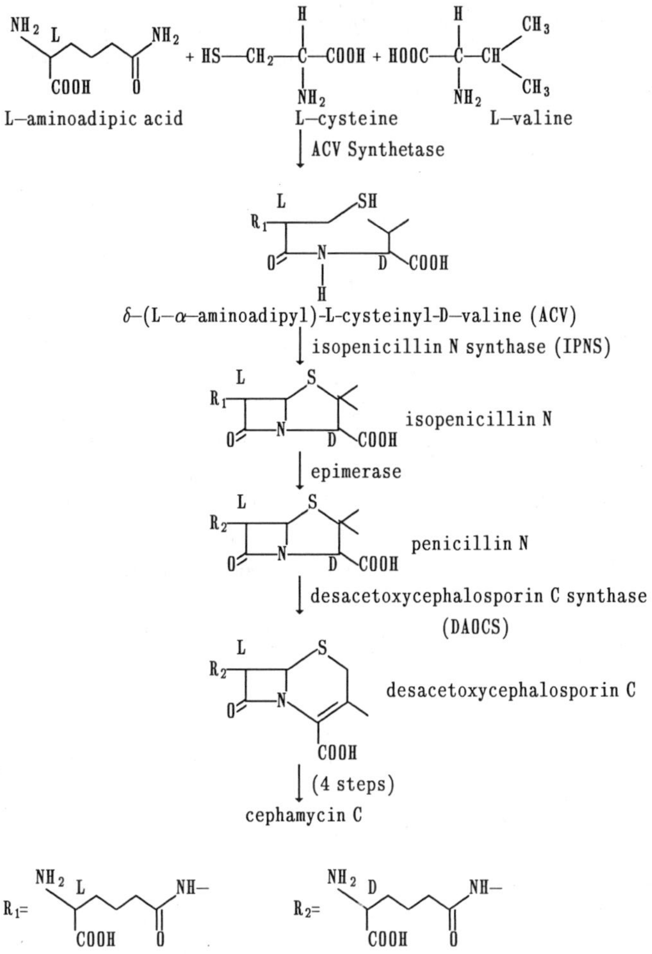

FIGURE 5.1. Biosynthetic pathway to penicillins and cephalosporins in *S. clavuligerus*.

oxygenase enzymes IPNS and DAOCS require ascorbate, ferrous ions, and oxygen, with DAOCS having an additional requirement for α-ketoglutarate (Jensen, 1986; Jensen et al., 1982a,b; Turner et al., 1978). There is, however, insufficient information in the literature on factors affecting levels of the last enzyme in the common pathway, desacetoxycephalospirin C hydroxylase (DACS), to include it in this modeling study. This lack of information is a result of the fact that most of the work on DAOCS has been carried out in fungi, which combine DAOCS and DACS activity in a single enzyme. While these two enzymes are separable in streptomyces, the assay is not readily adapted to physiological studies. Information in the literature indicates that the inhibition or repression of IPNS or DAOCS enzymes results in a buildup of ACV (Rollins et al., 1991) and penicillin N, respectively (Rollins et al., 1988a, 1990). The activity of these enzymes is lost during fermentation experiments and they are inactivated during catalytic turnover *in vitro* (Jensen et al., 1984), although in the case of DAOCS such inactivation may be reversible (Lubbe et al., 1985b).

3. REGULATION OF ANTIBIOTIC SYNTHESIS

When *S. clavuligerus* is grown in batch culture, cephamycin C formation begins in the late growth phase and continues through the stationary phase (Aharonowitz and Demain, 1978, 1979). Yield of cephamycin is reduced or eliminated if the growth medium contains high concentrations of readily metabolized carbon source, such as maltose (Aharonowitz and Demain, 1978), nitrogen source, such as ammonium (Aharonowitz and Demain, 1979; Lubbe et al., 1985c), or phosphate (Lubbe et al., 1985a). That is, cephamycin C formation is subject to carbon, nitrogen, and phosphate catabolite repression. Higher yields are observed when complex media are used, which serve to control the rate of release of utilizable carbon and nitrogen into the medium (Aharonowitz and Demain, 1978; Vining et al., 1987). These characteristics are similar to those found in other antibiotic-producing microorganisms such as *Penicillium chrysogenum* and *Cephalosporium acremonium*. *S. clavuligerus* produces cephamycin C when grown in chemostat culture under either nitrogen or carbon limited growth only at low dilution (i.e., growth) rates (Lebrihi et al., 1988b; Vining et al., 1987). That is, in either batch or chemostat culture optimal antibiotic production is only obtained under conditions where carbon and nitrogen catabolite repression does not occur.

During a batch fermentation carried out under very carefully controlled conditions, the appearance of biosynthetic enzymes occurs at similar times. The expression of the enzymes begins during exponential growth, apparently lagging behind the rapid increase in biomass but preceding any antibiotic formation (Fig. 5.2). Hence the enzyme system is in place when a nutrient becomes limiting and the pathway is switched on. Production of antibiotics in chemostat experiments with carbon, nitrogen, or phosphate sources as limiting nutrients (Lebrihi et al., 1988b; Vining et al., 1987) supports the conclusion that production of antibiotic is switched on under nutrient-limited conditions. During the stationary phase the activity pathway decays, so that production of antibiotics ceases at approximately the same time that the activity of the pathway enzymes is lost.

Regulation of antibiotic synthesis in actinomycetes can occur via three possible mechanisms (Vining and Doull, 1989):

1. *Availability of Precursors for the Biosynthetic Pathway*. This mode of control is likely operative during exponential growth when the enzymes have formed but production has not been initiated.
2. *Repression of Biosynthetic Enzymes*. The production of biosynthetic enzymes is regulated by catabolite levels.

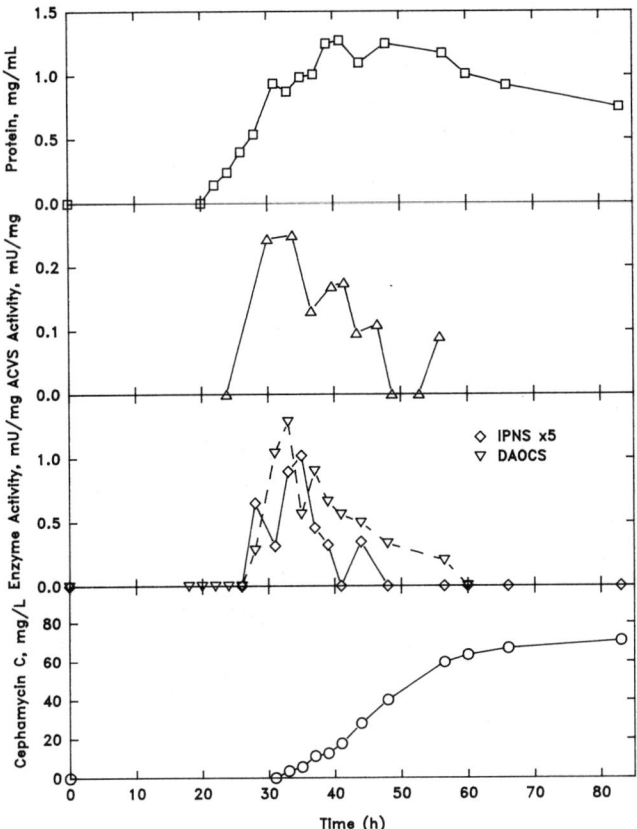

FIGURE 5.2. Time course of antibiotic fermentation. [Data for protein, IPNS, DAOCS, and cephamycin C are from Rollins et al. (1990); data for ACVS are from Rollins et al. (1991).]

3. *Inhibition of Biosynthetic Enzymes.* High concentrations of intermediates from the glycolytic pathways could inhibit the enzymes *in vivo*.

Each mechanism will be considered in turn.

3.1. Regulation of Amino Acid Precursors

Of the three amino acid precursors for the cephalosporin pathway, α-aminoadipic acid is unique in that it is derived from lysine. In prokaryotic cells this synthesis occurs via the diaminopimelic acid pathway from aspartic acid. As such, its synthesis has been considered as being a rate-limiting reaction controlling the amount of cephamycin C synthesized. Mendelovitz and Aharonowitz (1982) found that diaminopimelic acid and lysine stimulated antibiotic formation in *S. clavuligerus*. Aharonowitz et al. (1984) studied a regulatory mutant strain with an altered aspartokinase (the first enzyme of the common aspartate pathway), which was insensitive to feedback regulation. The mutant strain gave a 50% higher level of free intracellular amino acids than the wild type. In both strains the amino acid pool reached its maximal level during the first 12 h of growth, about 200 nmol/mg dry cell weight (DCW), during the lag time prior to rapid growth. The largest single difference between the mutant and the wild type was the intracellular concentration of diaminopimelic acid, which accounted for 0.5% of the amino acid pool in the wild type and 14.9% in the mutant after 24 h of growth. The mutant produced five times as much cephamycin C as the wild type.

Madduri et al. (1989) reported that lysine ϵ-aminotransferase (LAT) activity, the initial enzyme for converting lysine to α-aminoadipic acid, only occurred in actinomycetes that produced β-lactam antibiotics. Recently, Madduri et al. (1991a) reported that LAT was synthesized preferentially during early growth, and its formation was insensitive to induction by a precursor but was sensitive to carbon catabolite repression. That is, LAT behaved as a typical secondary metabolic pathway enzyme. Of particular interest was the observation that the intracellular concentration of its product (α-aminoadipic acid) in the amino acid pool in *S. clavuligerus* decreased during the late exponential phase, when cephamycin C was being produced. At the same time α-aminoadipic acid accumulated extracellularly. This observation suggests, at least under the conditions used, that the formation of α-aminoadipic acid was not a rate-limiting step in cephamycin C biosynthesis.

These results show that upstream regulation of amino acid synthesis, particularly in the aspartic acid pathway, can affect antibiotic production. The concentration of all the precursors would be affected by the overall level of protein synthesis in the cell. Synthesis of proteins during rapid growth would lower the concentration of the amino acid pool (Aharonowitz et al., 1984), which has two effects: lower antibiotic synthesis due to lack of precursors, and an overall stimulation of amino acid synthesis by positive feedback regulation.

3.2. Control of Enzyme Expression

As illustrated in Fig. 5.2, the expression of the biosynthetic enzymes in *S. clavuligerus* lags behind the initiation of rapid growth but precedes antibiotic formation. The latter is clearly controlled by the supply of precursors to the pathway, which is switched on when the nutrient supply to the cells becomes limiting. The signal for initiating enzyme expression is unknown. The level of activity of each enzyme was affected by catabolite repression. Hence IPNS was repressed by carbon and nitrogen source levels (Brana et al., 1985; Vining et al., 1987). DAOCS was repressed by the carbon source (Lebrihi et al., 1988a) and phosphate (Lebrihi et al., 1987) and derepressed (or promoted) by oxygen (Rollins et al., 1990). These observations indicated that the expression of IPNS and DAOCS was affected to different extents by the ambient conditions.

Experiments on *S. lactamdurans* showed that carbon catabolism repressed the first and last enzymes of the pathway, ACVS and DAOCS, but had little effect on IPNS (Cortes et al., 1986). For example, increasing the concentration of glucose from 55 to 165 mM reduced the level of DAOCS threefold and cut the yield of cephamycin C by an order of magnitude. This mode of control gave no buildup of intermediates in the pathway. By contrast, nitrogen catabolism repressed all significant enzymes in the pathway in a coordinated fashion (Castro et al., 1985).

The genes for the biosynthetic enzymes in streptomycetes are known to be clustered together, and the regulatory genes occur in the cluster or nearby (Chater, 1990). The regulation of these genes is positive. Madduri et al. (1991b) found that the genes for the entire biosynthetic pathway in *S. clavuligerus* were clustered together, from lysine ε-aminotransferase through to DAOCS. Although the exact level of expression of each biosynthetic enzyme may depend on carbon and nitrogen repression, the expression of enzymes in the pathway (which have been studied in detail) follow a similar time course. The combination of coordinated on/off control of expression with differential catabolite repression of each enzyme indicates a hierarchy of regulation. The top level enables synthesis of the enzymes for the entire pathway, while the lower level of catabolite repression modulates the level of individual enzymes.

The details of how catabolite repression operates are unknown. In the case of *S. antibioticus*, glucose repression acts at the transcriptional level (Jones, 1985). This mode of control is commonly observed, likely because of its efficiency in using cellular resources. When an enzyme is not required, the transcription of the gene is blocked and no resources are devoted to mRNA synthesis from this gene. Transcriptional regulation of the biosynthetic pathway in *S. clavuligerus* is supported by recent work on the mRNA for IPNS (A. Petrich and S. E. Jensen, Department of Microbiology, University of Alberta). When the culture was grown from spores in complex medium (as in Fig. 5.2), the mRNA for IPNS was not detected until 30 h into the fermentation. The concentration was maximal at 36–48 h and disappeared by 60 h. The time course for the mRNA, therefore, exactly matched the time course for the activity of IPNS (Fig. 5.2). When growing mycelia, however, were used as inoculum, the mRNA for IPNS was always detectable; that is, the cells did not revert and transcription continued at a steady rate in the new medium.

If we assume that regulation acts on transcription, what is the effector? Cortes et al. (1986) showed that the intracellular concentration of cyclic AMP did not correlate with the rate of enzyme synthesis, so that the regulation of the antibiotic pathway did not follow the same signals as the glycolytic pathway.

3.3. Inhibition of Biosynthetic Enzymes

The assays for the activity of the biosynthetic enzymes, as in Fig. 5.2, give an estimate of the amount of the enzyme at the time of sampling. The enzyme assays, however, do not indicate the actual activity *in vivo* because inhibitors such as phosphorylated glycolysis intermediates, the availability and concentration of substrate, and the fact that the enzyme is removed from the presence of other pathway enzymes could affect the rate of product formation. The main subjects for inhibition studies have been the interaction of intermediates of the glycolytic pathway with the biosynthetic enzymes. Cortes et al. (1986) found that 1 mM levels of glucose-6-phosphate inhibited the activity of IPNS and DAOCS by 73 and 53%, respectively. Fructose 1,6-diphosphate and fructose 2,6-diphosphate inhibited DAOCS activity by 46 and 65%, respectively, but did not inhibit IPNS. The phosphorylated derivatives of glycerol did not inhibit the enzymes, so that use of glycerol as a carbon source would not cause inhibition *in vivo*. The intracellular concentrations of cyclic AMP and glucose-6-phosphate declined as glucose was exhausted. This observation led to the conclusion that cyclic AMP did not mediate antibiotic biosynthesis in *S. lactamdurans*, in accord with previous studies on formation of secondary metabolites in streptomycetes (Chatterjee and Vining, 1981; Regan and Vining, 1978).

These data suggest that phosphorylated sugars may regulate biosynthesis to some degree. However, these compounds probably do not account for the intracellular repression of enzyme activity, because the level of inhibition associated with them seems too weak to account for the observed cephamycin profiles. The intracellular concentration of glucose-6-phosphate was at most 8 μmol/g dry cell weight (Cortes et al., 1986). With a water content of about 80% and a density near 1 g/mL, the concentration of glucose-6-phosphate would therefore be less than about 1.6 mM inside the cells. The inhibition of DAOCS activity by this concentration would be 30–40%, which was much too low to account for the complete lack of cephamycin synthesis by the cells. The data of Cortes et al. (1986) showed that cephamycin production started at the same time as the expression of DAOCS, suggesting that the regulation was via genetic control rather than mediated control of the enzymes.

Hu et al. (1984) studied catabolite inhibition in resting cells and found that addition of the same carbon source (glycerol or starch/maltose) that the cells had been grown on caused a reduction in antibiotic formation from the biosynthetic enzymes already in place (Fig. 5.3). When the carbon source was changed, that is, from glycerol to maltose, the new carbon source was not utilized during a 4-h experiment and did not inhibit antibiotic synthesis. Blockage of protein synthesis by the addition of chloramphenicol did not affect the formation of antibiotic in the control samples, but it completely eliminated the inhibitory effect of the glycerol (Fig. 5.3). Chloramphenicol did not block the uptake of the carbon source. Hu et al. (1984) suggested that chloramphenicol blocked the formation of the proteases responsible for the inactivation of the biosynthetic enzymes. This hypothesis will be considered later using the results of the mathematical model. A definite conclusion from Hu et al. (1984) was that although the carbon source was utilized, catabolite inhibition did not occur when protein synthesis was blocked. This observation supports the previous discussion on the weak role of phosphorylated intermediates.

The question of product inhibition in cephamycin C production, however, has received very little attention. Malik and Vining (1972) found that a streptomycete that produced chloramphenicol was only sensitive to the antibiotic prior to antibiotic synthesis; once antibiotic production started the cells became impermeable to the product in the medium. When this type of resistance is present, inhibition or feedback regulation by endogenously produced antibiotic should not be signif-

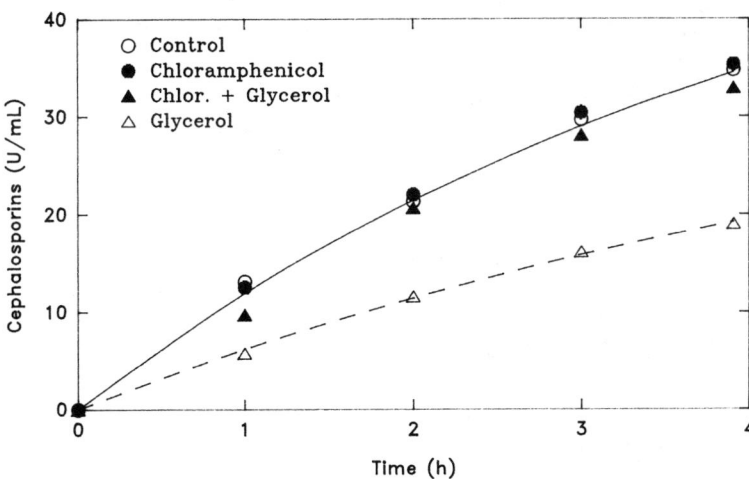

FIGURE 5.3. Antibiotic synthesis by resting cells of *S. clavuligerus* [data of Hu et al. (1984)].

icant. The mechanisms by which antibiotic-producing organisms, particularly *Streptomyces* spp., protect themselves from the antibiotics they produce were recently reviewed by Cundliffe (1989).

4. EFFECT OF DISSOLVED OXYGEN LEVELS

As a mechanism in studying the effect of environmental factors on the regulation of antibiotic production, the possible role of oxygen usually is not investigated. Recent investigations indicate that increased levels of oxygen, unlike carbon, nitrogen, and phosphorus, enhance the formation of the production of cephamycin C. Under controlled fermentation conditions with a soluble starch medium, Rollins et al. (1990) found that 100% oxygen saturation increased the expression of DAOCS, and to a lesser extent IPNS. Expression of ACV synthetase was not affected by oxygen levels, although the stability of the enzyme may have been decreased by oxygen saturation (Rollins et al., 1991). The addition of maltose to controlled-oxygen fermentations reduced the yield of cephamycin C by a factor of 2 and also reduced the activity of DAOCS threefold in comparison to controls (Rollins et al., 1988b).

In a defined glycerol–asparagine medium the stability of IPNS was improved by oxygen saturation (Rollins et al., 1989a), but 100% dissolved oxygen saturation gave lower yields of cephamycin C (Rollins et al., 1989b). This latter observation was surprising, in that the average activity of DAOCS was enhanced by oxygen saturation (Rollins et al., 1989a). The lack of sensitivity of yield to DAOCS activity, coupled with the fact that the penicillin N intermediate was unaffected, indicated that the main effect of oxygen in defined medium fermentation was on

ACVS activity. The role of oxygen concentration was therefore affected by the composition of the medium.

Given that all three enzymes (ACV, DAOCS, IPNS) were expressed at the same time and reached maximum activity at similar times, regardless of the oxygen level, we can conclude that the promotion or derepression effect of oxygen was secondary to the role of catabolite repression. This conclusion is further supported by the data from maltose feeding (Rollins et al., 1988b), where the effect of maltose on cephamycin C formation dominated over the oxygen concentration.

5. FRAMEWORK FOR THE KINETIC MODEL

From the previous discussion, the following phenomena are relevant to a model for the biosynthetic pathway:

1. Growth of biomass (as dry cell weight or protein) and substrate utilization.
2. Availability of the precursors for the pathway (α-aminoadipic acid, cysteine, and valine) and control over initiation of antibiotic synthesis.

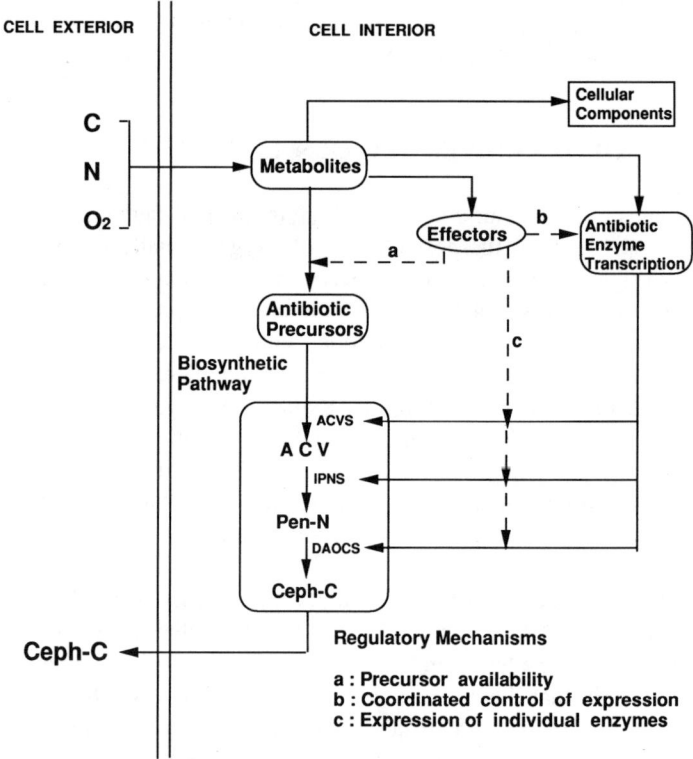

FIGURE 5.4. Schematic diagram of regulatory pathways for cephamycin C biosynthesis.

3. Kinetics of the biosynthetic enzymes, including initiation and regulation of enzyme expression, inactivation of enzymes, and inhibition.

The schematic diagram of Fig. 5.4 illustrates the key relationships in the biosynthesis of cephamycin C. Solid lines show fluxes of material in the cell, from the substrates for growth through to cephamycin C. Two main paths are shown: direct conversion of precursors to product through the biosynthetic pathway, and control of activity of the biosynthetic enzymes, which includes regulation of transcription. Dashed lines show the regulatory relationships, due to the action of the as yet unidentified effectors. As an approximation, we assume that the measurable extracellular concentration of utilizable nutrient is proportional to the intracellular concentration of effector. In the case of glycerol, for example, the regulation of intracellular events would be correlated to the measured extracellular concentrations. The regulation of enzyme transcription is approximated as a two-stage process. Catabolite repression exerts on/off control on transcription of the entire biosynthetic pathway. Carbon source, nitrogen source, and dissolved oxygen can more specifically regulate the levels of the individual enzymes (ACVS, IPNS, and DAOCS) during the time that transcription is active.

6. KINETICS OF TRANSCRIPTION AND INACTIVATION OF ENZYMES

The simplest model for intracellular enzyme concentrations was first verified by Price et al. (1962), to allow for synthesis and inactivation of enzyme:

$$\frac{d\mathrm{E}}{dt} = r_T - k_d \mathrm{E}, \tag{1}$$

where E is the concentration of enzyme, r_T is the rate of enzyme synthesis (equal to the rate of transcription, which is assumed to be the controlling step), and k_d is the rate constant for inactivation of enzyme. The level of enzyme at steady state, E_{ss}, is determined by the rates of transcription and inactivation:

$$\mathrm{E}_{ss} = r_T/k_d \tag{2}$$

If the rate of transcription is constant, and the inactivation rate constant does not change, then the buildup of enzyme is given by the integral of Eq. (1). For the conditions of $r_T > 0$, and no initial enzyme activity, $\mathrm{E}(0) = 0$:

$$\mathrm{E}(t) = r_T/k_d [1 - \exp(-k_d \{t - t_{\mathrm{lag}}\})], \tag{3}$$

where t_{lag} is the lag time before transcription is switched on. At a constant rate of transcription, Eq. (3) shows that the steady-state level of the enzyme is reached asymptotically, assuming that transcription continues long enough (Fig. 5.5). When transcription ceases, the enzyme activity decays exponentially:

$$E(t) = E_{max} \exp[-k_d(t - t_{max})], \quad \text{for } t > t_{max}, \quad (4)$$

where E_{max} is the enzyme level at time $t = t_{max}$ when transcription stops. If transcription ceases before the steady state is achieved, a very different profile of enzyme activity is observed, wherein the activity rises and falls very quickly (Fig. 5.5).

The time profiles for ACVS, IPNS, and DAOCS under different environmental conditions were available from the literature. The simplest analysis of these data was to use Eqs. (3) and (4) to estimate the rates of transcription and deactivation in different growth media. These estimated rates can then be used to determine whether a more sophisticated model is required.

The values of r_T, k_d, t_{max}, E_{max}, and t_{lag} from Eqs. (3) and (4) for the data on IPNS are listed in Table 5.1. The data for enzyme activities throughout this study have been converted to a common basis for unit activity; one unit of activity is an initial rate of conversion of 1 μmol substrate per minute. The transcription rate (r_T) was repressed by both carbon and nitrogen sources. Correlated with the change in r_T were the values of E_{max}, k_d, and t_{max}. The lag time was not correlated with the rate of transcription and was presumably sensitive to the details of preparing the inoculum. The time when transcription was active ($t_{max} - t_{lag}$) was also correlated with the transcription rate r_T. In general, larger values of r_T gave larger E_{max}, consistent with the limiting steady-state value of enzyme activity, which depends on r_T [Eq. (2)]. Richer media gave lower transcription rates, a longer time of active expression of enzyme, lower values of E_{max}, and lower values of the deactivation rate constant k_d. The latter observation is consistent with the lower

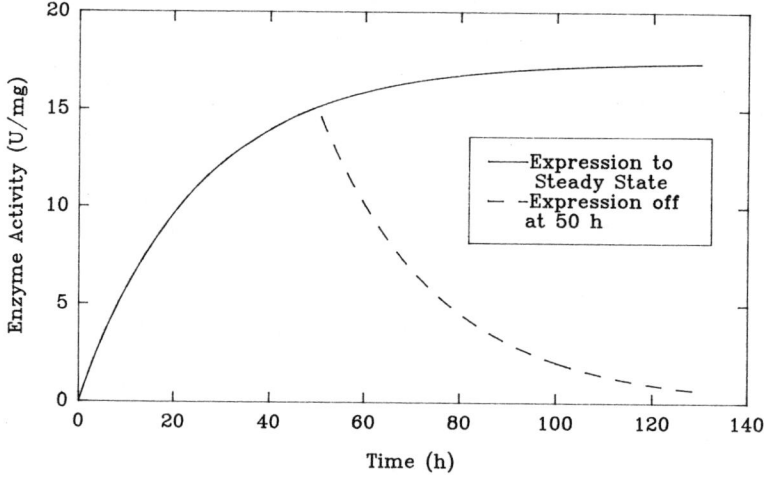

FIGURE 5.5. Kinetics of enzyme expression and inactivation [from Eqs. (3) and (4)].

TABLE 5.1. Expression and Inactivation of Isopenicillin N Synthase in *Streptomyces clavuligerus*

Medium	t_{lag} (h)	r_T (mU/mg cell protein/h)	Max IPNS^a (mU/mg cell protein)	t_{max} (h)	k_d^b (h^{-1})	Max Ceph C (mg/g cell protein)	Reference
Starch 10 g/L + TCS^c	12	0.78	4.7	23	0.089 (0.12)	83	Vining et al. (1987)
Starch 30 g/L + TCS	5.5	0.25	4.3	31	NA^d	NA	Vining et al. (1987)
Starch 10 g/L + 15 mM asparagine	15	0.44	3.7	28	0.042 (0.04)	47	Vining et al. (1987)
Glycerol 10 g/L + 15 mM asparagine	14	0.083	1.64	59	0.048	105^e	Brana et al. (1985)
Glycerol 10 g/L + 30 mM ammonium	8	0.015	0.77	68	0.0095	50^e	Brana et al. (1985)
Complex,^{f} no DO control	27	0.23	0.4	28	0.02	50	Rollins et al. (1990)
Complex,^{f} 100% DO	20	0.043	0.5	30	0.04	300	Rollins et al. (1990)

^{a}All activities were converted to a common basis of μmol product per minute.
^{b}Parameters are from a nonlinear least squares over the entire data set. The value in parentheses is the estimate based only on the decline phase.
^{c}TCS, trypticase soy broth, 30 g/L, equivalent to 160 mM total nitrogen source.
^{d}Insufficient data.
^{e}Total β-lactams, based on a disk diffusion assay calibrated against cephamycin C, converted to specific yield assuming 0.3 g cell protein/g dry cell weight.
^{f}Medium contained starch 10 g/L + 20 g/L tryptone and peptone (trypticase soy broth). Data for cephamycin C yields are from Rollins et al. (1988a).

rates of protein turnover in bacteria growing rapidly on balanced media (Switzer, 1977). The range in values for k_d varied by a factor of 9, from 0.0095 to 0.089 h^{-1}.

The same parameters are listed in Table 5.2 for the activity of DAOCS. The units for the data of Rollins et al. (1990) have been expressed as mU/mg of total cell protein based on their original activity definition (Rollins et al., 1988c). The repressive effect of the carbon source was similar to IPNS, but the enzyme activity was insensitive to the concentration of nitrogen source. The deactivation rate constant for DAOCS was much less sensitive to the metabolic state of the culture than was the parameter for IPNS; all but one value of k_d fell within a factor of 3, with most values falling in the vicinity of 0.04 h^{-1}. The only exception to this trend was the data of Rollins et al. (1989a), who observed a value of 0.14 h^{-1}. This value was inconsistent with previous experiments using the same defined medium and uncontrolled oxygen concentrations, that is, the studies of Brana et al. (1985) and Lebrihi et al. (1988a).

The time course of DAOCS activity is illustrated in Fig. 5.6, using the data of Lebrihi et al. (1988a), in a defined medium with different initial glycerol concentrations. The time when DAOCS transcription began was shifted from 27 to as much as 41 h by increasing the concentration of glycerol in the starting medium. The time when production of cephamycin C started also shifted, from 10 h at 4 g/L glycerol to 36 h at 15 g/L. The time lag between the appearance of DAOCS and the production of cephamycin C ranged from 0 to 12 h, showing that the biosynthetic pathway was controlled both by the expression of the enzymes and by the availability of precursors. The range of transcription rates with glycerol [0.015-0.035 mU/mg total cell protein/h from Lebrihi et al. (1988a)] showed that the expression of DAOCS was regulated by catabolism of the carbon source; as the concentration of glycerol increased, the average rate of transcription decreased.

The data of Fig. 5.6 show that the simple model [Eqs. (3) and (4)] provided a good fit to the activity data as long as the initial enzyme activity at time zero, for the inoculum, was greater than zero. The initial enzyme activity would depend on the details of how the inoculum was prepared. Lebrihi et al. reported a drop in activity from 11 to 24 h at 4 g/L glycerol, and similar declines were observed from 24 to 36 h at 10 g/L and 15 g/L. Such declines are consistent with an inoculum with a DAOCS activity of about 0.2-0.4 mU/mg total cell protein.

A very important experiment by Lebrihi et al. (1988a) was to add glycerol to the culture during the fermentation. When 10 g/L glycerol was added to a culture grown on starch at $t = 24$ h, when expression of DAOCS had just started, the synthesis of the enzyme was repressed almost immediately and the time course was similar to the experiment with the same total amount of glycerol and starch at time $t = 0$ (maximum activity of 5-6 mU/mg total cell protein at $t = 80$ h). This experiment showed that the lag between catabolism and regulation of expression of the enzymes was negligible, compared to the sampling time of about 8 h.

Rollins et al. (1990) reported IPNS and DAOCS activity data for *S. clavuligerus* grown on a complex medium with and without control of dissolved oxygen. Data

TABLE 5.2. Expression and Inactivation of Desacetoxycephalosporin C Synthase in *Streptomyces clavuligerus*

Medium	t_{lag} (h)	r_T (mU/mg cell protein/h)	Max DAOCS[a] (mU/mg cell protein)	t_{max} (h)	k_d^b (h^{-1})	Max Ceph C (mg/g cell protein)[c]	Reference
Glycerol 4 g/L + 15 mM asparagine	27	0.035	0.68	48	0.032 (0.04)	117	Lebrihi et al. (1988a)
Glycerol 10 g/L + 15 mM asparagine	39	0.020	0.39	72	0.041 (0.05)	33	Lebrihi et al. (1988a)
Glycerol 15 g/L + 15 mM asparagine	41	0.015	0.34	84	0.039 (0.03)	23	Lebrihi et al. (1988a)
Starch 10 g/L + 15 mM asparagine	21	0.030	0.68	73	0.043 (0.055)	120	Lebrihi et al. (1988a)
Starch 10 g/L + 10 g/L glycerol + 15 mM asparagine	6	0.007	0.28	87	0.020	32	Lebrihi et al. (1988a)
Glycerol 10 g/L + 15 mM asparagine	3	0.041	1.06	64	0.036	105	Brana et al. (1985)
Glycerol 10 g/L + 30 mM ammonium	27	0.036	0.82	88	0.04	50	Brana et al. (1985)
Glycerol 10 g/L + 15 mM asparagine, no DO control	38	0.096	0.9	55	0.14	100	Rollins et al. (1989)
Glycerol 10 g/L + 15 mM asparagine, 100% DO	18	0.233	0.9	23	0.05	45	Rollins et al. (1989)
Complex,[d] no DO control	27	0.21	1.2	33	0.07	50	Rollins et al. (1990)
Complex,[d] 100% DO	21	0.65	4.9	27.5	0.05	300	Rollins et al. (1990)

[a] All activities were converted to a common basis of μmol product per minute.
[b] Parameters are from a nonlinear least squares over the entire data set. The value in parentheses is the estimate based only on the decline phase.
[c] Data of Brana et al. (1985) and Lebrihi et al. (1988a) converted to specific yield assuming 0.3 g cell protein/g cell mass. Data from Brana et al. (1985) are for total β-lactams, based on a disk diffusion assay calibrated against cephamycin C.
[d] Medium contained starch 10 g/L + 20 g/L tryptone and peptone (trypticase soy broth). Data for cephamycin C yields are from Rollins et al. (1988a).

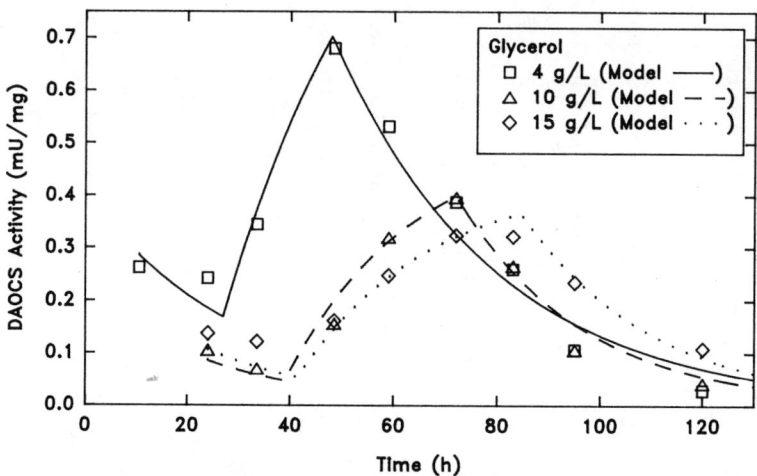

FIGURE 5.6. Effect of glycerol concentration on expression of DAOCS. The symbols shown in the figure represent the actual data points of Lebrihi et al. (1988a). The drawn lines represent results of respective models, which are the best fit of Eqs. (3) and (4), using the parameters listed in Table 5.2.

for DAOCS are illustrated in Fig. 5.7, along with the model fits from Eqs. (3) and (4). The parameters are listed in Table 2. In this case, the inoculum did not have any measurable DAOCS activity. No enzyme activity was observed until transcription was initiated at 20–27 h. A repeat experiment without DO control (not shown) gave a rise in activity at 20 h, as compared to 27 h for the data set in Fig. 5.7. Consequently, the apparent difference in lag time between high and low oxygen-

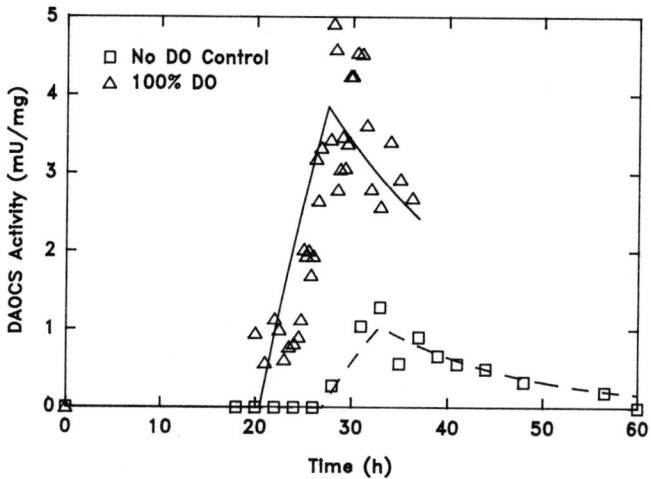

FIGURE 5.7. Effect of oxygen level on the expression of DAOCS. The symbols represent the actual data points of Rollins et al. (1990). The drawn lines represent results of respective models, which are the best fit of Eqs. (3) and (4), using the parameters listed in Table 5.2.

TABLE 5.3. Effect of Oxygen on Expression and Inactivation of ACV Synthetase[a]

Oxygen	Lag Time (h)	r_T (mU/mg cell protein/h)	Max ACVS (U/mg cell protein)	t_{off} (h)	k_d (h^{-1})
No control	27.5	0.089	0.25	32	0.10
100% DO	25	0.068	0.26	30	0.16

[a]From data of Rollins et al. (1991).

ation was not experimentally significant. In all cases the activity then increased rapidly to a maximum, then declined due to inactivation once transcription had ceased. The data of Table 5.2 for r_T suggest that oxygen has a significant effect on transcription in both defined and complex media.

Rollins et al. (1990) suggested that the data for DAOCS at 100% DO indicated oscillations in enzyme concentration with time. Analysis of two sets of DAOCS data for the period from 26 to 36 h [data from Rollins et al. (1990) without oxygen control (not shown) and the data in Fig. 6 with 100% DO] using a χ^2 test showed, however, that the relative variance for both data sets was statistically the same. In other words, there was no statistical evidence to show that the oscillations were occurring at 100% dissolved oxygen in comparison to the base case without oxygen control. The data of Fig. 5.7 simply show significant scatter of the data around the model curve and do not tend to support an argument for oscillations.

Very little data are available for the initial rate-limiting enzyme in the pathway, ACVS. The data of Rollins et al. (1990) were fitted to Eqs. (3) and (4) simultaneously, and the parameter values are listed in Table 5.3. An example data set is shown in Fig. 5.8. Although the transcription rates are not directly comparable,

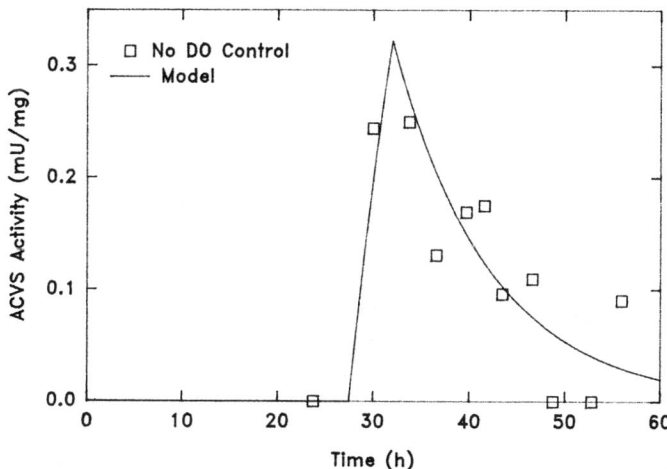

FIGURE 5.8. Kinetics of ACV synthetase expression. The symbols represent the actual data points of Rollins et al. (1991). The drawn line represents the results of the model, which was the best fit of Eqs. (3) and (4), using the parameters listed in Table 5.3.

the inactivation rates for ACVS were two to four times larger than for DAOCS or IPNS. As a result, the ACVS activity would tend to shut off first, leaving activity of IPNS and DAOCS to convert most of the residual ACV to antibiotic compounds.

7. REGULATION OF ENZYME TRANSCRIPTION

The transcription rate is dependent on the concentration of effectors and the concentration of the gene (G) in the cell (Lee and Bailey, 1984). For a given protein, the equations can be written as follows:

$$\frac{d[\text{mRNA}]}{dt} = k_p^0 \eta [G] - k_d [\text{mRNA}], \tag{5}$$

$$r_{tr} = k_q \zeta [\text{mRNA}], \tag{6}$$

where [mRNA] is the concentration of mRNA in the cells coding for the protein, k_p^0 is the transcription rate constant, k_d is the decay rate constant for mRNA, r_{tr} is the rate of translation of mRNA, and k_q is the rate constant for translation. The variable η, the efficiency of transcription of the gene, depends on the concentration of effectors for the gene. ζ is the efficiency of translation, which depends on the level of ribosomes, regulatory elements, and the binding of the mRNA to the ribosome. If we assume that expression of the proteins is controlled at the transcription step, and that the concentration of mRNA is in steady state, then the net rate of transcription of the gene to produce enzyme, r_T, can be written as follows:

$$r_T = (k_p^0 k_q \zeta / k_d) \eta [G] = (k_i^0 [G]) \eta, \tag{7}$$

where $k_i^0 [G]$ is the maximum rate of transcription, and η is the efficiency of transcription ($0 < \eta < 1$).

In order to use Eq. (7) in a model for *S. clavuligerus*, where the effectors are not understood in detail, we must use a simple empirical expression for the efficiency of transcription. A model for catabolite repression of enzyme expression, proposed by Imanaka and Aiba (1977), gave a simple mathematical expression for correlating changes in the rate of transcription to the changes in the measurable concentrations in nutrients:

$$\eta = \frac{1 + \alpha S^n}{1 + (1 + \beta)\alpha S^n}. \tag{8}$$

S is the concentration of the substrate that causes catabolite repression, and β, α, and n are adjustable parameters. Even though Eq. (8) was derived for catabolite repression by glucose following a cyclic AMP regulation as in *E. coli*, we adopt it for the model for *S. clavuligerus* as an empirical equation with reasonable functional dependence of transcription efficiency on repressor. Equation (8) gives an S-shaped curve, with $\eta = 1/(1 + \beta)$ at high substrate concentrations (fully repressed) and $\eta \to 1.0$ at very low substrate concentrations. This expression was used empirically by Matsumura et al. (1981) in a model for cephalosporin production, although no data for enzyme expression were actually reported. Their best fit parameter values for the repressive effect of glucose were $n = 5.1$, $\beta = 1.6$, and $\alpha = 4.6 \times 10^3$.

Lebrihi et al. (1988a) observed that the repressive effect of glycerol was relatively weak; an increase in concentration from 1 to 15 g/L lowered the maximal activity of DAOCS from 0.82 mU/mg total cell protein to 0.24 mU/mg total cell protein. In each fermentation, the glycerol was consumed with time, which might be expected to derepress DAOCS expression. The lack of sensitivity to glycerol suggested, however, that the initial low rates of expression of DAOCS could not be determined only by glycerol concentration. When Eq. (8) was fitted to the time course of DAOCS expression, using $S(t) = $ [glycerol] and $\eta(t) = r_T/r_{T,\max}$, the values of α and n varied dramatically with the initial concentration of glycerol.

The lack of success of Eq. (8) for continuously adjusting the rate of transcription suggested a different mode of regulation. Lebrihi et al. (1988a) observed that the effect of addition of glycerol to a culture grown on starch depended on timing; prior to 24 h the glycerol repressed DAOCS expression, while addition at later times had little effect. This observation, and the weak dependence of expression of DAOCS on glycerol concentration, suggests the following sequence of events:

Stage 1. Growth Without Expression of Biosynthetic Enzymes. Expression of DAOCS (and the other biosynthetic enzymes) is turned off, due to a lag between initiation of growth and initiation of expression. This lag would be due to repression by an effector during balanced exponential growth on an excess of all nutrients. The regulation of the lag phase by an effector is consistent with the change in lag time with the concentration of carbon source (Table 5.2 and Fig. 5.6). This mode of regulation is shown as pathway b in Fig. 5.4.

Stage 2. Initiation of Biosynthetic Enzyme Expression. At the end of the lag phase, the transcription of the biosynthetic enzymes is enabled. The transcription rate is set by the extent of catabolite repression at the time it is initiated, so that high carbon levels give a lower rate of transcription (regulatory pathway c in Fig. 5.4). Once the transcription begins it is relatively unaffected by the concentration of nutrient.

Stage 3. Initial Nutrient Limitation. Once a nutrient begins to limit growth, the precursors become available to the antibiotic pathway and production begins (regulatory pathway a in Fig. 5.4). This phase in batch growth corresponds to nutrient-limited growth in a chemistat.

Stage 4. Conversion of Precursors. Exhaustion of nutrients ends the expression of the biosynthetic enzymes. Conversion of precursors continues using the available enzyme activity.

Given this sequence of events, an expression like Eq. (8) can be used to correlate the dependence of the average rate of transcription of each biosynthetic enzyme with the concentration of carbon source, nitrogen source, or phosphate. A complete model of catabolite repression would require a description of the lag effect (Stage 1) and the functional dependence of transcription on each nutrient.

The data of Lebrihi et al. (1988a) provide a basis for modeling the effect of glycerol on DAOCS transcription. Maximum DAOCS activity was reported for glycerol concentrations in the range of 1–15 g/L (Fig. 5.9). Equation (2) and Fig. 5.5 showed that the maximum enzyme activity at steady state was r_T/k_d. Lebrihi et al. (1988a) did not observe the steady state, but if we assume that E_{max} was proportional to E_{ss} (i.e., the enzyme activity achieved a constant fraction of the ultimate steady-state value), then Eq. (8) can be used to analyze catabolite repression:

$$E_{max} = \bar{r}'_T/k_d = r_{T,max}\eta/k_d = \frac{r'_{T,max}(1 + \alpha S^n)}{k_d(1 + (1 + \beta)\alpha S^n)}, \qquad (9)$$

where $\bar{r}'_T$ is the effective mean rate of transcription and $r'_{T,max}$ is the maximum rate of transcription. If E_{max} differs from E_{ss} by a constant, then these effective transcription rates will differ from the true values by the same constant. A mean value of $k_d = 0.04$ h^{-1} was obtained from Fig. 5.6 and used to fit Eq. (9) to the data of Fig. 5.9, giving the solid curve in the figure. The optimal value of $n = 1.47$ was

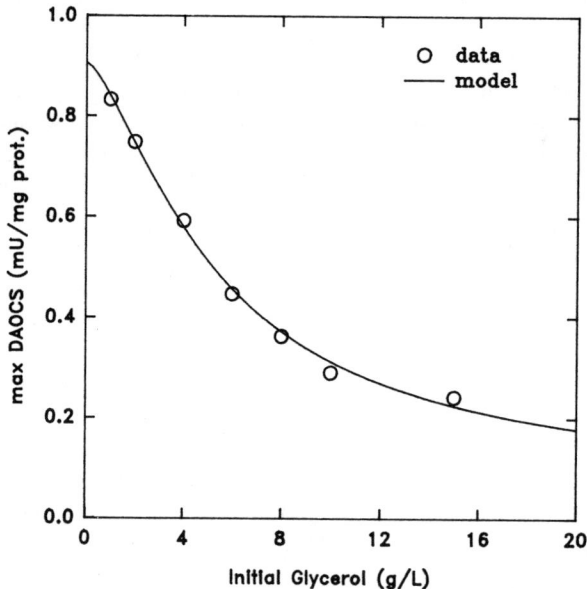

FIGURE 5.9. Repression of DAOCS expression by glycerol [data of Lebrihi et al. (1988a). The model curve is from Eq. (9) with $\alpha = 6.7 \times 10^{-3}$, $n = 1.47$, $\beta = 11.5$, $r'_{T,max} = 0.75$ mU/mg cell protein/h, $k_d = 0.04$ h^{-1}].

consistent with the weak repressive effect of glycerol, because strong repression would give a value $n \simeq 5$ (Matsumura et al., 1981).

The data of Table 5.2 can be used to check the assumption that E_{max} and E_{ss} differ by a constant factor. The values of r_T reported in the table decrease from 0.035 to 0.015 mU/mg cell protein/h as glycerol concentration increases from 4 to 15 g/L. These values for r_T are 1.5–1.63 times larger than the estimates of $\bar{r}'_T$ calculated from the values of E_{max} in Fig. 9 [via Eq. (9) with $k_d = 0.04$ h^{-1}]. Consequently, the assumption of constant proportionality between E_{ss} and E_{max} was accurate over this range of glycerol concentration, with an average ratio of 1.59. The value of $r_{T,max}$ therefore was 1.59 times the best fit value of $r'_{T,max}$ from Fig. 5.9, giving a value of 0.057 mU/mg cell protein/h.

The same procedure would also be used to model the effects of nitrogen source or phosphate on the biosynthetic enzymes, for example, using the data of Lebrihi et al. (1987) for phosphate repression. The functional form for the promoting (or derepressing) effect of dissolved oxygen cannot be tested, because Rollins et al. (1990) reported data at only two sets of conditions (and dissolved oxygen was not constant in the run without control of DO). As Fig. 5.9 indicates, two measurements at arbitrary concentrations cannot determine the maximum and minimum rates of transcription; rather, a range of experimental conditions are required.

8. KINETICS OF CEPHAMYCIN C SYNTHESIS

Given the model scheme in Fig. 5.4, the kinetics of product formation at each step of the biosynthetic pathway are controlled by three factors:

1. Enzyme activity *in vivo*, controlled by the amount of enzyme and any inhibition.
2. Concentration of precursor.
3. Rate expression for the enzyme.

In order to study the kinetics of a given enzyme, therefore, we require detailed data on activity, product, and precursor concentrations as a function of time.

If we assume that the biosynthetic reactions follow Michaelis–Menten kinetics, then the following rate expression would apply for formation of product P:

$$\frac{d[P]}{dt} = \frac{k_{cat}[E][X][S]}{(K_M + K_i[I] + [S])}, \qquad (10)$$

where k_{cat} and K_M are the reaction rate constant and Michaelis constant, [E] is the concentration of active enzyme as mU/mg total cell protein, [X] is the concentration of total cell protein in mg/mL, [S] is the concentration of the limiting substrate, K_i is the inhibition constant, and [I] is the lumped concentration of any inhibitors. If more than one substrate is required, the expression would contain additional terms. The most efficient utilization of precursors will occur when [S] ≪ K_M, and with the effect of inhibitors minimized, so that the reaction is first order:

$$\frac{d[P]}{dt} = k_p[E][X][S], \qquad (11)$$

where $k_p = k_{cat}/K_M$. In terms of modeling, we can take the approach of using this simple rate expression to determine whether it can reconcile the data for product, precursor, and enzyme activity.

Rollins et al. (1988a, 1990) gave data on DAOCS activity, penicillin N concentration, and cephamycin C concentration as a function of time. Since the conversion of penicillin N (Pen-N) to desacetoxycephalosporin C (DOC) is the rate-limiting step for this portion of the biosynthetic pathway, the rate of formation of DOC will equal the rate of formation of the ultimate product, cephamycin C (Ceph-C):

$$\frac{d[DOC]}{dt} \simeq 0 = k_p[DAOCS][X][Pen-N] - \frac{d[Ceph-C]}{dt}, \qquad (12)$$

with [DOC], [DAOCS], [Pen-N], and [Ceph-C] being the respective concentrations of the four compounds. The simplest rate expression for the formation of cephamycin C then becomes

$$\frac{d[Ceph-C]}{dt} = k_p[DAOCS][X][Pen-N] \qquad (13)$$

$$[Ceph-C] = \int_0^t \left(\frac{d[Ceph-C]}{dt}\right) dt$$

$$= \int_0^t k_p[DAOCS][X][Pen-N]\, dt. \qquad (14)$$

Data for DAOCS, penicillin N, protein content, and cephamycin C were given by Rollins et al. (1990) as a function of time for fermentations without oxygen control (see Figs. 5.2 and 5.7). Using the model curve for DAOCS (Fig. 5.7), a regression of Eq. (14) on the data of Rollins et al. (1990) gave an excellent fit with a value of $k_p = 0.147$ mL/(mU)/h. As illustrated in Fig. 5.10, the model apparently underestimated the yield of product at 34–42 h, when the concentrations of both cepahmycin C and penicillin N were small. This discrepancy indicates that either the simple rate expression is inadequate or the error in measuring the concentrations was more significant at low concentrations.

Two factors may influence the activity of DAOCS *in vivo*: inhibition due to phosphorylated sugars and oxygen concentration (since DAOCS is a dioxygenase).

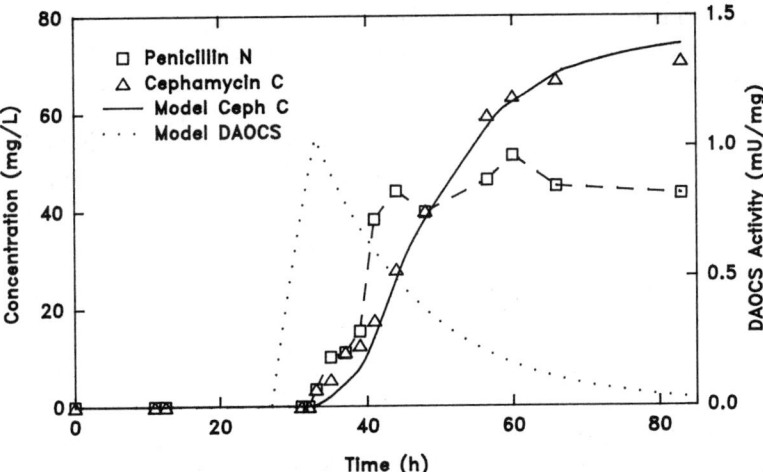

FIGURE 5.10. Kinetics of cephamycin C formation without control of dissolved oxygen. The symbols represent the actual data points of Rollins et al. (1990). The drawn line for DAOCS (· · ·) represents the results of the model [Eqs. (3) and (4) with parameters from Table 5.2]. The drawn line for cephamycin C (——) represents the results of model Eq. (14) with $k_p = 0.147$ mL/(mU·h). The drawn line for penicillin N (- - -) shows the trend of the data but was not derived from a model.

Both factors would tend to suppress enzyme activity most during the active growth phase (20–38 h) as compared to later in the fermentation (38–80 h). Carbon catabolism would be most active during active growth, giving maximal inhibition due to phosphorylated intermediates, and dissolved oxygen concentration would be minimal. Rollins et al. (1990) reported that DO fell to about 15% for the period 34–40 h, then rose gradually rapidly to 60–70% saturation.

If either inhibition or lack of oxygen were important during the period from 32 to 42 h, the equation without these mechanisms [Eq. (14)] would overpredict the rates of formation of cephamycin C. Rather than overpredicting the yield of cephamycin C during the period from 32 to 42 h, Eq. (14) underpredicts the yield (Fig. 5.10). Any attempt to include inhibition or oxygen concentration terms in Eq. (14) would only increase the discrepancy between the model and the data.

9. KINETICS OF ACV FORMATION AND CONVERSION

As intermediates in the biosynthetic pathway, ACV and penicillin N are formed and converted simultaneously. If we consider the case of penicillin N, the rate expression can be written

$$\frac{d[\text{Pen-N}]}{dt} = k_a [\text{IPNS}][X][\text{ACV}] - \frac{d[\text{Ceph-C}]}{dt}, \qquad (15)$$

where [ACV] represents the concentration of that compound, and the first term on the right is the rate of conversion of ACV by IPNS, assuming first-order kinetics with a rate constant k_a, and the second term is the rate of conversion of penicillin N to cephamycin C. If the first term is larger than the second, penicillin N will accumulate as in Fig. 5.10. If the two terms are of similar magnitude, then little penicillin N will accumulate. Rearranging Eq. (15) to solve for [ACV] gives

$$[ACV] = (d\{[Pen-N] + [Ceph-C]\}/dt)/(k_a[IPNS][X]). \qquad (16)$$

The variables on the right-hand side can be evaluated from the data of Rollins et al. (1990). The total rate of change of penicillin N plus cephamycin C can be calculated from measured values (Fig. 5.10), and the activity of IPNS and the total cell protein concentration, [X], were measured. Equation (16) gives the concentration of ACV to within a constant, k_a.

Figure 5.11 shows the data for ACV from Rollins et al. (1991) as a function of time. The levels of intracellular ACV were measured by disk diffusion assay and were not converted to true concentration units. The values here follow Rollins et al. (1991) and are given as "nmol," that is, nanomoles of ACV from cell extract prepared under the experimental conditions. The data for penicillin N, cephamycin C, IPNS, and protein were used to calculate the levels of ACV that would have existed, using Eq. (16). The solid line in Fig. 5.11 shows the predicted values using a value of $k_a = 0.1$ mmol/(mU · nmol · h). The solid line approximates the experimental data both in magnitude and overall time course, which indicates

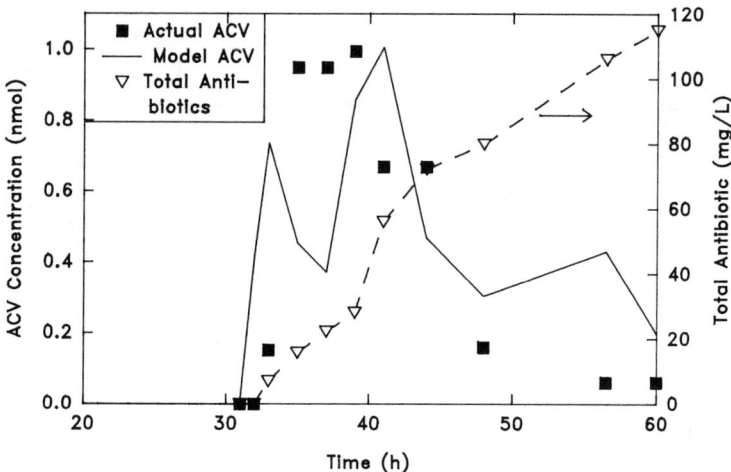

FIGURE 5.11. ACV concentrations during fermentation. The symbols represent the actual data points of Rollins et al. (1990, 1991). The drawn line for the predicted ACV concentration (——) is from the results of the respective model [Eq. (16) with $k_a = 0.1$ mmol/(mU · nmol · h)]. The drawn line for total antibiotic concentration (– – –) shows the trend of the data but was not derived from a model.

that Eq. (15) is a good description of the enzyme kinetics given the limited data available. This qualitative agreement leads to the conclusion that both DAOCS and IPNS performance are governed by linear kinetics, and that the biosynthesis of antibiotics is determined by the expression of enzyme activity and the supply of precursors.

A similar expression can be written for the formation of ACV, as a function of the rate-limiting precursor, which is given here as α-aminoadipic acid (α-AAA):

$$\frac{d[ACV]}{dt} = k_i[ACVS][X][\alpha\text{-AAA}] - k_a[IPNS][X][ACV]. \qquad (17)$$

[ACVS] and [α-AAA], respectively, represent the concentrations of those compounds. In order for ACV to form, precursor must be present. The data for ACV in Fig. 5.11 show that the level of ACV lags behind the formation of ACVS (Fig. 5.2), indicating a delay between expression of the enzymes and supply of the precursors. If ACVS follows simple enzyme kinetics, as assumed in Eq. (17), then the time course of the rate-limiting precursor would have to follow the concentration of ACV. The available data on amino acid pools (Aharonowitz et al., 1984) and α-AAA (Madduri et al., 1991a) indicate a rapid rise to maximal levels early in the fermentation, followed by a gradual decline. We can therefore infer that the *in vivo* kinetics of all the biosynthetic enzymes probably follow the *in vitro* activity measurements, and that the sequence of reactions follows approximately linear kinetics.

10. PREDICTIONS OF THE KINETIC MODEL

The model developed so far consists of two parts: the conceptual framework embodied in Fig. 5.4 and the quantitative expressions for the individual pathways and linkages. The available literature data have been used to develop mathematical expressions for a number of steps, including transcription and inactivation of enzymes, catabolite repression, and enzymatic conversion of substrates in the biosynthetic pathway. In order to test and validate the model, two sets of data are considered. First, the model is used to consider the results of Hu et al. (1984) for resting cells. Second, the kinetic model for DAOCS is used to predict yields of cephamycin C at 100% dissolved oxygen saturation. The model equations are then used to analyze options for enhancing antibiotic production.

10.1. Addition of Carbon Source to Resting Cells

The data of Hu et al. (1984) (Fig. 5.3) showed that the carbon source inhibited antibiotic formation only when protein synthesis was active, leading the authors to suggest that synthesis of specific proteases was involved in the observed inhibition. The simple model of Eq. (1) can be used to investigate this hypothesis. If transcription is blocked, then the activity of the enzyme will decay exponentially:

$$\frac{dE}{dt} = -k_d E. \tag{18}$$

If $k_d \simeq$ constant,

$$E = E_0 \exp(-k_d t) \tag{19}$$

If specific proteases are expressed, the rate of decay will increase; that is, the value of k_d will change. The rate of formation of antibiotic, in simplified form, is given by

$$\frac{d[\text{Ceph}]}{dt} = k_{\text{ceph}} E [\text{Precursors}], \tag{20}$$

where [Ceph] represents the total concentration of produced cephalosporins, k_{ceph} is the overall rate constant for their formation, and [Precursors] is the total supply of precursors for the biosynthetic pathway. If the supply of precursors is similar in both cases (glycerol with and without chloramphenicol), the ratio of the antibiotic production from the two experiments will depend on the rates of inactivation:

$$d[\text{Ceph}]_1/d[\text{Ceph}]_2 \simeq E_1/E_2 \simeq \exp[-(k_{d,1} - k_{d,2})t], \tag{21}$$

where subscript 1 is for glycerol alone and subscript 2 is for glycerol with chloramphenicol. Integration of this expression shows that the ratio of the antibiotic concentrations must increase exponentially with time if the deactivation rate constants differ [i.e., $k_{d,1} > k_{d,2}$ as suggested by Hu et al. (1984)]. Inspection of the data of Fig. 3 shows, however, that the ratio was constant at all times at approximately 0.5, so that differential inactivation could not account for the results.

Equation (21) holds the key to the only possible explanation. The rate constants cannot change because the carbon source was only inhibitory when protein synthesis was active. The data are inconsistent with a change in the rate of inactivation due to protein synthesis, so the enzyme activities must be comparable. The observed reduction in yield was therefore due to a reduction in the supply of precursors (i.e., regulatory pathway a in Fig. 5.4). Protein synthesis would be extremely active in resting cells under these conditions (Cortes et al., 1986). Given that the antibiotic pathway branches off from amino acid synthesis, we would expect the supply of precursors to diminish if resting cells are supplied with carbon source because they would be diverted for use in new proteins. The data of Hu et al. (1984) suggest that this diversion can reduce antibiotic synthesis in resting cells by 50%.

10.2. Prediction of Cephamycin C Formation at 100% DO

The simple rate expression for the conversion of penicillin N to cephamycin C [Eq. (13)] indicates that the biosynthetic pathway is regulated by two mechanisms: supply of precursors and the quantity of active enzyme. This hypothesis can be tested by using data for cephamycin C formation at 100% dissolved oxygen from Rollins et al. (1988a, 1990).

DAOCS was expressed to higher concentrations at 100% DO, as illustrated in Fig. 5.7. Penicillin N apparently oscillated between 55 and 0 mg/mL (Rollins et al., 1988a). Such an oscillation would require rapid, coordinated changes in the *in vivo* activity of IPNS and DAOCS by some unknown mechanism. An alternative explanation is that the apparent oscillations were the result of the experimental method used to determine the concentration of penicillin N. Total antibiotic concentration was measured by a diffusion assay, where the ring of inhibition was measured on a plate with a supersensitive strain of *E. coli*. Cephamycin C was measured by adding a β-lactamase to remove the penicillins. The concentration of penicillin N was determined by the difference between the two assays. This approach would give reasonable results if penicillin N and cephamycin C were present in similar amounts (Fig. 10). If the cephamycin C concentration were much larger than the penicillin N, then the error in the difference would be significant. Rollins et al. (1988a) reported cephamycin C levels up to 320 mg/L at 100% oxygen. Penicillin N at <50 mg/L cannot be determined accurately by the difference method under these conditions.

The approach to modeling was to use Eq. (14) for cephamycin C formation. Data for cephamycin C and protein were obtained from Rollins et al. (1988a). DAOCS activities were extrapolated for the period 36–96 h using Eq. (4) and parameters from Table 5.2 because no data were reported for the latter stages of fermentation. Given the statistical arguments regarding penicillin N, a constant concentration of 18 mg/L was used [the mean of the data of Rollins et al. (1988a) was 16.8 mg/L with a standard deviation of 16.5 mg/L]. A value of 0.147 mL/U/h for k_p (identical to the uncontrolled dissolved oxygen case) gave excellent agreement with the data, as illustrated in Fig. 5.12.

This result shows that the model is predictive for changes in cephamycin C yield in response to differences in enzyme level and precursor concentration. The increase in DAOCS expression due to increased oxygen gave more efficient conversion of penicillin N to cephamycin C. Higher oxygen must also have increased the overall supply of the ACV precursor, as indicated by the six- to sevenfold increase in final antibiotic titer. The increase in cephamycin C level suggests that product inhibition was not significant over this range of concentration. The simple model for DAOCS kinetics was also consistent with a study by Skatrud et al. (1989) on cloning of DAOCS in *C. acremonium*. Insertion of a plasmid bearing the gene for DAOCS increased the assayed enzyme activity by a factor of 2 relative to the unmodified strain. The effect of the higher enzyme activity was to enhance the conversion of penicillin N to cephamycin C, which demonstrated that in the nontransformed recipient the activity of DAOCS was rate limiting.

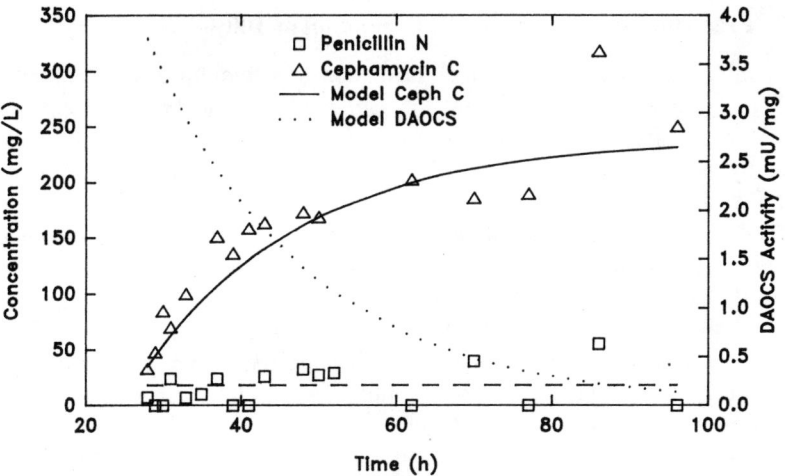

FIGURE 5.12. Kinetics of cephamycin C formation with 100% dissolved oxygen saturation. The symbols represent the actual data points of Rollins et al. (1990). The drawn line for DAOCS (· · ·) represents the results of the model [Eq. (4) with parameters from Table 5.2]. The drawn line for cephamycin C (——) represents the results of model Eq. (14) with $k_p = 0.147$ mL/(mU · h). The drawn line for penicillin N (- - -) shows its average concentration during the fermentation.

10.3. Enhancement of Enzyme Activity by Prolonged Transcription

A significant observation in all the experiments was that a steady-state activity of enzyme was not achieved or maintained because transcription was enabled for a relatively short period in these batch experiments. The steady-state activity of enzyme is given by Eq. (2), so that the parameters in Tables 5.1–5.3 can be used to calculate the enzyme levels that could be obtained if transcription remained on longer. These predicted values are given in Table 5.4 for ACVS, IPNS, and DAOCS.

The estimated values of E_{ss} show considerable variation depending on the medium, which is expected from the earlier discussion on how r_T and k_d tend to be related, particularly for IPNS. As a general comment, most of the experiments using defined media achieved a high fraction of the potential steady-state enzyme activity; in the range of 35–99%, with the majority of values in the range of 80–95%. The fermentations in complex media, on the other hand, gave both better enzyme activities and a much higher ultimate potential activity. The experiments in complex media only achieved 3–60% of the potential enzyme activity, so that even better yields could be achieved if transcription were prolonged.

On the basis of this argument, the appropriate design of fermentation conditions to prolong transcription would yield significant improvements in enzyme activity. Because the on/off regulatory mechanism for the cluster of biosynthetic genes is not known, the exact conditions cannot be identified quantitatively.

TABLE 5.4. Predicted Steady-State Enzyme Activities in *Streptomyces clavuligerus*

Enzyme	Medium	Predicted E_{ss} (mU/mg cell protein)	Percentage of E_{ss} Observed
ACVS	Complex, no DO control	0.89	28
	Complex, 100% DO	0.43	60
IPNS	Starch + TCS	8.76	54
	Starch + asparagine	10.5	35
	Glycerol + asparagine	1.73	95
	Glycerol + NH_4	1.58	49
	Complex, no DO control	11.5	3
	Complex, 100% DO	1.08	46
DAOCS	Glycerol + asparagine	0.38–1.09	62–94
	Starch + asparagine	0.69	99
	Starch + glycerol + asparagine	0.35	80
	Complex, no DO control	3.0	40
	Complex, 100% DO	13.0	38

10.4. Optimization of Fermentation Conditions

The kinetics of cephamycin C formation and ACV conversion indicate that two factors control the yield of antibiotic: activity of biosynthetic enzymes and supply of precursors. Any optimization scheme must seek to enhance both requirements simultaneously. Optimization of the precursors is difficult because few data are available on the kinetics of the free amino concentrations, but the available data do show that the pool of intracellular amino acids should be maintained. The conditions required to achieve this aim are the same as for prolonging transcription; growth must continue at a low rate. The condition in a batch fermentation that favors both transcription and the availability of precursors is the restricted but nonzero supply of a key nutrient, which occurs in the late growth phase. Fed-batch operation would provide a straightforward method of maintaining cellular metabolism in this regime, wherein the rate of feeding is adjusted to keep the nutrient at levels that support slow growth, that is, levels that support secondary metabolite production. Further design of the feeding strategy will require a better knowledge of regulatory pathways a and b in Fig. 5.4.

10.5. Modification of Cellular Regulation

The previous discussion on the stimulatory effect of cloning extra copies of the DAOCS gene, which resulted in an increase in enzyme level, illustrates the point that an increase in a rate-limiting enzyme improved yield only to the point where the available precursor, penicillin N, was consumed. A similar experiment with the IPNS enzyme (Skatrud et al., 1989) was unsuccessful in increasing antibiotic yield, indicating that this activity was not limiting the conversion of pathway intermediates to cephalosporin C. Cloning of the entire cluster of biosynthetic genes would suffer from the same limitation in that more enzyme will convert the available precursor faster. The yield improvement will be modest unless steps are also taken to supply more amino acids for conversion. A combination of increasing the content of rate-limiting enzymes through cloning together with appropriate fermentation conditions could enhance the conversion of the amino acid pool into antibiotic because both sides of the reaction requirements would be met. The results of Rollins et al. (1990), illustrated in Fig. 5.11, showed that enhancing the enzyme activity and the supply of precursors could give a dramatic increase in yield (about four times as much cephamycin C was produced at 100% DO relative to the uncontrolled experiments). A few hours more of transcription and precursor supply could give dramatic improvements in yield.

Whatever the means of achieving both prolonged transcription and continued supply of amino acid precursors, the simple model of Section 8 can predict the resulting yields of cephamycin C. The combination of Eqs. (3), (4), and (14) gives the predicted yield for any combination of transcription time and supply of precursors. To illustrate, the following cases were considered:

Base Case. Yields of cephamycin C as observed by Rollins et al. (1990), with transcription of DAOCS turned on from 20.5 to 27.5 h (duration 7 h) and penicillin N available at an average concentration of 18 mg/L.

Case 1—Prolonged Transcription. The duration of DAOCS transcription was doubled, from 7 to 14 h. The concentration of penicillin N was also maintained at 18 mg/L, which implies continued activity of the earlier enzymes in the biosynthetic pathway, particularly ACVS.

Case 2—Enhanced Supply of Precursors. The duration of DAOCS transcription was 7 h, as in the base case, but the supply of precursors was doubled, giving an average penicillin N concentration of 36 mg/L.

Case 3—Prolonged Transcription and Enhanced Supply of Precursors. Both the duration of transcription of DAOCS and the concentration of penicillin N were doubled, to 14 h and 36 mg/L, respectively.

The activity of DAOCS and the yields of cephamycin C for these cases are illustrated in Fig. 5.13. The two strategies of either prolonged transcription or enhanced supply of precursors give equivalent yields of cephamycin C, about 400 mg/L. A combination of the two strategies doubles the yield to about 850 mg/L after 100 h of fermentation. Although Fig. 13 only shows the activity of DAOCS, similar activities of ACVS and IPNS must be sustained in order to achieve the yields of cephamycin C illustrated. For example, if DAOCS transcription were prolonged but the supply of penicillin N were not maintained, then the 18 mg/L of penicillin N intermediate would be consumed and product formation would cease, as in the case of DAOCS inserted on plasmids in *C. acremonium* (Skatrud et al., 1989).

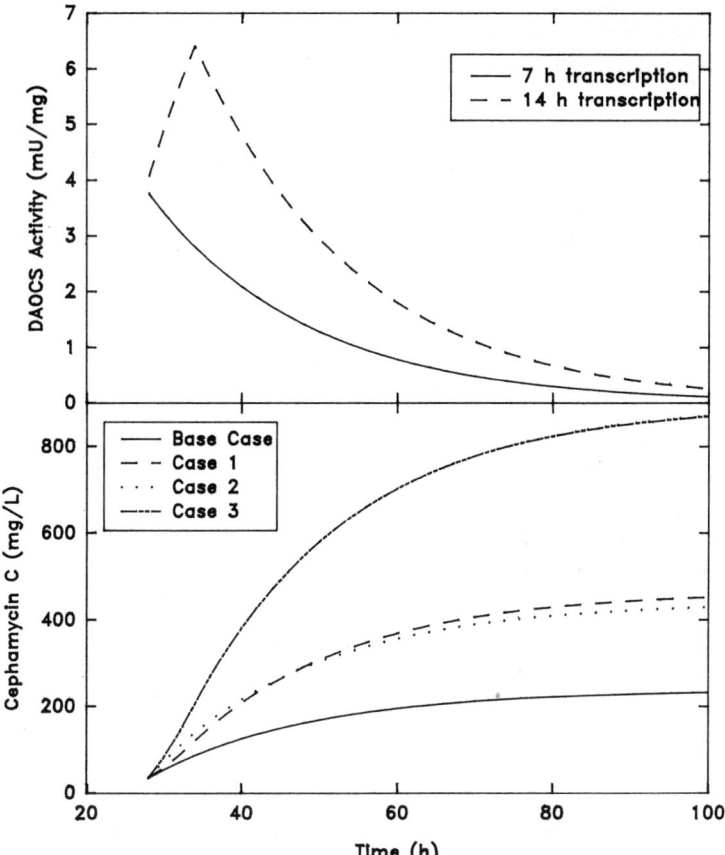

FIGURE 5.13. Predicted kinetics of cephamycin C formation with prolonged transcription and enhanced supply of precursors. Base Case is from Rollins et al. (1990) and Fig. 5.12. Case 1—doubling the duration of DAOCS transcription to 14 h; Case 2—doubling the supply of penicillin N to 36 mg/L; Case 3—doubling both the duration of transcription and the supply of penicillin N.

11. CONCLUSION

The kinetics of cephamycin C production by *S. clavuligerus* can be modeled using the activities of the biosynthetic enzymes and the concentration of precursors. The time course of the biosynthetic enzyme activities follows a pattern consistent with a burst of transcription followed by gradual inactivation. Catabolite repression can be modeled empirically given the initial concentration of nutrient, but more information is required on the global regulation of the antibiotic pathway in order to model the relationship between cell metabolism and production of secondary metabolites. The kinetic model showed that the most effective strategy to increase the yield of cephamycin C was to prolong transcription of the biosynthetic enzymes and to maintain the supply of amino acid precursors.

REFERENCES

Aharonowitz, Y., and Demain, A. L. (1978), *Antimicrob. Agents. Chemother.*, **145**, 159–164.
Aharonowitz, Y., and Demain, A. L. (1979), *Can. J. Microbiol.*, **25**, 61–67.
Aharonowitz, Y., Mendelovitz, S., Kirenberg, F., and Kuper, V. (1984), *J. Bacteriol.*, **157**, 337–340.
Bajpai, R. K., and Reuss, M. (1980), *J. Chem. Technol. Biotechnol.*, **30**, 332–344.
Brana, A. F., Wolfe, S., and Demain, A. L. (1985), *Can. J. Microbiol.*, **31**, 736–743.
Castro, J. M., Liras, P., Cortes, J., and Martin, J. F. (1985), *Appl. Microbiol. Biotechnol.*, **22**, 32–40.
Chater, K. F. (1990), *Bio/Technol.*, **8**, 115–121.
Chatterjee, S., and Vining, L. C. (1981), *Can. J. Microbiol.*, **28**, 311–317.
Cortes, J., Liras, P., Castro, J. M., and Martin, J. F. (1986), *J. Gen. Microbiol.*, **132**, 1805–1814.
Cundliffe, E. (1989), *Annu. Rev. Microbiol.*, **43**, 207–233.
Hu, W.-S., Brana, A. F., and Demain, A. L. (1984), *Enzyme Microb. Technol.*, **6**, 155–160.
Imanaka, T., and Aiba, S. (1977), *Biotechnol. Bioeng.*, **19**, 757–764.
Jensen, S. E. (1986), *Crit. Rev. Biotechnol.*, **3**, 277–301.
Jensen, S. E., Westlake, D. W. S., and Wolfe, S. (1982a), *J. Antibiot.*, **35**, 483–490.
Jensen, S. E., Westlake, D. W. S., and Wolfe, S. (1982b), *J. Antibiot.*, **35**, 1351–1360.
Jensen, S. E., Westlake, D. W. S., and Wolfe, S. (1984), *Appl. Microbiol. Biotechnol.*, **20**, 155–160.
Jones, G. H. (1985), *J. Bacteriol.*, **163**, 1215–1221.
Lebrihi, A., Germain, P., and Gerard, L. (1987), *Appl. Microbiol. Biotechnol.*, **26**, 130–135.
Lebrihi, A., Lefebvre, G., and Germain, P. (1988a), *Appl. Microbiol. Biotechnol.*, **28**, 44–51.
Lebrihi, A., Lefebvre, G., and Germain, P. (1988b), *Appl. Microbiol. Biotechnol.*, **28**, 39–43.
Lee, S. B., and Bailey, J. E. (1984), *Biotechnol. Bioeng.*, **24**, 1372–1382.
Lubbe, C., Wolfe, S., and Demain, A. L. (1985a), *Arch. Microbiol.*, **140**, 317–320.
Lubbe, C., Wolfe, S., and Demain, A. L. (1985b), *Enzyme Microb. Technol.*, **7**, 353–356.
Lubbe, C., Demain, A. L., and Bergman, K. (1985c), *Appl. Microbiol. Biotechnol.*, **22**, 424–427.
Madduri, K., Stuttard, C., and Vining, L. C. (1989), *J. Bacteriol.*, **171**, 299–302.

Madduri, K., Shapiro, S., DeMarco, A. C., White, R. L., Stuttard, C., and Vining, L. C. (1991a), *Appl. Microbiol. Biotechnol.*, **35**, 358–363.

Madduri, K., Stuttard, C., and Vining, L. C. (1991b), *J. Bacteriol.*, **173**, 985–988.

Malik, V. S., and Vining, L. C. (1972), *Can. J. Microbiol.*, **18**, 583–590.

Matsumura, M., Imanaka, T., Yoshida, T., and Taguchi, H. (1981), *J. Ferment. Technol.*, **59**, 115–123.

Mendelovitz, S., and Aharonowitz, Y. (1982), *Antimicrob. Agents Chemother.*, **21**, 74–84.

Nestaas, E., and Wang, D. I. C. (1983), *Biotechnol. Bioeng.*, **30**, 781–796.

Price, V. E., Sterling, W. R., Tarantola, V. A., Hartley, R. W., and Reichcigl, M. (1962), *J. Biol. Chem.*, **237**, 3468–3475.

Regan, C. M., and Vining, L. C. (1978), *Can. J. Microbiol.*, **24**, 1012–1015.

Rollins, M. J., Jensen, S. E., and Westlake, D. W. S. (1988a), *J. Ind. Microbiol.*, **3**, 357–364.

Rollins, M. J., Jensen, S. E., Wolfe, S., and Westlake, D. W. S. (1988b), *Biotechnol. Lett.*, **10**, 295–300.

Rollins, M. J., Westlake, D. W. S., Wolfe, S., and Jensen, S. E. (1988c), *Can. J. Microbiol.*, **34**, 1196–1202.

Rollins, M. J., Jensen, S. E., and Westlake, D. W. S. (1989a), *Can. J. Microbiol.*, **35**, 1111–1117.

Rollins, M. J., Jensen, S. E., and Westlake, D. W. S. (1989b), *Appl. Microbiol. Biotechnol.*, **31**, 390–396.

Rollins, M. J., Jensen, S. E., Wolfe, S., and Westlake, D. W. S. (1990), *Enzyme Microbiol. Technol.*, **12**, 40–45.

Rollins, M. J., Jensen, S. E., and Westlake, D. W. S. (1991), *Appl. Microbiol. Biotechnol.*, **35**, 83–88.

Skatrud, P. L., Tietz, A. J., Ingolia, T. D., Cantwell, C. A., Fisher, D. L., Chapman, J. L., Queener, S. W. (1989), *Bio/Technol.*, **7**, 477–485.

Switzer, R. L. (1977), *Annu. Rev. Microbiol.*, **31**, 135–157.

Turner, M. K., Farthing, J. E., Brewer, S. J. (1978), *Biochem. J.*, **173**, 839–850.

Vining, L. C., and Doull, J. L. (1989), in *Proceedings of the 7th International Symposium on the Biology of Actinomycetes*, pp. 406–411.

Vining, L. C., Jensen, S. E., Westlake, D. W. S., Aharonowitz, Y., and Wolfe, S. (1987), *Appl. Microbiol. Biotechnol.*, **27**, 240–246.

Wolfe, S., Demain, A. L., Jensen, S. E., and Westlake, D. W. S. (1984), *Science*, **226**, 1386–1392.

6 Comparing the Accuracy of Equation Formats for Modeling Microbial Population Decay Rates

CHRISTON J. HURST
U.S. Environmental Protection Agency

DEANNA K. WILD
National Institute for Occupational Safety and Health

ROBERT M. CLARK
U.S. Environmental Protection Agency

1. Introduction 149
2. Statistical design 150
   2.1. Additive error format 150
   2.2. Multiplicative error formats 151
   2.3. Logistic format 152
   2.4. Exponential formats 153
3. Comparison of equation formats 156
   3.1. Development of models and regression of predicted versus observed values 156
   3.2. Visual test for comparing the accuracy of equation formats 157
   3.3. Numerical test for comparing the accuracy of equation formats 171
   3.4. Full models developed using the multiplicative error II equation format 173
4. Summary 174
   References 175

1. INTRODUCTION

A microbial population can be described as being in either a state of active growth, stasis, or senescence at any specified point in time, assuming the absence of predation. When in senescence, the number of viable organisms within a microbial population decreases with the passage of time due to the death or loss of viability of individual members of that population. Unfortunately, the small physical size of microorganisms usually makes it impossible to follow the loss of viability for each individual member within a microbial population. It is for this reason that microbial die-off is commonly discussed in terms of the fate of the microbial population taken as a whole. The die-off of a microbial population can be described as being analogous to the decay rates associated with radioisotopes, in that the frequency of individual death or decay events is assumed to occur at a statistically calculable rate. Because of this analogy, the rapidity with which the members of a microbial population die off is often expressed as a rate function, and the results can be termed "microbial population decay rates."

The study presented in this chapter addresses viral environmental stability. Survival of viruses in natural environments is an important topic relevant to both human health and the health of other animals and plants. Viral stability is also a point of interest with regard to food chain studies and to research on the possible role of viruses in microbial population dynamics (Bergh et al., 1989; Børsheim et al., 1990; Suttle et al., 1990; Williams et al., 1987). Furthermore, accurate models for microbial stability are required for understanding the environmental transport of microorganisms.

Viral inactivation is an exponential function and under environmental conditions viral inactivation is predominantly influenced by deleterious temperature effects. Additional detrimental environmental influences can include aerobic microbial activity, low relative humidity in relation to desiccation, solar radiation, and in some cases identifiable ions or classes of chemicals present in the surrounding environmental medium. Viral adsorption onto surfaces provides a protective effect. No standard format exists for modeling environmental virus stability. The present study was intended to help fill this void and compared the relative accuracy of eight regression equation formats for empirically modeling the influence of environmental variables on viral population decay rates. Based on these equation formats, sets of models were produced using a database generated during a descriptive study on the stability of seeded human viruses in natural surface freshwaters (Hurst et al., 1989). Accuracy of the equation formats was assessed by then using the developed models to predict the outcome of that same virus-seeding study, and comparing sets of predicted values generated by the individual equation formats against their corresponding experimentally observed values. That comparison involved a series of simple linear regressions for which Y and X, respectively, represented the predicted and observed values and included a visual comparison of the plotted linear regressions. During the first part of this analysis project, when the model equations were generated, Y represented the actual experimentally observed values. During the second part of the project, when subsequently using those model equations to predict the outcome of the study, the generated predicted values represented estimates of the expected value of Y: $E(Y)$.

2. STATISTICAL DESIGN

2.1. Additive Error Format

Eight different equation formats were used to develop models for this study. The first equation format examined was a multiple linear regression based on the following equation:

$$Y = \beta_0 + \beta_1 X_1 + \beta_2 X_2 + \cdots + \beta_n X_n, \qquad (1)$$

which employed as its dependent variable (Y) predetermined viral inactivation rates, expressed in terms of $[\log_{10}(N_t/N_0)]/t$. These rates were slope values that had been calculated by linear regression of daily values for viral titer change [expressed in terms of $\log_{10}(N_t/N_0)$] as the dependent variable versus the length of incubation time (t, expressed in days) since the outset of a defined study period (Hurst et al., 1989). This approach yielded the following equation format, which was used for actual construction of models:

$$\frac{\log_{10}(N_t/N_0)}{t} = \beta_0 + \beta_1 X_1 + \beta_2 X_2 + \cdots + \beta_n X_n. \qquad (2)$$

Models based on this equation format were termed "additive error." Selection of this equation format for use in the study was based on its previous application for modeling the influence of soil characteristics on viral stability in soil (Hurst et al., 1980). For all the equation formats that are presented in this chapter, N_0 represents the titer of a microbial population at time 0, the outset of a defined study period. N_t represents the titer of that same population after the passage of any time period t. Time was always expressed in terms of days for the comparative modeling study that is presented in this chapter. The eight equation formats used for constructing the models presented in this chapter differ with respect to whether time appears as an independent variable or is incorporated into the dependent variable. For this chapter the symbols X_1 through X_n represent a series of independent variables, which included water characteristics that previously had been shown to correlate individually with the inactivation rates measured for viral populations incubated in surface freshwaters (Hurst et al., 1989). Incubation temperature, expressed in Kelvin and represented by T, was included among these independent variables in some of the models. The latter models were termed "across temperature" or "full range" models, as will be discussed later. For this chapter, β_0 always represents a Y-axis intercept value. The values β_1 through β_n represent the respective coefficients assigned to the independent variables X_1 through X_n. As mentioned above, some of the equation formats that are presented in this chapter employ incubation time, represented by t, as an independent variable. β_t always represents the coefficient assigned to time when time was used as an independent variable.

2.2. Multiplicative Error Formats

Two types of "multiplicative error" equation formats were used in this comparative modeling study. Both of these were based on the following equation:

$$Y = \beta_0 X_1^{\beta_1} X_2^{\beta_2} \cdots X_n^{\beta_n}, \qquad (3)$$

which, in order to construct models, was $\log_{10}$ transformed to

$$\log_{10} Y = \log_{10} \beta_0 + \beta_1 \log_{10} X_1$$
$$+ \beta_2 \log_{10} X_2 + \cdots + \beta_n \log_{10} X_n. \quad (4)$$

The two multiplicative error formats differed with respect to their dependent variable. A multiplicative error format employing the survival ratio (N_t/N_0) as dependent variable (Y) has previously been used to model chemical disinfection studies (Hom, 1972) and roughly corresponds to the equation format that we term "multiplicative error II." The final equation for the multiplicative error II format, following its $\log_{10}$ transformation, was

$$\log_{10} \frac{N_t}{N_0} = \log_{10} \beta_0 + \beta_1 \log_{10} X_1 + \beta_2 \log_{10} X_2$$
$$+ \cdots + \beta_n \log_{10} X_n + \beta_t \log_{10} t. \quad (5)$$

Our "multiplicative error I" equation format uses as its dependent variable (Y) the same rate values $\{[\log_{10}(N_t/N_0)]/t\}$ as employed for dependent variable in the additive error format and was selected to provide an intermediate equation form between the additive error and multiplicative error II formats. Thus the equation for the multiplicative error I format following its $\log_{10}$ transformation was

$$\log_{10}\left[\frac{\log_{10}(N_t/N_0)}{t}\right] = \log_{10}\beta_0 + \beta_1 \log_{10} X_1 + \beta_2 \log_{10} X_2$$
$$+ \cdots + \beta_n \log_{10} X_n. \quad (6)$$

It is important to note that time is a part of the dependent variable for both the additive error and multiplicative error I equation formats. In contrast to this, the multiplicative error II equation format employs time (t) as one of the independent variables and thus allows time to have a coefficient (β_t).

2.3. Logistic Format

The fourth type of equation format used in this modeling comparison study was termed "logistic" and was based on the following equation:

$$Y = \frac{1}{1 + \exp[-(\beta_0 + \beta_1 X_1 + \beta_2 X_2 + \cdots + \beta_n X_n + \beta_t t)]}, \quad (7)$$

which, in order to construct models, was "logit" transformed to

$$\log_n\left[\frac{Y}{1-Y}\right] = \beta_0 + \beta_1 X_1 + \beta_2 X_2 + \cdots + \beta_n X_n + \beta_t t. \quad (8)$$

The survival ratio (N_t/N_0) was used as dependent variable (Y) in this equation format. Thus the final form of the logistic equation format was

$$\log_n \left[\frac{N_t/N_0}{1 - (N_t/N_0)} \right] = \beta_0 + \beta_1 X_1 + \beta_2 X_2 + \cdots + \beta_n X_n + \beta_t t. \quad (9)$$

Use of the logistic equation format was based on its classical application for studying survival of macroorganisms such as humans, although in that form a differential assessment of "live versus dead" for each individual organism is used as the dependent variable. Such a differential assessment cannot be made for individual members of a virus population. This is the reason why the survival ratio was employed as dependent variable in the logistic models. As with the multiplicative error II equation format, our logistic equation format incorporates time (t) among the independent variables and thus allows time to have a coefficient (β_t).

2.4. Exponential Formats

A series of four equation formats termed "exponential" were also evaluated in this modeling study. Their use was prompted by the proposal of Chick (1908), and later examinations by Casolari (1981), that the decrease in titer of a microbial population due either to disinfectant activity (Chick, 1908) or some other inactivating factor (Casolari, 1981) was an exponential function corresponding to a general exponential decay equation:

$$\frac{dN}{dt} = -kN. \quad (10)$$

According to this equation, the amount of titer decrease within a given time period is dependent on the number of organisms present at the start of that time period. Integration of Eq. (10) yields

$$\frac{N_t}{N_0} = e^{-kt}, \quad (11)$$

with k being a rate constant. The additive error and multiplicative error equation formats defined above are closely related to Eq. (11) but are not identical to it. The logistic equation format is related to Eq. (11) only in a very restricted sense. The four different equation formats that we have termed "exponential" for the purpose of this study are based directly on Eqs. (10) and (11). In order to derive these exponential equation formats, Eq. (11) was first taken through the following intermediate equation:

$$\log_n \frac{N_0}{N_t} = kt. \quad (12)$$

Both sides of Eq. (12) were then divided by t to yield:

$$\frac{\log_n (N_0/N_t)}{t} = k. \qquad (13)$$

Based on Eq. (13), two equation formats were then generated, which model k as a function of multiple independent variables. The two equation formats generated from Eq. (13) were termed "exponential IA" [Eq. (14)] and "exponential IB" [Eq. (15), which was subjected to a natural logarithmic transformation to yield Eq. (16) prior to its actual use in constructing models]. The exponential IA equation format uses an additive error structure to model the influence of the independent variables. In contrast, the exponential IB equation format models the influence of the independent variables using a multiplicative error structure. Both of these two exponential I equation formats, as with the additive error and multiplicative error I equation formats, employ time (t) as part of the dependent variable. Therefore neither of these exponential I equation formats, the additive error equation format nor the multiplicative error I equation format, allows the effect of time to be assigned a "beta" coefficient (which would be labeled as β_t).

$$\frac{\log_n (N_0/N_t)}{t} = \beta_0 + \beta_1 X_1 + \beta_2 X_2 + \cdots + \beta_n X_n, \qquad (14)$$

$$\frac{\log_n (N_0/N_t)}{t} = \beta_0 X_1^{\beta_1} X_2^{\beta_2} \cdots X_n^{\beta_n}, \qquad (15)$$

$$\log_n \left[\frac{\log_n (N_0/N_t)}{t} \right] = \log_n \beta_0 + \beta_1 \log_n X_1 + \beta_2 \log_n X_2$$
$$+ \cdots + \beta_n \log_n X_n. \qquad (16)$$

Two additional "exponential" equation formats, termed "exponential II," were also based on Eq. (12). The exponential II equation formats differ from the exponential I equation formats in that the "exponential II" formats incorporate time (t) among the independent variables. The exponential II equation formats thus allow time to have a coefficient (β_t). This, of course, is the same manner of difference that exists between the multiplicative I and multiplicative II equation formats. The exponential IIA equation format uses an additive error structure to model the influence of the independent variables. The exponential IIB equation format models the influence of the independent variables by means of a multiplicative error structure. The exponential IIA equation format is represented by Eq. (17). The exponential IIB equation format is represented by Eq. (18), which, in order to construct models, was first logarithmically transformed to Eq. (19):

$$\log_n \frac{N_0}{N_t} = \beta_0 + \beta_1 X_1 + \beta_2 X_2 + \cdots + \beta_n X_n + \beta_t t, \qquad (17)$$

$$\log_n \frac{N_0}{N_t} = \beta_0 X_1^{\beta_1} X_2^{\beta_2} \cdots X_n^{\beta_n} t^{\beta_t}, \tag{18}$$

$$\log_n \left[\log_n \frac{N_0}{N_t} \right] = \log_n \beta_0 + \beta_1 \log_n X_1 + \beta_2 \log_n X_2$$
$$+ \cdots + \beta_n \log_n X_n + \beta_t \log_n t. \tag{19}$$

The eight equation formats used for constructing the regression models are further described in Table 6.1. These additive error, multiplicative error, logistic,

TABLE 6.1. Equation Formats Used for Regression Models

Equation Format	Dependent Variable (Y)	Independent Variables[a] (X's)
Additive error	$[\log_{10} (N_t/N_0)]/\text{day}$[b]	Conductivity, turbidity, ability to support bacterial growth
Multiplicative error I	$[\log_{10} (N_t/N_0)]/\text{day}$[b]	Conductivity, turbidity, ability to support bacterial growth
Multiplicative error II	N_t/N_0	Conductivity, turbidity, ability to support bacterial growth, incubation time in days
Logistic	N_t/N_0	Conductivity, turbidity, ability to support bacterial growth, incubation time in days
Exponential IA	$[\log_n (N_0/N_t)]/\text{day}$	Conductivity, turbidity, ability to support bacterial growth
Exponential IB	$[\log_n (N_0/N_t)]/\text{day}$	Conductivity, turbidity, ability to support bacterial growth
Exponential IIA	$\log_n (N_0/N_t)$	Conductivity, turbidity, ability to support bacterial growth, incubation time in days
Exponential IIB	$\log_n (N_0/N_t)$	Conductivity, turbidity, ability to support bacterial growth, incubation time in days

[a] Conductivity expressed as μmho/cm, turbidity as nephelometric turbidity units, and ability to support bacterial growth expressed as the mean number of generations of growth that independently seeded populations of *Pseudomonas fluorescens*, *Enterobacter cloacae*, and *Klebsiella oxytoca* could undergo in sterilized but otherwise identical samples of the same waters. Incubation temperature, expressed in Kelvin, was used as an additional independent variable in the models developed across temperature. Models developed for individual incubation temperatures did not include temperature as an independent variable.
[b] Each of the inactivation rate values used as dependent variable for the additive error and multiplicative error I formats represents the extent of viral inactivation observed for any given combination of water source, virus type, and incubation temperature over a period of 8 weeks for incubation at 22°C or 12 weeks for incubation at either 1°C or −20°C. Rate values were calculated independently by linear regression.

and exponential equation formats examine the nature of the relationship between the survival ratio, time (t), and their interaction with temperature and the water characteristics.

3. COMPARISON OF EQUATION FORMATS

3.1. Development of Models and Regression of Predicted Versus Observed Values

For the purpose of these modeling efforts, missing values were ignored. Weights were not assigned to the independent variables. Temperature was modeled using the Kelvin scale to prevent loss of data for temperatures below 0°C during log transformations. Inherent variability in viral assays occasionally results in survival ratio values greater than 1.0. For logistic modeling, any survival ratio values greater than 1.0 were recoded as 1.0 to prevent their rejection during logit transformation.

The three water characteristics of conductivity, turbidity, and the ability of a water sample to support bacterial growth were selected for use as independent variables in this modeling study because each correlated in a statistically significant manner with virus survival during the study that generated the database being modeled (Hurst et al., 1989). These three variables did not strongly cross-correlate with one another (Hurst et al., 1989). The study by Hurst et al. (1989) employed a factorial design in which purified viruses were added to water collected from five different surface water sources, and replicate portions of each seeded water stored, respectively, for prolonged time at −20, 1, or 22°C. The seeded waters were stored in the dark under aerobic conditions with the water containers kept loosely capped to allow gas exchange with the atmosphere. Viral titers for the seeded waters were determined both at the outset of storage (N_0) and periodically during the course of storage. Data for four of these waters were considered suitable for this modeling comparison. Data for the fifth water were eliminated from consideration because that water could not be assessed for one of the independent variables: ability to support bacterial growth.

A set of four models was developed for each of the eight equation formats, one apiece using the data from the individual incubation temperatures, plus one termed an ''across temperature'' model, which utilized the data from all three incubation temperatures and thus necessarily included temperature as an additional independent variable. The resulting 32 models were then used to predict the outcome of that same virus survival study. Relative accuracy of the eight different equation formats was determined by comparing the sets of predicted values generated by the models versus the actual experimentally observed values. Each model was examined for ability to accurately predict the range of values for which it was developed (i.e., using a model developed for 1°C to predict outcome for 1°C, or using an across temperature model developed for the full range of −20 to 22°C to predict outcome for that same full range). Also examined was whether or not a model developed for a broad temperature range (−20 to 22°C) was accurate at predicting values for a subset of that range (−20, 1, or 22°C). The latter was accomplished by segregating the values predicted by each across temperature model with respect to incubation temperature, prior to regressing those sets of predicted values against their corresponding experimentally observed values. Comparison of corresponding sets of predicted and observed values was done by simple linear regression [Eq. (20)]. The equations that resulted from these regressions of predicted (Y) and observed (X) values are listed in Table 6.2 along with their respective values for n, r^2, and p:

$$Y = \beta_0 + \beta_1 X. \tag{20}$$

3.2. Visual Test for Comparing the Accuracy of Equation Formats

Figures 6.1 through 6.9 present the use of a visual test employed for assessing the comparative accuracy of models based on the different equation formats. A visual test was used for this comparison because of the differences between the dependent variables that were employed by the equation formats. Parts (a) of these figures compare the values predicted by the across temperature models generated from these different equation formats versus the experimentally observed values for the full temperature range. The most important things to note for parts (a) of the individual Figs. 6.1–6.9 are, in each case, how closely the line generated by regressing the predicted and observed values (solid line) comes in terms of overlaying the ideal line (dashed line representing $Y = X$, i.e., predicted = observed) and how well the individual triangles, which correspond to the coordinates of the paired predicted and observed values, follow the path of these two lines. If a model equation fits the experimental data (observed values) perfectly, then the set of values subsequently predicted by that model should exactly duplicate the set of observed values. In such a case, the slope and Y-axis intercept resulting from linear regression of the predicted versus observed values should produce the ideal line.

Theoretically, an ideal equation format should be able to generate a model for a broad temperature range that can perfectly predict any subset of that same temperature range. In addition, an ideal equation format should be able to generate individual temperature models that are perfectly extendable to any larger temperature range. Of course, no such ideal equation format is likely to be found. The unlikelihood of finding a perfect equation format is further compounded by both an inability to recognize and include in a model all the factors that influence microbial survival, and an inability to model the variation that naturally exists in any body of experimental data. Nevertheless, these two topics—ability of a broad range model to accurately predict a subset of its range and extendability of models—are examined in parts (b) of Figs. 6.1–6.9.

In order to assess the ability of each across temperature model to accurately predict values for an individual part of its temperature range, the predicted values generated by that across temperature model were segregated according to the incubation temperature that they represented, before being linearly regressed against

TABLE 6.2. Comparison of Values Predicted by Model Equations Versus Experimentally Observed Values[a]

Equation Format of Model Used to Generate Predicted Values for the Comparison	Form of the Compared Values		Model Developed Across All Incubation Temperatures with Predicted and Observed Values then Compared for:[b]				Models Developed for Individual Incubation Temperature and the Predicted Values then Compared with Observed Values for that Same Temperature of:		
		−20 to 22°C		−20°C	1°C	22°C	−20°C	1°C	22°C
Additive error	$\log_{10}(N_t/N_0)$/day	Slope	0.856	Slope −0.731	Slope 0.323	Slope 0.165	Slope 0.328	Slope 0.514	Slope 0.481
		Y-intercept	−0.011	Y-intercept 0.004	Y-intercept −0.059	Y-intercept −0.129	Y-intercept −0.003	Y-intercept −0.024	Y-intercept −0.088
		n	36	n 12	n 12	n 12	n 12	n 12	n 12
		r^2	0.856	r^2 0.030	r^2 0.055	r^2 0.473	r^2 0.328	r^2 0.514	r^2 0.481
		p	≤0.0001	p 0.589	p 0.462	p 0.013	p 0.052	p 0.009	p 0.012
Multiplicative error I	$\log_{10}(N_t/N_0)$/day	Slope	1.089	Slope 0.068	Slope −0.149	Slope 0.105	Slope 0.367	Slope 0.509	Slope 0.504
		Y-intercept	0.002	Y-intercept −0.005	Y-intercept −0.042	Y-intercept −0.179	Y-intercept −0.003	Y-intercept −0.024	Y-intercept −0.082
		n	36	n 12	n 12	n 12	n 12	n 12	n 12
		r^2	0.897	r^2 0.175	r^2 0.187	r^2 0.096	r^2 0.320	r^2 0.514	r^2 0.480
		p	≤0.0001	p 0.176	p 0.161	p 0.327	p 0.055	p 0.009	p 0.012
Multiplicative error II	$\log_{10}(N_t/N_0)$	Slope	0.889	Slope 1.004	Slope 0.578	Slope 0.545	Slope 0.166	Slope 0.901	Slope 0.908
		Y-intercept	−0.275	Y-intercept 0.179	Y-intercept −1.529	Y-intercept −2.064	Y-intercept −0.241	Y-intercept −0.272	Y-intercept −0.471
		n	392	n 143	n 144	n 105	n 143	n 144	n 105
		r^2	0.889	r^2 0.157	r^2 0.892	r^2 0.894	r^2 0.166	r^2 0.901	r^2 0.908
		p	≤0.0001	p ≤0.0001	p ≤0.0001	p ≤0.0001	p ≤0.0001	p ≤0.0001	p ≤0.0001
Logistic	$\log_e\left[\dfrac{N_t/N_0}{1-(N_t/N_0)}\right]$	Slope	0.855	Slope 0.560	Slope 0.613	Slope 0.222	Slope 0.193	Slope 0.845	Slope 0.691
		Y-intercept	−0.839	Y-intercept 0.246	Y-intercept −2.799	Y-intercept −8.952	Y-intercept 0.128	Y-intercept −0.998	Y-intercept −3.639
		n	365	n 120	n 140	n 105	n 120	n 140	n 105
		r^2	0.855	r^2 0.178	r^2 0.832	r^2 0.577	r^2 0.193	r^2 0.845	r^2 0.691
		p	≤0.0001	p ≤0.0001	p ≤0.0001	p ≤0.0001	p ≤0.0001	p ≤0.0001	p ≤0.0001

Model	Statistic							
Logistic $\log_{10}(N_t/N_0)$	Slope	0.776	0.605	0.620	0.164	0.227	0.835	0.524
	Y-intercept	-0.722	-0.175	-1.196	-4.210	-0.201	-0.476	-2.507
	n	428	155	156	117	155	156	117
	r^2	0.778	0.185	0.880	0.560	0.217	0.884	0.686
	p	≤0.0001	≤0.0001	≤0.0001	≤0.0001	≤0.0001	≤0.0001	≤0.0001
Exponential IA $\dfrac{\log_n(N_0/N_t)}{\text{day}}$	Slope	0.554	0.311	0.264	0.014	0.066	0.212	0.040
	Y-intercept	0.143	-0.097	0.325	0.806	0.015	0.118	0.929
	n	392	143	144	105	143	144	105
	r^2	0.554	0.032	0.196	0.039	0.066	0.212	0.040
	p	≤0.0001	0.031	≤0.0001	0.044	0.002	≤0.0001	0.042
Exponential IB $\log_n\left[\dfrac{\log_n(N_0/N_t)}{\text{day}}\right]$	Slope	0.829	0.034	0.165	0.059	0.061	0.208	0.043
	Y-intercept	-0.377	-4.072	-1.746	-0.069	-4.026	-1.541	-0.156
	n	365	120	140	105	120	140	105
	r^2	0.829	0.047	0.186	0.025	0.060	0.208	0.043
	p	≤0.0001	0.017	≤0.0001	0.107	0.007	≤0.0001	0.034
Exponential IIA $\log_n(N_0/N_t)$	Slope	0.735	1.406	0.598	0.158	0.224	0.887	0.689
	Y-intercept	1.386	-0.493	2.836	8.566	0.476	0.660	3.283
	n	428	155	156	117	155	156	117
	r^2	0.735	0.216	0.878	0.614	0.224	0.887	0.689
	p	≤0.0001	≤0.0001	≤0.0001	≤0.0001	≤0.0001	≤0.0001	≤0.0001
Exponential IIB $\log_n[\log_n(N_0/N_t)]$	Slope	0.781	0.195	0.580	1.503	0.197	0.725	0.881
	Y-intercept	0.248	-0.386	0.508	-1.067	-0.499	0.466	0.286
	n	365	120	140	105	120	140	105
	r^2	0.781	0.181	0.716	0.858	0.197	0.725	0.881
	p	≤0.0001	≤0.0001	≤0.0001	≤0.0001	≤0.0001	≤0.0001	≤0.0001

[a] These linear regression equations describe the relationships between predicted values representing the different models (as dependent variable) versus corresponding experimentally observed values (as independent variable).

[b] Predicted values from the models developed across all temperatures were regressed against their corresponding observed values both without (−20 to 22°C) and with (−20, 1, and 22°C) the predicted values first having been segregated according to temperature.

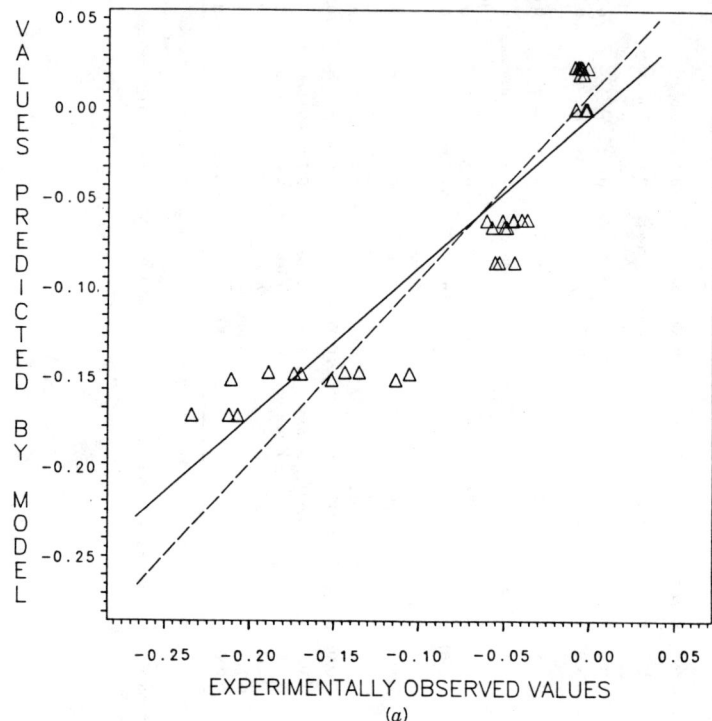

FIGURE 6.1. Graphs of linear regression equations* comparing values predicted by models generated using the *additive error* equation format versus corresponding experimentally observed values, with the compared values represented as $[\log_{10}(N_t/N_0)]/\text{day}$.

*The equations representing regression of predicted and corresponding observed values are presented as solid lines. The appropriate lines of identity ($Y = X$) appear as broken lines. Part (*a*) of this figure represents the model derived for the full temperature range, using linear regression to compare the "predicted" values generated by that model versus the actual "experimentally observed" values from all three incubation temperatures (−20, 1, and 22°C). The coordinate positions of the corresponding predicted and observed values for that comparison of the "across temperature model" are plotted as open triangles. Part (*b*) of this figure shows the graphs from linear regression of predicted versus corresponding observed values both for the across temperature model, with the values predicted by that model having been segregated according to incubation temperature (−20, 1, or 22°C) prior to their being linearly regressed against the corresponding observed values; and graphs of predicted versus observed values representing "individual temperature models," those developed independently for the individual incubation temperatures. The symbols used to identify linear regression lines for predicted versus observed values in part (*b*) of this figure are: *open circle*, across temperature model applied to −20°C; *open square*, across temperature model applied to 1°C; *open hexagon*, across temperature model applied to 22°C; *filled circle*, individual temperature model developed for −20°C; *filled square*, individual temperature model developed for 1°C; and *filled hexagon*, individual temperature model developed for 22°C.

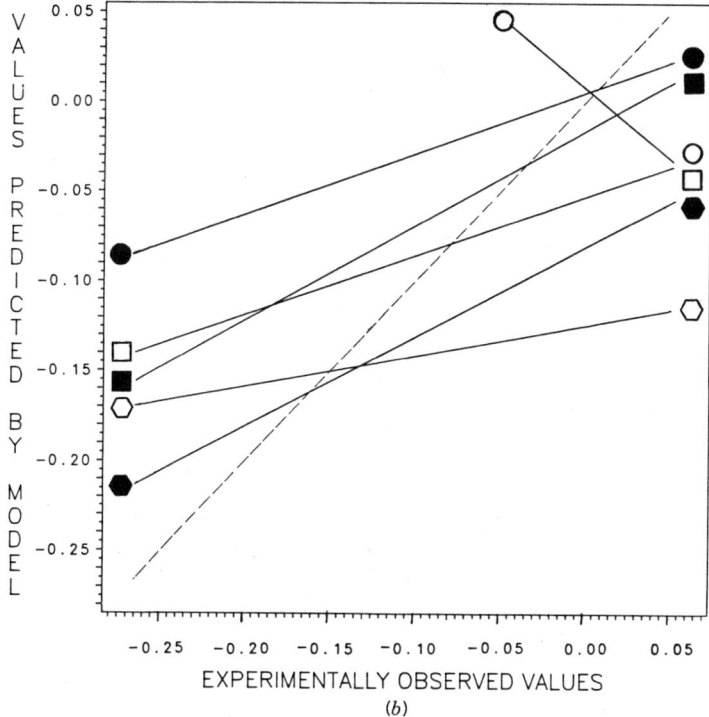

FIGURE 6.1. (*Continued*)

their corresponding experimentally observed values. The drawn lines representing these regressions are presented as solid lines in parts (*b*) of Figs. 6.1–6.9 and have been extended across the full width of the graphs (extension of regression lines was done for the sake of illustration and statistically is not recommended). Parts (*b*) of Figs. 6.1–6.9 also examine the effect that occurred when models developed for individual temperatures were extrapolated to the full temperature range. This was done by regressing predicted values generated from individual temperature models (which, respectively, represented the different equation formats) versus the corresponding observed values for that same temperature, and extending the drawn regression lines (also represented as solid lines) across the full width of the graph. The most important thing to note when examining parts (*b*) of Figs. 6.1–6.9 is, in each case, how closely the group of six calculated regression lines (solid lines) fit in terms of overlaying the ideal line (dashed line, representing $Y = X$).

The most obvious modeling failures observed in this study were those occasions when lines representing the regression of predicted and observed values had slopes that were close to zero, such that the plotted regression lines crossed the graphs horizontally rather than as they should, diagonally with a positive slope value.

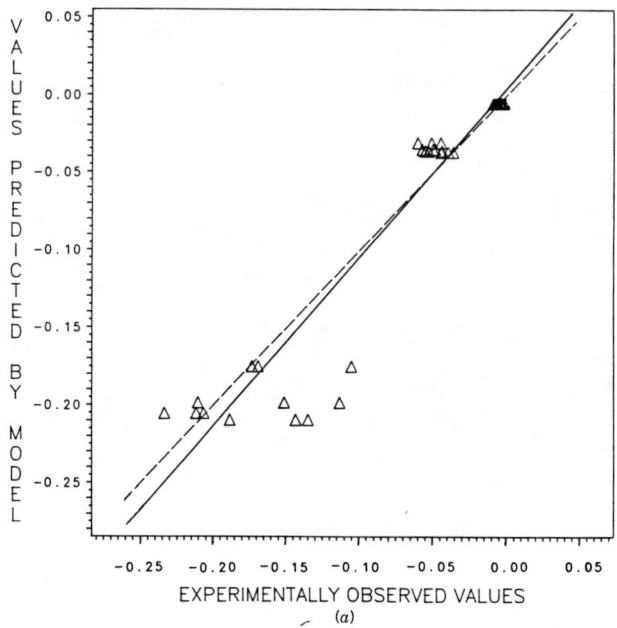

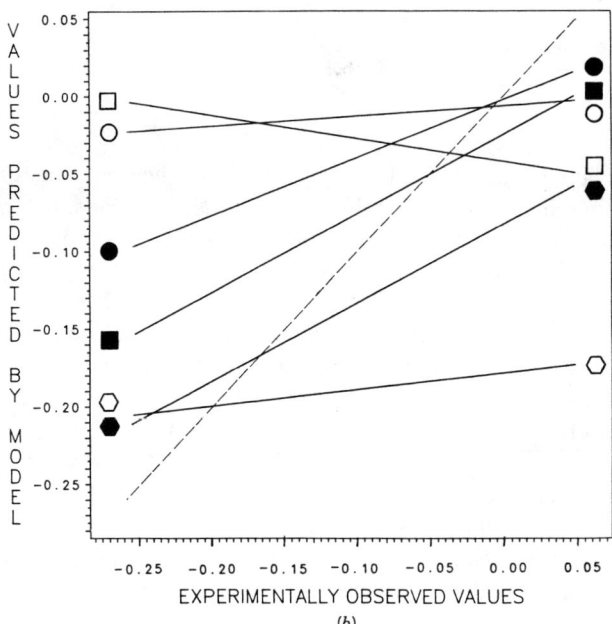

FIGURE 6.2. Graphs of linear regression equations* comparing values predicted by models generated using the *multiplicative error I* equation format versus corresponding experimentally observed values, with the compared values represented as $[\log_{10}(N_t/N_0)]/$day.

3. COMPARISON OF EQUATION FORMATS

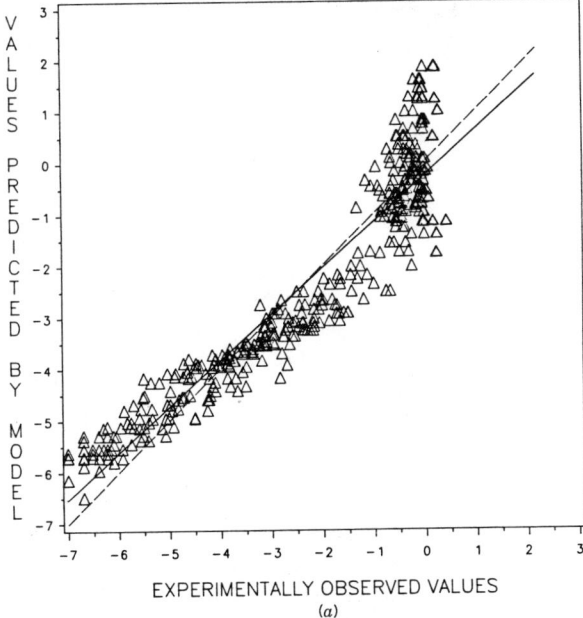

(a)

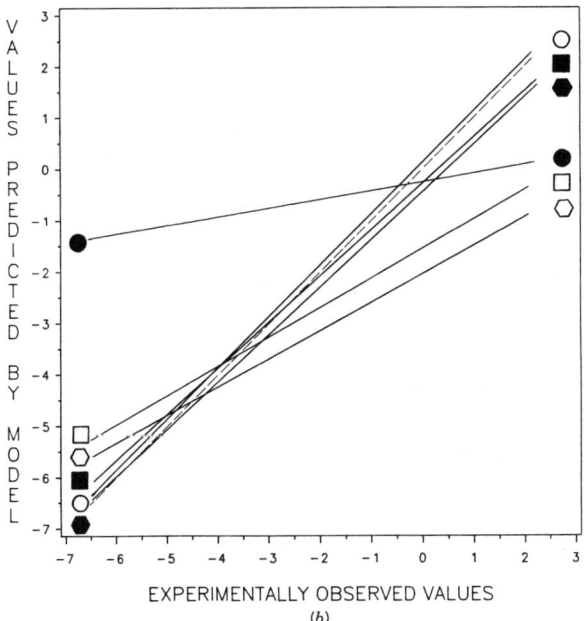

(b)

FIGURE 6.3. Graphs of linear regression equations* comparing values predicted by models generated using the *multiplicative error II* equation format versus corresponding experimentally observed values, with the compared values represented as $\log_{10}(N_t/N_0)$.

164 MODELING MICROBIAL POPULATION DECAY RATES

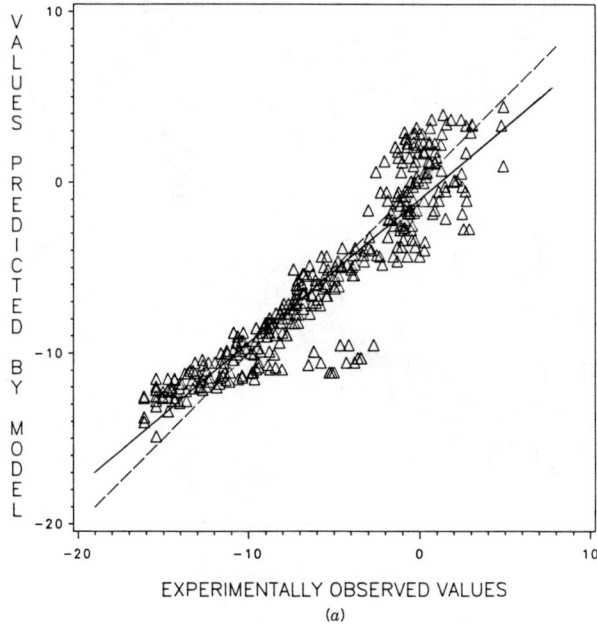

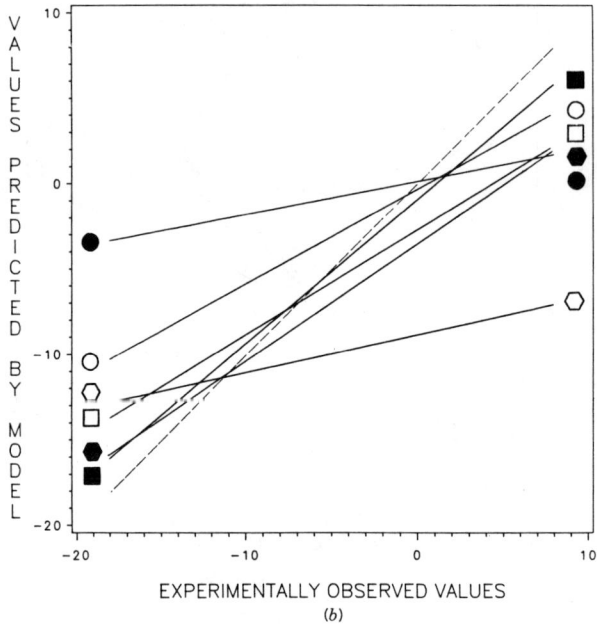

FIGURE 6.4. Graphs of linear regression equations* comparing values predicted by models generated using the *logistic* equation format versus corresponding experimentally observed values, with the compared values represented as $\log_n \{[N_t/N_0]/[1 - (N_t/N_0)]\}$.

3. COMPARISON OF EQUATION FORMATS

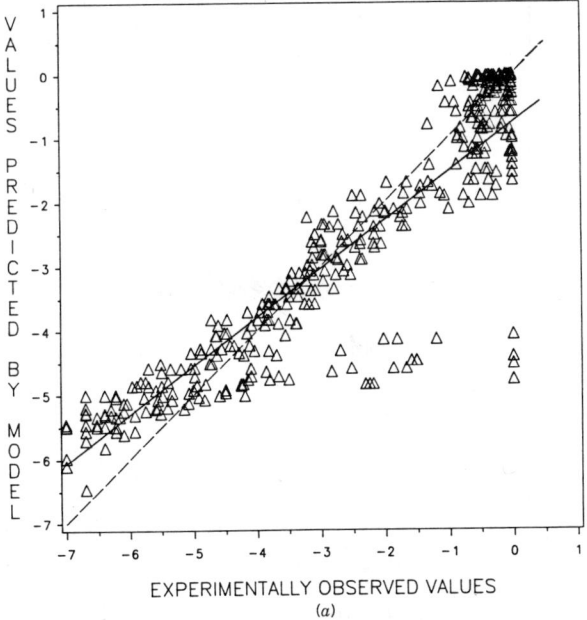

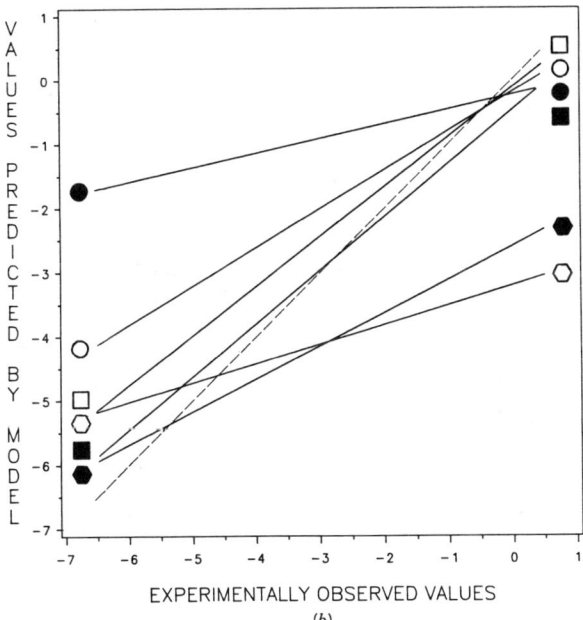

FIGURE 6.5. Graphs of linear regression equations* comparing values predicted by models generated using the *logistic* equation format versus corresponding experimentally observed values, with the compared values represented as $\log_{10}(N_t/N_0)$.

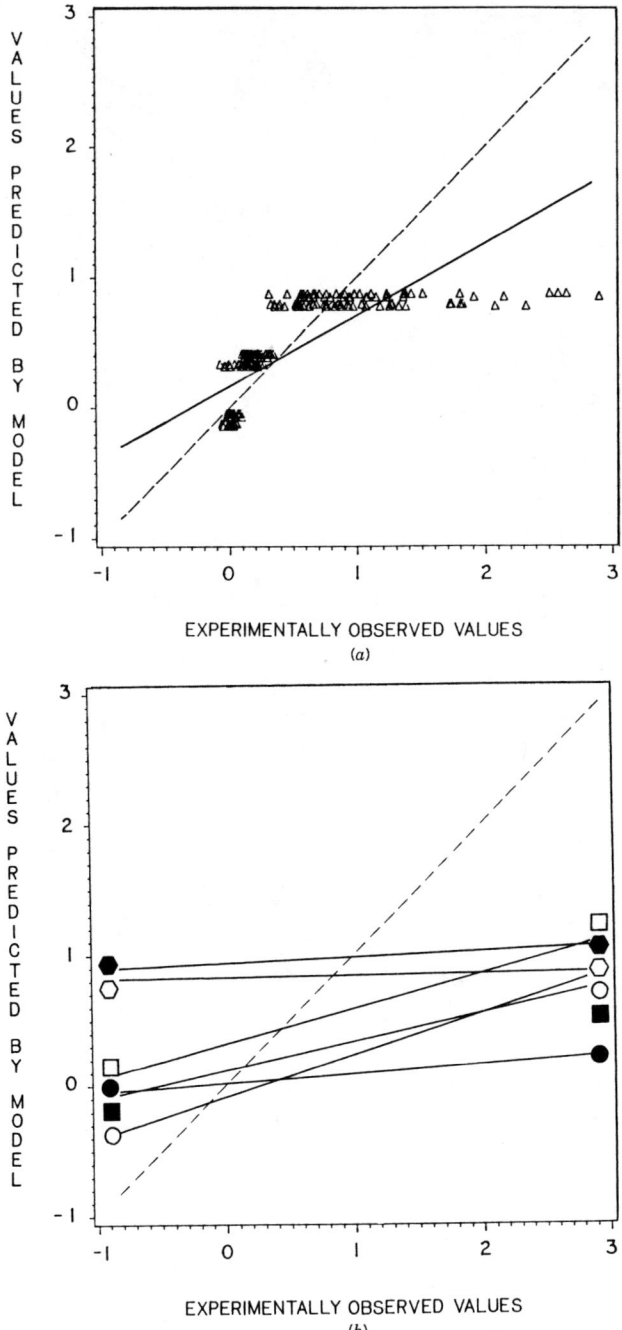

FIGURE 6.6. Graphs of linear regression equations comparing values predicted by models generated using the *exponential IA* equation format versus corresponding experimentally observed values, with the compared values represented as $[\log_n (N_0/N_t)]/\text{day}$.

3. COMPARISON OF EQUATION FORMATS 167

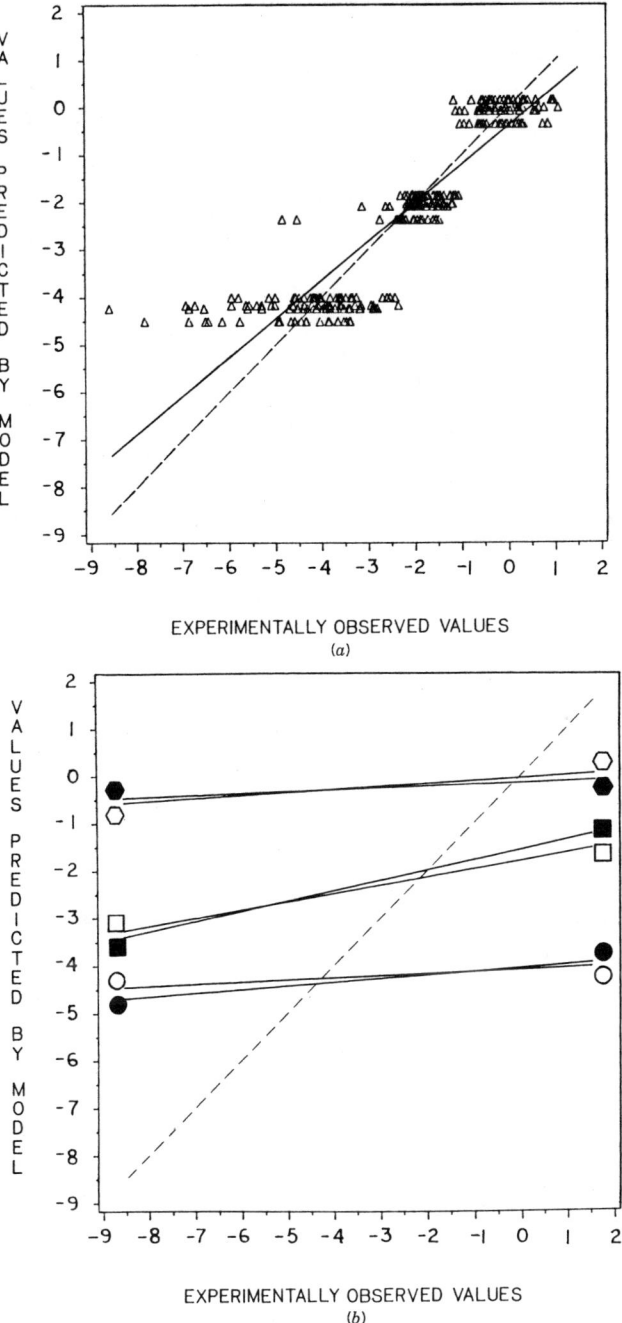

FIGURE 6.7. Graphs of linear regression equations* comparing values predicted by models generated using the *exponential IB* equation format versus corresponding experimentally observed values, with the compared values represented as $\log_n \{[\log_n (N_0/N_t)]/\text{day}\}$.

168 MODELING MICROBIAL POPULATION DECAY RATES

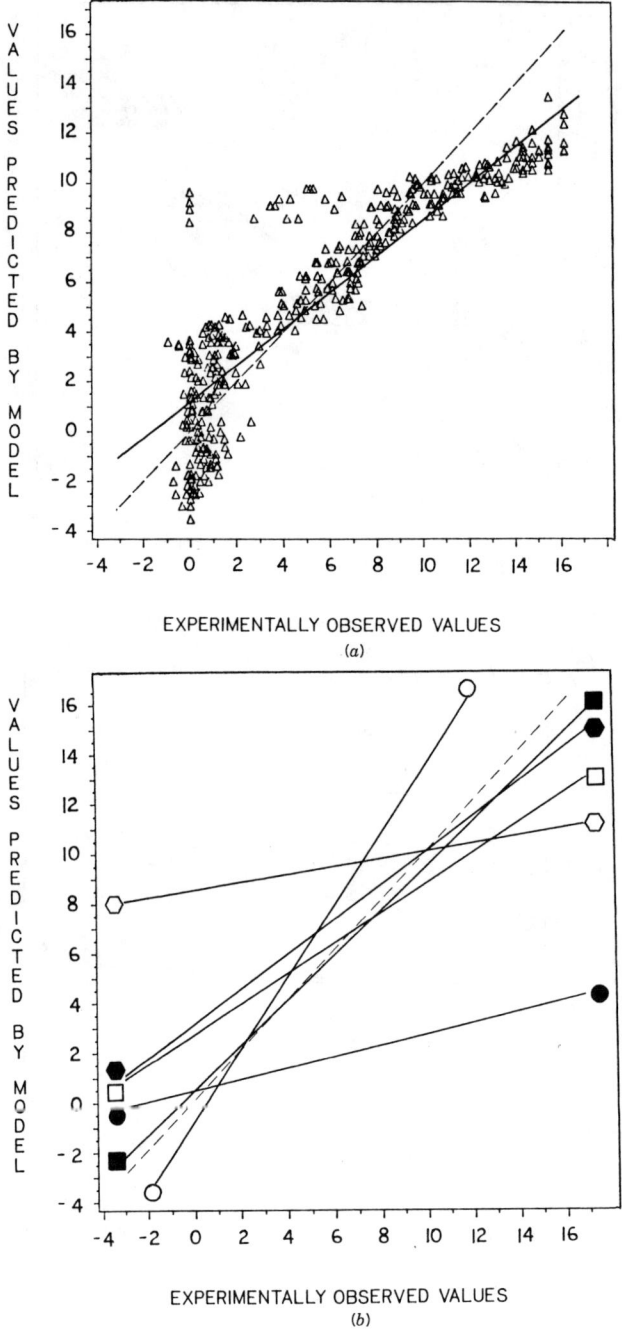

FIGURE 6.8. Graphs of linear regression equations* comparing values predicted by models generated using the *exponential IIA* equation format versus corresponding experimentally observed values, with the compared values represented as $\log_n (N_0/N_t)$.

3. COMPARISON OF EQUATION FORMATS

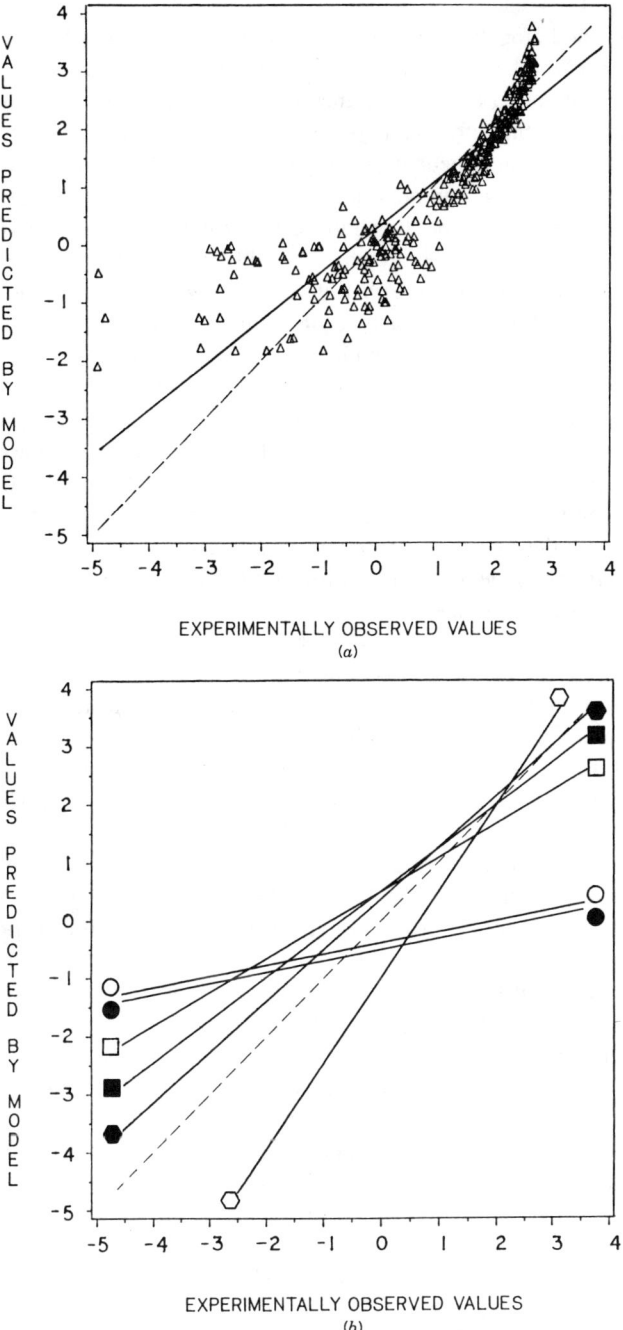

FIGURE 6.9. Graphs of linear regression equations* comparing values predicted by models generated using the *exponential IIB* equation format versus corresponding experimentally observed values, with the compared values represented as $\log_n [\log_n (N_0/N_t)]$.

This type of modeling failure was most prominently noted when examining the extendability of models generated by the exponential IB equation format (Fig. 6.7b). These horizontal lines reflect the fact that incorporation of time as a part of the dependent variable effectively yields a very narrow range of predicted values for each of the individual incubation temperatures. This effect would not have been noted if those incubation temperatures selected for this study were more closely spaced. Another type of modeling failure is represented in Fig. 6.1b, where the across temperature model generated by the additive error equation format predicted that viral titer would increase with time at the lowest ($-20°C$) incubation temperature (open circle). The latter type of failure is evidenced by the fact that the regression line connecting the open circles in that figure has a negative slope value and indeed it crosses the ideal line almost perpendicularly.

In Fig. 6.1b it can be seen that when using the Additive error equation format, the individual temperature model generated for $-20°C$ (filled circle) came closer to approximating the ideal line than did the corresponding prediction from the across temperature model (same figure, open circle). This suggests that the presence of data for the two higher temperatures (1 and $22°C$) in that additive error across temperature model caused misdirection of the model with respect to the low end of the temperature range.

In contrast to this problem, Fig. 6.3b reveals that when using the multiplicative error II equation format, the across temperature model did a better job of predicting the low end of the temperature range (open circle) than did its corresponding individual temperature model (filled circle). This is evidenced by the fact that in Fig. 6.3b the solid line connecting the open circles comes closer to approximating the ideal line than does the solid line that connects the filled circles. The suggestion drawn from these observations regarding Fig. 6.3b is that when using the multiplicative error II equation format, including the data for the two higher temperatures was beneficial in terms of predicting what happened at the low temperature.

Examining graphs of the various regression lines (solid lines) shown in parts (a), and more noticeably in parts (b), of the different figures for their closeness of fit to the drawn ideal lines (dashed lines representing identity, $Y = X$), suggests that models developed using the multiplicative error II equation format had the greatest overall accuracy. This closeness of fit is particularly noticeable in Fig. 6.3b, which represents the multiplicative error II equation format and wherein three of the six solid lines (open circle, filled square, and filled hexagon) almost replicate the dashed line, and two (open squares and open hexagon) of the other three solid lines come very close to matching the slope of the dashed line. Models developed using the multiplicative error I equation format (Fig. 6.2) appeared least accurate. Models developed using the logistic equation format are represented both in terms of their customary logit transformed dependent variable ($\log_n \{[N_t/N_0]/[1 - (N_t/N_0)]\}$) (Fig. 6.4) and as $\log_{10}$ transformed survival ratio values [$\log_{10} (N_t/N_0)$] (Fig. 6.5). The latter is a construct, which, in this case, was obtained by first undoing the logit transformation of the predicted values to yield N_t/N_0, followed by a $\log_{10}$ transformation. This is somewhat artificial but allowed us to use more similar scales on the figure axes when visually comparing results from the logistic and multiplicative error II models.

Based on this visual test, the accuracy of the additive error, logistic, and exponential equation formats appeared to be intermediate between those of the two multiplicative error equation formats.

There were some instances, such as shown in Fig. 6.7b, where the ranges of predicted values generated for individual temperatures by an across temperature model (open symbols) were nearly identical to those values generated by individual temperature models based on the same equation format (same figure, filled symbols). This is shown by the fact that the solid lines connecting the open symbols in that figure nearly overlay those solid lines that connect the corresponding filled symbols (i.e., open hexagon and filled hexagon). This ability of the across temperature model to replicate predictions of the individual temperature models suggests good precision, and yet the fact that each of the solid lines shown in Fig. 6.7b is approximately 45° away from overlaying the ideal (dashed) line indicates that this equation format is basically inaccurate.

3.3. Numerical Test for Comparing the Accuracy of Equation Formats

If an ideal equation format existed, then sets of predicted values generated by models based on that equation format should exactly replicate their corresponding sets of experimentally observed values. Therefore linearly regressing any individual set of predicted values based on that ideal equation format, versus its corresponding set of experimentally observed values, should yield a slope value of 1.0 and a Y-axis intercept value of 0.0. As explained earlier in this chapter, we examined eight different equation formats for their relative accuracy in modeling the same data on virus survival in water. Table 6.2 lists the individual equations that represent linear regressions of sets of predicted versus experimentally observed values for models based on the equation formats that we examined in this study. Any individual differences between the calculated slope and Y-intercept values for those regressions listed in Table 6.2 and the ideal line, whose slope value in each case should be equal to 1.0 with a Y-axis intercept value of 0.0, are assumed to represent inaccuracy on the part of the equation formats. The linear regressions listed in Table 6.2 are graphed as solid lines in Figs. 6.1–6.9. Each figure also shows a dashed line, which represents the ideal regression line, whose slope is 1.0 and whose Y-axis intercept is 0.0. The visual test of modeling accuracy that is represented by Figs. 6.1–6.9 relies on both the slope and Y-axis intercept values for the linear regressions that are listed in Table 6.2.

It would be difficult to use the Y-axis intercept values listed in Table 6.2 as the basis for a numerical test to compare the accuracy of the different equation formats that we examined in this study. The reason for this is because the equation formats differed with respect to both the content and form of their dependent variable. This fact also caused the units of the axes for Figs. 6.1–6.9 to vary considerably, although all figures were drawn to comparative scale. However, it is possible to use absolute numerical differences between the individual slope values listed in Table 6.2 and the ideal slope value of 1.0 as the basis for a numerical test to compare the accuracy of the different equation formats. Our approach for performing an accuracy test based on these slope values consisted of calculating the absolute numerical difference between each of the individual slope values listed in Table 6.2 and the ideal value of 1.0, and then grouping those values of numerical difference according to the equation format that they represented. Table 6.2 lists seven linear regressions apiece for each of the additive error, multiplicative error I, multiplicative error II, exponential IA, exponential IB, exponential IIA, and exponential IIB equation formats. Thus for each of these equation formats there were seven "difference" values. Table 6.2 lists a total of 14 linear regressions for the logistic equation format; seven representing the normal form of its dependent variable, $\log_n \{[N_t/N_0]/[1 - N_t/N_0]\}$, and seven representing the compared values in terms of $\log_{10} (N_t/N_0)$. Thus there were 14 "difference" values for the logistic equation format, seven apiece for each of the two forms of the dependent variable. A mean was calculated for each set of seven "difference" values, and those mean values are listed in Table 6.3. As an example of how these calculations were performed, the slope values listed in Table 6.2 for the additive error equation format are 0.856, −0.731, 0.323, 0.165, 0.328, 0.514, and 0.481. The corresponding absolute numerical difference values for the additive error equation format were thus 0.144, 1.731, 0.677, 0.835, 0.672, 0.486, and 0.519; with the mean of these difference values being 0.723.

According to this numerical test, the most accurate equation format would be that which yielded the smallest value for mean of individual differences between the calculated slope values given in Table 6.2 and the ideal slope value of 1.0. The multiplicative error II equation format proved best in this regard, with a mean difference value of 0.288. The exponential IB equation format proved to be the worst, with a mean difference value of 0.800. As expected, using the normal form of the dependent variable for the logistic equation format, $\log_n \{[N_t/N_0]/[1 - (N_t/N_0)]\}$, proved more accurate in this test (with a mean of differences value equal to 0.432) than did comparing the values from the logistic format in terms of $\log_{10} (N_t/N_0)$ (mean of differences value equal to 0.464). If we had based this numerical test on the values for median of the individual differences between the calculated slope values given in Table 6.2 and the ideal slope value of 1.0 (these median values are listed in Table 6.3), rather than using the mean values as described above, there would have been only a single position change in ranking the comparative accuracy of the different equation formats. That change would have been to rank the results for expressing the dependent variable from the logistic format in terms of $\log_{10} (N_t/N_0)$ (median equal to 0.395), immediately below the results from expressing the dependent variable for the logistic format in its normal $\log_n \{[N_t/N_0]/[1 - (N_t/N_0)]\}$ form (median equal to 0.387).

It is interesting to note that of the eight equation formats that we examined in this study, those four that employed time as part of the dependent variable ranked lowest according to this numerical test (Table 6.3). Those four equation formats were the multiplicative error I, additive error, exponential IA, and exponential IB.

3. COMPARISON OF EQUATION FORMATS

TABLE 6.3. Results from Using a Numerical Test to Compare Accuracy of the Different Equation Formats

Equation Format[a]	Mean of Individual Differences Between Calculated Regression Slope Values Listed in Table 2 and Ideal Slope Value of 1.0
Multiplicative error II	0.288 ± 0.297 (0.111)[d]
Logistic[b]	0.432 ± 0.270 (0.387)
Exponential IIA	0.445 ± 0.268 (0.402)
Exponential IIB	0.449 ± 0.273 (0.420)
Logistic[c]	0.464 ± 0.256 (0.395)
Multiplicative error I	0.669 ± 0.354 (0.633)
Additive error	0.723 ± 0.494 (0.672)
Exponential IA	0.791 ± 0.191 (0.788)
Exponential IB	0.800 ± 0.285 (0.939)

[a] The equation formats are listed in ranked order according to the mean of absolute numerical differences between the slope values listed in Table 2 (that represent the individual linear regressions of predicted versus observed values) and, in each case, the defined ideal slope value of 1.0. According to this test an ideal equation format would have generated sets of predicted values, which identically matched the actual experimentally observed values. Hence linearly regressing sets of predicted values generated by an ideal equation format versus their corresponding sets of experimentally observed values should, in each case, produce a slope value equal to 1.0 and result in a mean of differences equal to 0.0.
[b] This comparison represents the normal form of the dependent variable for the logistic equation format, $\log_n\{[N_t/N_0]/[1 - (N_t/N_0)]\}$.
[c] This comparison represents expression of results from the logistic equation format in terms of $\log_{10}(N_t/N_0)$.
[d] Values are presented as mean ± standard deviation, followed by the median in parentheses.

This suggests that for the set of data that we examined, employing time as an independent variable allowed us to obtain better modeling accuracy.

3.4. Full Models Developed Using the Multiplicative Error II Equation Format

The full models developed using the multiplicative error II equation format including incorporation of log transformation, which in this study appeared to be the most accurate format for modeling microbial population decay rates, are the following:

incubation at $-20°C$,

$$\log_{10} \frac{\widehat{N_t}}{N_0} = 0.188 - 0.362 \log_{10} t - 0.0971 \log_{10} C + 0.178 \log_{10} \text{NTU}$$

$$-0.384 \log_{10} \text{PEK}; \quad (21)$$

incubation at 1°C,

$$\log_{10} \widehat{\frac{N_t}{N_0}} = 8.157 - 3.587 \log_{10} t - 0.546 \log_{10} C + 0.875 \log_{10} \text{NTU}$$
$$- 4.910 \log_{10} \text{PEK}; \qquad (22)$$

incubation at 22°C,

$$\log_{10} \widehat{\frac{N_t}{N_0}} = -0.219 - 3.914 \log_{10} t + 2.141 \log_{10} C + 0.451 \log_{10} \text{NTU}$$
$$- 6.676 \log_{10} \text{PEK}; \qquad (23)$$

across temperature,

$$\log_{10} \widehat{\frac{N_t}{N_0}} = 215 - 2.349 \log_{10} t + 0.462 \log_{10} C + 0.534 \log_{10} \text{NTU}$$
$$- 3.562 \log_{10} \text{PEK} - 87.302 \log_{10} T; \qquad (24)$$

with t, time in days; C, conductivity; NTU, turbidity; PEK, measure of growth of bacteria seeded in the water as defined in the footnote for Table 6.1; and T, temperature in Kelvin units. Undoing their log transformation would yield the starting form of the dependent variable that was used for this equation format (N_t/N_0). These models represented by Eqs. (21)–(24) may be useful for predicting viral stability in environmental freshwaters. A circumflex has been placed over the dependent variable for Eqs. (21)–(24) because these equations have been supplied for use in a predictive mode, and appropriate numerical values have been supplied for the coefficients associated with the independent variables.

Discovery that the multiplicative error II equation format appeared to have been the most accurate may reflect a greater ability on the part of this format to accommodate nonlinear interactions in comparison to the other equation formats that were examined in this study. The multiplicative error I equation format, which in this comparative evaluation proved relatively inaccurate for modeling viral population decay rates, does not allow time to have a coefficient. The multiplicative error II equation format, which appeared to be the most accurate, differs from the multiplicative error I format only in that it allows time to have a coefficient, which takes the form of an exponent.

4. SUMMARY

The health significance of human or animal viral contaminants in air, soil, or water largely depends on the virus' ability to persist in these environments. Viral stability is also a consideration in studies on the role of natural virus populations in plant and microbial ecology. No standard formats exist for modeling environmental virus stability. This chapter presents a study that was intended to help fill this void and compares the relative accuracy of eight regression equation formats for empirically modeling virus survival in surface freshwaters using a human virus database. Greatest modeling accuracy was achieved by an equation format that represented multiplicative error regression, which we termed "multiplicative error II." This equation format used a survival ratio (N_t/N_0; N_t representing population titer at elapsed time t, N_0 representing population titer at the outset of an incubation period) as the dependent variable. The independent variables used in that regression equation format were water conductivity expressed as micromhos per centimeter, water turbidity expressed as nephelometric turbidity units, a laboratory measured value for ability of the different tested waters to support growth of aerobic microorganisms, and incubation time expressed in days. Incubation temperature expressed in Kelvin units was used as an additional independent variable when modeling data that represented more than a single incubation temperature.

REFERENCES

Bergh, Ø., Børsheim, K. Y., Bratbak, G., and Heldal, M. (1989), *Nature*, **340**, 467–468.

Børsheim, K. Y., Bratbak, G., and Heldal, M. (1990), *Appl. Environ. Microbiol.*, **56**, 352–356.

Casolari, A. (1981), *J. Theor. Biol.*, **88**, 1–34.

Chick, H. (1908), *J. Hygiene*, **8**, 92–158.

Hom, L. M. (1972), *J. Sanit. Eng. Div. Am. Soc. Civil. Eng.*, **98**, 183–194.

Hurst, C. J., Gerba, C. P., and Cech, I. (1980), *Appl. Environ. Microbiol.*, **40**, 1067–1079.

Hurst, C. J., Benton, W. H., and McClellan, K. A. (1989), *Can. J. Microbiol.*, **35**, 474–480.

Suttle, C. A., Chan, A. M., and Cottrell, M. T. (1990), *Nature*, **347**, 467–469.

Williams, S. T., Mortimer, A. M., and Manchester, L. (1987), in *Phage Ecology* (S. M. Goyal, C. P. Gerba, and G. Bitton, Eds.), Wiley, New York, pp. 157–179.

7 Mathematical Modeling and Quantitative Characterization of Bacterial Motility and Chemotaxis

ROSEANNE M. FORD
Department of Chemical Engineering, Thornton Hall, University of Virginia

1. Introduction 177
   1.1. Bacterial chemotaxis in microbial ecology 178
   1.2. General description of bacterial chemotaxis 179
   1.3. The chemosensory mechanism 181
2. Survey of mathematical models 183
   2.1. Population balance equation and constitutive relation 184
   2.2. Subpopulation cell balance equations and constitutive relations 184
3. Evaluation of cell transport coefficients from experimental data 192
   3.1. Description of population experiments 193
      3.1.1. Capillary assay 193
      3.1.2. Laser densitometry assay 194
      3.1.3. Stopped-flow diffusion chamber assay 194
   3.2. Modeling equations 196
   3.3. Determination of random motility coefficients from population assays 197
      3.3.1. Capillary assay 199
      3.3.2. Stopped-flow diffusion chamber assay 200
   3.4. Determination of chemotactic sensitivity coefficients from population
      assays 200
      3.4.1. Capillary assay 201
      3.4.2. Laser densitometry assay 205
      3.4.3. Stopped-flow diffusion chamber assay 209
   3.5. Validation of mathematical model 211
4. Conclusions 212
   References 213

1. INTRODUCTION

Bacteria in their natural habitats are continually exposed to changing environmental conditions. Their survival depends on their capacity to respond favorably to adverse circumstances. Since their ability to modify their surroundings is limited because of their small size (1-2 μm) and simple structure, they respond either by migration to a more desirable location or by adaptation of their internal metabolic processes (Macnab, 1980). Adaptation occurs naturally through genetic modification, but this process is relatively slow. Motile bacteria can respond much more quickly by moving to a more favorable environment; they swim toward increasing concentrations of nutrients and away from increasing concentrations of harmful substances. This ability of bacterial populations to direct their migration in response to chemical gradients is known as chemotaxis.

The focus of this chapter is a mathematical model to describe bacterial chemotaxis and its use in characterizing experimental data in terms of fundamental transport properties of bacterial populations. The importance of chemotaxis in microbial processes, the general features of the phenomenon, and the sensory mechanism employed by bacteria to move chemotactically are described in Sections 1.1-1.3. Following this introduction to bacterial chemotaxis, the remainder of this chapter is divided into three sections. In Section 2, three- and one-dimensional cell balance equations and the relationships between them are reviewed. The constitutive equations, which relate the population parameters in the cell balance equations to the single-cell behavior (described in Section 1), are also given in Section 2. Section 3 contains descriptions of three experimental assays for measuring chemotactic responses in bacterial populations, the general equations for modeling such assays (obtained by combining the cell balance and constitutive equations from Section 2), and the boundary conditions appropriate to each assay. Values for bacterial transport coefficients (the random motility and the chemotactic sensitivity) obtained from the assays are presented and summarized in Table 7.1. Finally, in Section 4, we present our conclusions, including a brief discussion of the use of the experimentally determined bacterial transport coefficients in population dynamics models proposed by Lauffenburger and co-workers (Kelly et al., 1988; Lauffenburger, 1988).

1.1. Bacterial Chemotaxis in Microbial Ecology

This migratory behavior of bacterial populations is believed to play a significant role in many microbial ecological processes (Chet and Mitchell, 1976). For example, bacteria are intimately involved in the nitrogen cycle, a vitally important concern in agricultural ecology. *Rhizobia meliloti*, one particular bacterial strain involved in this process, has been shown to respond chemotactically to root exudates isolated from the soil of leguminous plants (Gulash et al., 1984). The implication is that bacteria in the surrounding soil are guided to nodules on the roots of nitrogen-fixing plants by chemical gradients. Denitrification, although a process that robs the soil of essential nutrients, is nevertheless an important step in the nitrogen cycle. Certain species, typically *Pseudomonads*, exhibit chemotaxis toward high concentrations of fixed nitrogen and reduce it to elemental nitrogen (Kennedy and Lawless, 1985). A delicate balance between these two processes is critical for optimal plant growth. Other microbial processes involving chemotaxis include the pathogenesis of infection (Drake and Montie, 1987, 1988; Freter et al., 1979) and the development of biofilms (Chet et al., 1975; Gristina, 1987).

Two areas of current interest for which bacterial migration and dispersion play a significant role are bioremediation and the release of genetically engineered organisms into the environment. Bioremediation involves the treatment of chemical and hazardous waste with microorganisms to facilitate its degradation to innocuous compounds. Using microorganisms to clean up toxic waste sites is appealing because essentially all organic chemicals are susceptible to biodegradation and are ultimately reduced to carbon dioxide and water. This technology is superior to more conventional treatment schemes because the chemical waste is actually transformed instead of concentrated or contained. One major difficulty with large-scale implementation is getting the bacteria in contact with the toxin. The ability of bacteria to direct their random, diffusive motion toward increasing concentrations of a particular chemical provides one mechanism by which bacteria can direct their migration toward toxins in the environment. Recent studies show that certain species prevalent in the soil move in response to concentration gradients of benzene, toluene, trichloroethylene, and chlorinated benzoates (Harwood et al., 1990; King et al., 1990; López-de-Victoria, 1989). Exploitation of this behavior should provide a basis for enhancing the effectiveness of bioremediation.

The responsible release of genetically engineered organisms for use in the environment requires reliable predictions concerning the spread of these organisms and their interactions with native species. To be effective and ecologically safe they must be able to compete with native species, but not overpopulate them. Native species are already optimally adapted to their surroundings, therefore new species may require a competitive advantage to survive in environments where the availability of nutrients is limited. Lauffenburger and co-workers (Kelly et al., 1988; Lauffenburger, 1988) have shown that for certain parameter ranges, chemotaxis can provide such an advantage. Ford and co-workers (1991) have recently developed an assay to measure these parameter values characterizing bacterial migration, which in combination with population dynamics models provides a powerful tool for predicting bacterial distributions in the environment.

1.2. General Description of Bacterial Chemotaxis

Bacteria sense and respond to a wide variety of chemical stimuli including sugars, amino acids, oxygen, hydrocarbons, metal ions, and alcohols. Stimuli are generally grouped into one of two categories either as attractants or repellents according to the nature of the response they induce in bacteria. Attractants tend to be nutrients or carbon sources such as amino acids and sugars, which are essential for growth. For aerobic species oxygen is also an attractant, but anaerobic bacteria such as *Clostridium* are repelled by it. Some chemicals that are not directly beneficial to bacteria are attractants by virtue of their structural similarity to food sources. Fucose, for example, a structural analog of galactose, is an attractant although it cannot be metabolized by *Escherichia coli*. Repellents include chemicals toxic to bacteria such as acetate and the metal ions, Ni^{2+} and Co^{2+}, or metabolites that indicate overly populated conditions. Extremes of pH can also be detected and avoided through the bacterial sensory system. Discussion of mathematical models in this chapter will be restricted to chemical attractants, although bacteria have also been observed to direct their motion in response to certain wavelengths of light (phototaxis), magnetic fields (magnetotaxis), temperature gradients (thermotaxis), and osmotic potentials (osmotaxis) (Macnab, 1987).

Bacterial populations have been observed to travel in bands of high cell density in response to temporally changing chemical gradients. This behavior was noted by Adler (1966) when a plug of bacteria was introduced at the mouth of a tube containing a metabolizable chemical attractant in soft agar. As bacteria consumed the attractant a gradient was created and a band of high cell density formed and traveled along the capillary as more and more of the attractant was consumed. Similar behavior was observed for bacteria migrating on semisolid agar plates (Adler, 1966; Wolfe and Berg, 1989). In this case rings formed, which migrated toward the edge of the plate as nutrients in the agar were consumed by the bacterial population.

Peritrichious bacteria such as *Escherichia coli* and *Salmonella typhimurium* propel themselves through the surrounding media by coordinated rotation of flagella 5–10 μm in length, which are randomly spaced around the perimeter of the cell (Stewart and Dalquist, 1987). The path that a single cell traces as it swims through the surrounding medium consists of a series of alternating runs and tumbles (Berg and Brown, 1972; Macnab and Koshland, 1972; Spudich and Koshland, 1975) as shown schematically in Fig. 7.1. The runs are relatively straight lines of smooth swimming that last for a few seconds. The tumbles last only a fraction of a second

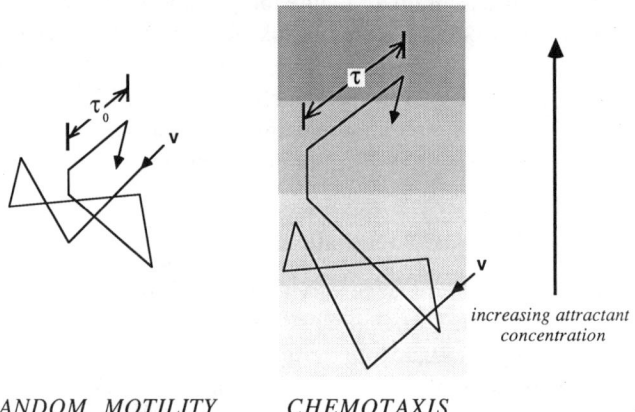

FIGURE 7.1. Observed single-cell behavior in an isotropic medium (left) resembles a random walk with basal mean run length $\langle \tau_0 \rangle$ and swimming speed v. In the presence of an attractant gradient (right) run lengths are increased when moving toward increasing concentrations of attractant, yielding a mean run length $\langle \tau \rangle$ greater than $\langle \tau_0 \rangle$. [Figure adapted from Macnab (1980).]

and result in a random change of direction. Changes between modes of smooth swimming and tumbling behavior are controlled by the direction in which the flagellar motors rotate. When the motors operate in a counterclockwise manner, the flagella coordinate their rotation to form a bundle and smooth swimming results. Clockwise rotation causes the flagellar bundle to unravel, generating a tumble. In the absence of any chemical gradients, the swimming pattern resembles a three-dimensional random walk similar to Brownian motion in molecular diffusion, except that changes in direction result from tumbles instead of molecular collisions. The term random motility is used to describe this behavior. In the presence of an attractant gradient, bacteria are able to bias their random walk by altering their basal tumbling frequency. Decreasing their tumbling frequency when moving toward higher attractant concentrations extends the run lengths in that direction, resulting in a net bias of movement toward more favorable conditions. Thus the individual cell technically exhibits chemoklinokinesis (the ability to modulate tumbling frequency in the presence of chemical gradients), but the overall result for the entire population is directed motion, hence the designation chemotaxis (Nossal, 1979).

1.3. The Chemosensory Mechanism

As bacteria move about exploring their surroundings, they monitor changes in chemical concentrations through special proteins called receptors located near the cell surface. These receptors, like enzymes, have specific binding sites to which only a narrow range of structurally similar chemical substrates can bind. In *E. coli* and *S. typhimurium* about 20 different receptors have been identified for various attractants and repellents (Koshland, 1979). It is the change in the number of bound receptors over time that provides information to the cell regarding chemical gradients (Macnab and Koshland, 1972).

Larger eukaryotic cells such as polymorphonuclear leukocytes are known to detect spatial gradients across the surface of the cell (Zigmond et al., 1985), but the size of a bacterium is an order of magnitude smaller, 1–2 μm compared to 10–15 μm for a white blood cell; therefore a spatial mechanism seemed unlikely to account for the sensitivity of the bacterial response to very shallow gradients. Macnab and Koshland (1972) designed a clever temporal gradient assay in which they elegantly demonstrated that bacteria sense temporal as opposed to spatial changes in concentration. Two bacterial suspensions differing only in stimulus concentration were rapidly mixed together in a chamber and the swimming behavior of individual cells was observed microscopically. When exposed to a rapid increase in attractant concentration, bacteria exhibited smooth swimming for several minutes before recovering their basal tumbling frequency. No spatial gradients were present in this assay; therefore it was concluded that cells must sense temporal changes. Sensing of spatial gradients is possible because the bacterium is moving through the spatial gradient at a given velocity, making it appear to the bacterium that the concentration is changing with time. The gradual return to normal swimming patterns by cells exposed to a large temporal gradient suggests that bacteria are able to adapt to new concentration levels. In other words, they have a short-term and long-term memory. The short-term memory compares changes in the number of bound receptors between tumbles while the long-term memory recalls conditions from several minutes into the past, enabling cells to adapt their response and remain in a favorable environment once they have located it.

Much has been learned about the chemosensory signaling mechanism by studying the behavior of *E. coli* and *S. typhimurium* mutants that are defective in chemotaxis. [Some reviews include Koshland (1988), Parkinson (1988), Stewart and Dalquist (1987) and Stock and Stock (1987).] Although the complete biochemical pathways communicating concentration gradient information to the flagellar motors have not been elucidated, the initial steps in the pathway are well known. These are illustrated in the diagram shown in Fig. 7.2 (Borkovich and Simon, 1990). Four methyl-accepting chemotaxis proteins (MCPs) identified in *E. coli* span the cytoplasmic membrane and act as chemotactic signal transducers: Tsr (*t*axis to *s*erine and from *r*epellents), Tar (*t*axis to *a*spartate and from *r*epellents), Trg (*t*axis to *r*ibose and *g*alactose), and Tap (*t*axis *a*ssociated *p*rotein) for dipeptides. These transmembrane proteins convey information from the exterior of the cell across the cytoplasmic membrane to the interior. Located on the periplasmic side of the transducer proteins are specific binding sites for chemoattractants and repellents. Amino acids such as serine and aspartate bind directly to Tsr and Tar,

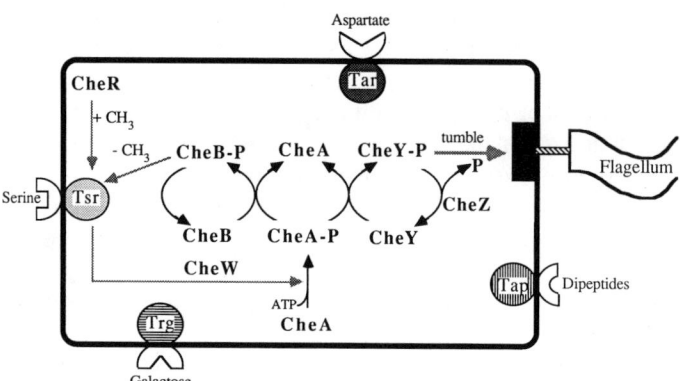

FIGURE 7.2. Schematic representation of the sensory transduction mechanism of *E. coli*. Chemical stimuli such as sugars and amino acids bind to transmembrane proteins labeled Tar, Tsr, Tap, and Trg, which act as receptors. Once bound, these proteins undergo a conformational change on the interior of the cell membrane, which permits interaction with CheA and CheW. CheA is necessary for the transfer of phosphoryl groups to CheY and CheB. CheB and CheR are enzymes necessary for the methylation and demethylation of the transmembrane proteins, which allow the cell to adapt to high levels of stimulus concentration. The accumulation of CheY-P results in a tumble. CheZ reduces the amount of CheY-P. [Figure adapted from Borkovich and Simon (1990).]

respectively, while sugars such as ribose and galactose bind first to periplasmic binding proteins, which in turn interact with Trg. Periplasmic binding proteins are also involved in the transport of sugars into the cell. The binding event on the periplasmic portion of the transducer induces a structural change on the cytoplasmic side, which permits interaction with a signal protein (CheA) necessary for the transfer of information to the flagellar motors. The cytoplasmic portion of the transducer also has several reversible methylation sites, which allow the bacteria to adapt to high levels of stimulus concentration and enable them to gradually desensitize and return to their basal tumbling frequency (Stock and Stock, 1987). Six key gene products involved in the internal signal transduction have been identified (reviewed by Stewart and Dalquist, 1987): methylation is regulated by CheR and CheB; CheA and CheW are involved in signal integration; and the phosphorylation of CheY, which is regulated by CheZ, conveys signal information to the switch controlling the direction of flagellar rotation. This signal transduction pathway has generated a great deal of interest because of its similarity to many other bacterial processes that are regulated by protein phosphorylation, such as nitrogen regulation, osmoregulation, and phosphorous regulation (Stock et al., 1989). Analogies exist between the regulation of bacterial chemotaxis and systems in eukaryotic organisms such as hormone function (Borkovich and Simon, 1990). Therefore bacterial chemotaxis may be a useful model for understanding signal processing in higher-order organisms (Koshland, 1980).

2. SURVEY OF MATHEMATICAL MODELS

Our interest in this chapter lies in modeling experimental assays of bacterial migration (described in Section 3) in order to extract fundamental bacterial transport properties that can be incorporated into population dynamics models. The assays that we consider are performed at the cell population level by measuring spatially and temporally resolved cell densities. These assays are to be distinguished from measurements of individual cell behavior performed by Berg and co-workers (Block et al., 1983; Segall et al., 1985, 1986). The challenge in developing mathematical models for the experimental assays is to determine the population level transport coefficients (random motility and chemotactic sensitivity coefficients) using information reported about mechanisms at the individual cell level. (An analogy can be drawn with the development of macroscopic transport coefficients for fluids in terms of molecular parameters via kinetic theory.) Between the extremes of the population level and the individual cell level is the subpopulation level, at which we differentiate cells according to the direction of their motion.

At each level, two classes of equations are required to produce a complete mathematical model. The balance equations, which are exact and account for the evolution in time and space of a conserved quantity, comprise the first class of equations. The second class of required equations are the constitutive equations, which usually involve some level of approximation and are based on empirical observation and/or are used to define phenomenological coefficients. It is in the constitutive equations that the approximations are introduced that define the model.

In Section 2.1, we present the population level balance equation and constitutive relation of Keller and Segel (1971a). Section 2.2 contains several three-dimensional and one-dimensional subpopulation cell balance equations that describe the evolution in time and space of the densities of cells moving in specific directions. The one-dimensional cell balance equations of Segel (1977) are combined with the constitutive relations of Rivero et al. (1989) to yield expressions for random motility and chemotactic velocity for use in the population level models.

2.1. Population Balance Equation and Constitutive Relation

We consider the cell density c to be a function of position $\vec{r}$ and time t. The time evolution of the cell density is given by the conservation equation

$$\frac{\partial c}{\partial t} = -\vec{\nabla}_{\vec{r}} \cdot \vec{J}_c \qquad (1)$$

where $\vec{J}_c$ is the cell flux and $\vec{\nabla}_{\vec{r}}$ is the spatial divergence. The subscript $\vec{r}$ on the divergence operator is included to emphasize that the divergence is with respect to the spatial coordinate $\vec{r}$. Equation (1) is an exact statement in the absence of cell death or reproduction. In their original phenomenological model to describe population behavior, Keller and Segel (1971a) proposed the following constitutive relation for the bacterial flux, which includes one term to describe diffusion-like or random motility behavior and a second term for convection-like or chemotactic motion:

$$\vec{J}_c = -\mu \vec{\nabla}_{\vec{r}} c + \vec{V}_c c \qquad (2)$$

where μ is the random motility coefficient, analogous to a molecular diffusion coefficient, and $\vec{V}_c$ is the chemotactic velocity. The fundamental transport properties, μ and $\vec{V}_c$, are both functions of the attractant concentration; expressions for their specific dependence on the attractant gradient will be described in the following section.

2.2. Subpopulation Cell Balance Equations and Constitutive Relations

In the mathematical modeling of the motion of chemotactic bacteria, several authors (Alt, 1980; Othmer et al., 1988; Patlack, 1953; Segel, 1977; Stroock, 1974) have developed cell balance equations of varying degrees of rigor. The most general and rigorous are the three-dimensional cell balance equations derived by Alt (1980) that take into account explicitly the position, direction, and elapsed run time of a bacterial population. A simplified three-dimensional cell balance equation, known as Stroock's equation (Stroock, 1974), can be derived from Alt's when one assumes that the probability of a bacterium tumbling is independent of run time. The Segel equations (1977) are one-dimensional phenomenological equations that do not explicitly include the dependence on run time.

In Alt's cell balance equation, the cells are assumed to have piecewise linear paths (runs), where the mean speed, v, depends on position and time; that is, $v = v(\vec{r}, t)$. The quantity $\sigma(\vec{r}, \hat{s}, \tau, t)$ is the number density of cells per unit volume at point $\vec{r}$ moving in direction $\hat{s}$ (a unit vector) with run time τ at time t. The probability per unit time that a cell moving in direction $\hat{s}$ at $\vec{r}$ with run time τ (counted from the beginning of the run) at time t tumbles is given by $\beta(\vec{r}, \hat{s}, \tau, t)$. If a cell tumbles at $\vec{r}$ at time t after a run with direction $\hat{s}_1$, the probability that after tumbling it chooses the direction $\hat{s}_2$ as its new direction is $k(\vec{r}, \hat{s}_1, t; \hat{s}_2)$, where k is referred to as the turning probability. Alt's fundamental equation (1980) is then for $\tau > 0$

$$\frac{\partial \sigma(\vec{r}, \hat{s}, \tau, t)}{\partial t} = -\frac{\partial \sigma(\vec{r}, \hat{s}, \tau, t)}{\partial \tau} - \hat{s} \cdot \vec{\nabla}_{\vec{r}}[v(\vec{r}, t)\sigma(\vec{r}, \hat{s}, \tau, t)]$$
$$- \beta(\vec{r}, \hat{s}, \tau, t)\sigma(\vec{r}, \hat{s}, \tau, t) \qquad (3)$$

and for $\tau = 0$, we have

$$\sigma(\vec{r}, \hat{s}, 0, t) = \int_0^\infty \int \beta(\vec{r}, \hat{s}', \tau, t)\sigma(\vec{r}, \hat{s}', \tau, t)k(\vec{r}, \hat{s}', t; \hat{s}) \, d\hat{s}' \, d\tau. \qquad (4)$$

Physically, Eq. (3) states that the rate of change in the population of cells at $\vec{r}$ at time t moving in direction $\hat{s}$ with run time τ is given by a term that takes into account the change in the run time of the cell population, a convective term that accounts for the net motion of the cells away from the point $\vec{r}$, and a loss term due to cells tumbling (with probability β, so density $\beta\sigma$). Note that implicit in this cell balance equation is the notion that the swimming motion of the cells is piecewise linear and that changes in direction are assumed to occur *only* through the tumbling mechanism [taken into account by the $\beta\sigma$ term on the right-hand side of Eq. (3)]. Equation (4) states that the way in which one obtains an initial (i.e., run time = 0) population of cells moving in direction $\hat{s}$ is by considering cell populations that were moving in another direction $\hat{s}'$, which tumbled at time t with run time τ, $\beta(\vec{r}, \hat{s}', \tau, t)\sigma(\vec{r}, \hat{s}', \tau, t)$, multiplying by the probability that the cell after tumbling moves in the direction $\hat{s}$, given by $k(\vec{r}, \hat{s}', t; \hat{s})$. One needs to add up over all such populations: hence the integration over all directions $\hat{s}'$ and all run times τ.

Stroock's three-dimensional equation for the transport of chemotactic cell populations implicitly assumes that the tumbling probability β is independent of run time, yielding the following relationship:

$$\frac{\partial n(\vec{r}, \hat{s}, t)}{\partial t} = -\hat{s} \cdot \vec{\nabla}_{\vec{r}} v(\vec{r}, t) n(\vec{r}, \hat{s}, t) + \int p(\vec{r}, \hat{s}', \hat{s}, t) n(\vec{r}, \hat{s}', t) \, d\hat{s}'$$
$$- \int p(\vec{r}, \hat{s}, \hat{s}', t) n(\vec{r}, \hat{s}, t) \, d\hat{s}' \quad (5)$$

to describe the time rate of change of the cell number density $n(\vec{r}, \hat{s}, t)$, where $p(\vec{r}, \hat{s}', \hat{s}, t)$ is the probability density for a cell to change direction at point $\vec{r}$ at time t from direction $\hat{s}'$ to direction $\hat{s}$. The first term on the right-hand side is for net convective flow of cells at the point $\vec{r}$ having direction $\hat{s}$, the second term gives the rate at which cells change direction from $\hat{s}'$ to $\hat{s}$ [and is thus a gain term for $n(\vec{r}, \hat{s}, t)$], and the third term gives the rate at which cells change from $\hat{s}$ to $\hat{s}'$ [and is thus a loss term for $n(\vec{r}, \hat{s}, t)$]. Stroock's cell balance equation, Eq. (5), can be derived from the form of Alt's equations, which is obtained by integrating over τ (Alt, 1980; Ford and Cummings, 1992),

$$\frac{\partial n(\vec{r}, \hat{s}, t)}{\partial t} = -\hat{s} \cdot \vec{\nabla}_{\vec{r}} [v(\vec{r}, t) n(\vec{r}, \hat{s}, t)] - \int_0^\infty \beta(\vec{r}, \hat{s}, \tau, t) \sigma(\vec{r}, \hat{s}, \tau, t) \, d\tau$$
$$+ \int_0^\infty \int \beta(\vec{r}, \hat{s}', \tau, t) \sigma(\vec{r}, \hat{s}', \tau, t) k(\vec{r}, \hat{s}', t; \hat{s}) \, d\hat{s}' \, d\tau. \quad (6)$$

Under the assumption that $\beta(\vec{r}, \hat{s}, \tau, t)$ is independent of τ (valid for a Poisson process), Eq. (6) simplifies to

$$\frac{\partial n(\vec{r}, \hat{s}, t)}{\partial t} = -\hat{s} \cdot \vec{\nabla}_{\vec{r}} [v(\vec{r}, t) n(\vec{r}, \hat{s}, t) - \beta(\vec{r}, \hat{s}, t) n(\vec{r}, \hat{s}, t)$$
$$+ \int \beta(\vec{r}, \hat{s}', t) n(\vec{r}, \hat{s}', t) k(\vec{r}, \hat{s}', t; \hat{s}) \, d\hat{s}'. \quad (7)$$

Then for Eqs. (5) and (7) to be consistent requires that

$$p(\vec{r}, \hat{s}', \hat{s}, t) = \beta(\vec{r}, \hat{s}', t) k(\vec{r}, \hat{s}', t; \hat{s}), \quad (8)$$

$$\beta(\vec{r}, \hat{s}, t) = \int p(\vec{r}, \hat{s}, \hat{s}', t) \, d\hat{s}'. \quad (9)$$

Segel (1977) considered the motion of bacteria in one dimension (z) only and wrote down cell balance equations accordingly. If we define $n^+(z, t)$ as the density of cells at point z at time t moving in the positive z direction and $n^-(z, t)$ as the density of cells at point z at time t moving in the negative z direction, then Segel's equations are

$$\frac{\partial n^+}{\partial t} = -\frac{\partial}{\partial z}(vn^+) + p^- n^- - p^+ n^+, \tag{10}$$

$$\frac{\partial n^-}{\partial t} = \frac{\partial}{\partial z}(vn^-) + p^+ n^+ - p^- n^-. \tag{11}$$

In these equations, $p^+ = p^+(z, t)$ is the probability per unit time that a cell moving in the positive z direction becomes a cell moving in the negative direction (by tumbling), and $p^- = p^-(z, t)$ is the probability per unit time that a cell moving in the negative z direction becomes a cell moving in the positive direction. The one-dimensional cell density conservation equation is obtained by adding together Eqs. (10) and (11) to obtain

$$\frac{\partial c(z, t)}{\partial t} = -\frac{\partial}{\partial z} J_{cz}, \tag{12}$$

where

$$c(z, t) = n^+(z, t) + n^-(z, t) \tag{13}$$

and

$$J_{cz}(z, t) = v[n^+(z, t) - n^-(z, t)]. \tag{14}$$

The subscript on J contains a z to emphasize that it is a one-dimensional flux in the z direction. The cell flux evolution equation is obtained by subtracting Eqs. (10) and (11) to give

$$\frac{\partial J_{cz}}{\partial t} - \frac{1}{v} J_{cz} \frac{\partial v}{\partial t} = -v(n^+ - n^-)(p^+ + p^-) - \frac{\partial}{\partial z}(vc) - vc(p^+ - p^-)$$

$$= -J_{cz}(p^+ + p^-) - \frac{\partial}{\partial z}(vc) - vc(p^+ - p^-). \tag{15}$$

Until recently, the relationship between the three-dimensional cell balance equations of Alt and Stroock and the one-dimensional Segel equations has been obscure. Ford and Cummings (1992) have shown that the reduction of Alt's rigorous three-dimensional cell balance equations to the Segel one-dimension equations requires assuming that the cell motion is restricted to one dimension and that run times follow a Poisson distribution.

Assuming the cell flux has reached its equilibrium value [which is appropriate when the persistence time, $(p^+ + p^-)^{-1}$, is small compared to the experimental observation time period] and cell speed is not a function of attractant concentration

(Berg and Brown, 1972; Nossal and Chen, 1973), Eq. (15) can be rearranged in the form of the one-dimensional version of the phenomenological expression of Keller and Segel, given previously in Eq. (2):

$$J_{cz} = -\left[\frac{v^2}{p^+ + p^-}\right]\frac{\partial c}{\partial z} + \left[\frac{v(p^- - p^+)}{p^+ + p^-}\right]c. \qquad (16)$$

A comparison of the one-dimensional version of Eq. (2) with Eq. (16) leads to the following relationships between the population level transport properties (random motility and chemotactic velocity) and the individual cell properties (cell swimming speed v and tumbling probabilities p^+ and p^-):

$$\mu = \frac{v^2}{(p^+ + p^-)}, \qquad (17)$$

$$V_c = \frac{v(p^- - p^+)}{p^+ + p^-}. \qquad (18)$$

The model of Rivero et al. (1989), hereafter referred to as the RTBL model, follows from the use of Eqs. (10) and (11) with p^+ and p^- given as $p^\pm = p_t^\pm p_r$, where p_r is the probability that a cell reverses direction after tumbling [which, on the basis of experimental observations (Macnab, 1980), is independent of the presence of a chemical stimulus] and $p_t^\pm = p_t^\pm(z, t)$ is the probability of tumbling (i.e., ending a run) for a cell moving in the $\pm$ direction. In the RTBL model the tumbling probability is related to the mean run time according to

$$p_t^\pm = \frac{1}{\langle\tau^\pm\rangle}, \qquad (19)$$

where $\langle\tau^+\rangle$ and $\langle\tau^-\rangle$ are the average run times for cells moving in the positive and negative z directions, respectively. Equation (19) is valid provided that tumbling can be described as a Poisson process, a conclusion that is supported by the experimental data of Berg and co-workers (Berg and Brown, 1972). Experimental observations of individual-cell paths by Berg and Brown (1972) indicated that the reorientation after a tumble may not be completely random. In other words, bacteria may exhibit directional persistence. Analysis of these experimental observations (Alt, 1980) yielded $p_r = (1 - \psi)/2$, where ψ accounts for the directional persistence. For *E. coli* responding to amino acid attractants, the value of ψ is approximately 0.3 (Alt, 1980).

Berg and Brown (1972) also studied the effect of a chemical stimulus on individual cell paths and observed that the mean run times, $\langle\tau\rangle$, increased exponentially with the change in the number of receptor–attractant complexes, N_b, over mean run times measured in the absence of a chemical gradient, $\langle\tau_0\rangle$. (For the assumed Poisson process for tumbling, run time is related to Alt's general cell balance equations through $\beta = 1/\langle \tau \rangle$.) Based on this observation, the RTBL model for mean run time is given by

$$\langle \tau \rangle = \langle \tau_0 \rangle \exp\left(\nu \frac{DN_b}{Dt}\right) = \langle \tau_0 \rangle \exp\left(\nu \frac{\partial N_b}{\partial t} + \nu \vec{v} \cdot \vec{\nabla}_{\vec{r}} N_b\right), \quad (20)$$

where N_b is the number of bound receptors in the cell population, ν is the differential tumbling frequency, and D/Dt is the material (or substantial) derivative. The material derivative in Eq. (20) is necessary to account for both temporal and spatial changes in attractant concentration sensed by cells swimming at speed v (Berg and Brown, 1972; Macnab and Koshland, 1972; Spudich and Koshland, 1975). The quantity N_b is related to the chemical attractant concentration, a, by the following expression for receptor/ligand binding equilibrium:

$$N_b = \frac{N_T a}{(K_d + a)}, \quad (21)$$

where K_d is the dissociation constant for chemoattractant/receptor binding and N_T is the total number of receptors.

For substitution into the one-dimensional Segel equations through Eq. (19), Rivero et al. (1989) made use of a simplified version of Eq. (20) given by

$$\langle \tau^{\pm} \rangle = \langle \tau_0 \rangle \exp\left(\nu \frac{\partial N_b}{\partial t} \pm \nu v \frac{\partial N_b}{\partial z}\right). \quad (22)$$

Using these constitutive relationships [Eqs. (17), (18), (19), (21), and (22)], Rivero et al. (1989) derived expressions for the random motility coefficient and the chemotactic velocity, which are presented below.

Previous analyses have treated the random motility coefficient, μ, as a constant although the RTBL model (1989) predicts the following dependence on the spatial and temporal attractant concentration gradients:

$$\mu(a) = \mu_0 \exp\left(\nu \frac{N_T K_d}{(K_d + a)^2} \frac{\partial a}{\partial t}\right) \text{sech}\left(\nu \frac{N_T K_d}{(K_d + a)^2} v \frac{\partial a}{\partial z}\right). \quad (23)$$

Since bacterial random motility is characterized by the mean run length of an individual bacterium and the presence of a chemical gradient is known to alter the mean run length, it is not surprising that μ is dependent on the attractant gradient. In the absence of chemical gradients, the random motility coefficient reduces to a constant value, the gradient-independent random motility coefficient μ_0, defined in terms of individual cell properties as

$$\mu_0 = \frac{1}{2} \frac{v^2}{p_t(1-\psi)/2} = \frac{v^2}{p_t(1-\psi)}. \tag{24}$$

The individual-cell properties of swimming speed (v), tumbling probability in the absence of a gradient (p_t), and directional persistence (ψ) can be measured independently in individual cell studies. An analogy can be drawn to the relationship between the macroscopic transport quantity for the molecular diffusivity of a gas (D) and the individual molecular properties of average molecular speed ($\bar{u}$) and mean free path length (λ, sec^{-1}) given by $D = 0.5\,\bar{u}^2/\lambda$. The *derivation* of the formula for the random motility coefficient in terms of single-cell properties is also analogous to the derivation of the formula for the diffusion coefficient. In the latter case one begins with a molecular level kinetic theory (the Boltzmann equation) and by integrating over all possible directions of the molecular motion derives the macroscopic diffusion equation with the diffusivity defined above (Chapman and Cowling, 1960).

The driving force for the chemotactic velocity V_c is the attractant concentration gradient. A great amount of modeling effort has therefore been directed at determining an appropriate functional dependence on the attractant gradient. Keller and Segel (1971b) first proposed a linear dependence on the attractant gradient:

$$V_c(a) = \chi(a) \frac{\partial a}{\partial z}, \tag{25}$$

where $\chi(a)$ is the chemotaxis coefficient with unspecified dependence on attractant concentration, which only varies in the z direction. Several expressions have been suggested for $\chi(a)$. Keller and Segel (1971b) proposed

$$\chi(a) = \frac{\Delta}{a}, \tag{26}$$

where Δ is a proportionality constant. This particular form was chosen because it generated a traveling wave solution that could describe the traveling bands experimentally observed by Adler (1966, 1969). For an exponential attractant gradient this relationship predicts a constant chemotactic velocity. However, experiments performed in an exponential gradient did not yield a constant chemotactic velocity as expected; the velocity varied by a factor of 8 over the concentration range studied (Dalquist et al., 1972). (See Section 3.4.2 for details.) Lapidus and Schiller (1976) proposed a different functional form,

$$\chi(a) = \delta \frac{k}{(k+a)^2}, \tag{27}$$

to qualitatively satisfy the behavior of a dose–response curve in the capillary accumulation assay, as reported by Mesibov et al. (1973). The chemotactic response was proportional to the chemotactic mobility δ and had a maximum sensitivity at a particular concentration k, which they set equal to the initial attractant concentration. Segel (1977) later derived the following expression:

$$\chi(a) = \chi_M \frac{a^{n-1} K^n}{(K^n + a^n)^2}, \tag{28}$$

based on experimental observations of individual cell behavior by Brown and Berg (1972). The Hill cooperativity number n was included to match the sensitivity of response observed by Dalquist et al. (1972) over a wide range of attractant concentrations. For $n = 1$, the Lapidus and Schiller (1976) form is recovered with the sensitivity constant now defined as the dissociation constant K and the chemotactic mobility as the constant χ_M.

In both the capillary (Adler, 1973) and stopped-flow diffusion chamber (SFDC) assays, very steep gradients are initially present. According to Eq. (25), this would predict very large chemotactic velocities approaching infinity at short times. However, this is physically unrealistic since the chemotactic velocity is necessarily bounded by the swimming speed of an individual bacterium. Therefore Rivero et al. (1989), following an approach similar to Segel's (1977), derived a saturating expression for the dependence of the chemotactic velocity on the gradient. The resulting velocity,

$$V_c(a) = v \tanh\left(\nu \frac{N_T K_d}{(K_d + a)^2} v \frac{\partial a}{\partial z}\right), \tag{29}$$

depends on the hyperbolic tangent of the gradient in contrast to the linear dependence proposed by Keller and Segel (1971b). The individual cell properties that govern the chemotactic velocity include the one-dimensional swimming speed (v), the total number of receptors (N_T), the dissociation constant for the receptor–attractant complex (K_d), and the differential tumbling frequency (ν) representing the fractional change in cell run time per unit temporal change in receptor occupancy. The unit of the differential tumbling frequency is expressed as (receptors/time)$^{-1}$. Values for the cell speed and differential tumbling frequency can be measured independently in single-cell experiments such as the three-dimensional tracking studies by Berg and Brown (1972). The constitutive equations given in Eqs. (23) and (29) provide relationships between individual cell properties and macroscopic transport properties, which can be used to predict the dispersion and migration of bacterial populations in the environment.

Note that for shallow gradients Eq. (29) simplifies to a linear dependence on the gradient. The chemotactic coefficient $\chi(a)$ is defined in the following manner:

$$\chi(a) = \nu v^2 \frac{N_T K_d}{(K_d + a)^2} = \frac{\chi_0}{K_d} \frac{1}{(1 + a/K_d)^2}, \tag{30}$$

with χ_0 designated as the chemotactic sensitivity coefficient,

$$\chi_0 = \nu v^2 N_T, \tag{31}$$

representing a fractional change in dispersal capability per unit fractional change in receptor occupancy with units of distance^2/time. The population experiments performed in the SFDC enable us to test the validity of this functional form for the chemotaxis coefficient $\chi(a)$ over a wide range of attractant concentrations (Ford and Lauffenburger, 1991a).

3. EVALUATION OF CELL TRANSPORT COEFFICIENTS FROM EXPERIMENTAL DATA

We now turn to one of the main applications of the mathematical modeling described in this chapter, the determination of macroscopic cell transport coefficients, μ and V_c, from experimental measurements. In general, the coefficients μ and V_c depend on the chemoattractant gradient. Therefore it is much more useful for predictive modeling to determine the gradient-independent parameters μ_0, the random motility coefficient, and χ_0, the chemotactic sensitivity coefficient, that arise from the RTBL model and are given by Eqs. (24), (30), and (31). Provided these equations are valid (see Section 3.5), this allows μ and V_c to be calculated for arbitrary chemoattractant gradients. Thus we focus our attention on the determination of μ_0 and χ_0 from the experimental assays.

We begin by describing the three experimental assays—the capillary assay, the laser densitometry assay, and the SFDC assay—in Section 3.1. In order to extract the transport coefficients from each experiment, it is necessary to relate them to the quantities measured in each assay. We do this through the cell population balance and constitutive equations of Keller and Segel, Eqs. (1) and (2), supplemented with material balances for the chemoattractant, and utilizing Eqs. (23) and (29). The balance equations that are common to all three assays are derived in Section 3.2. Section 3.3 contains the analysis of the capillary assay and the SFDC necessary to determine the gradient-independent random motility coefficient μ_0. In Section 3.4, we consider the calculation of the chemotactic sensitivity coefficient for each of the assays in turn. Each assay is distinguished by different boundary conditions for the cell balance equations. This section also contains a table of μ_0 and χ_0 values obtained using these methods of analysis. Section 3.5 describes the use of the experimental data to verify the functional form of Eq. (30). Another test of the model equations is presented in which the RTBL equations are used to deduce the differential tumbling frequency from experimental data. This value is then compared to the value obtained from individual cell properties according to Eq. (31).

3.1. Description of Population Experiments

3.1.1. Capillary Assay. Adler's capillary technique (1973) continues to be the most widely used assay in the study of bacterial chemotaxis because of its experimental simplicity. A thin capillary (1 μL) filled with a chemical stimulus is placed into a chamber (0.2 mL) containing a dilute bacterial suspension; the chamber is formed by a U-shaped tube between a microscope slide and cover slip. As the stimulus diffuses out of the capillary into the surrounding suspension, a transient gradient is established within the capillary. Bacteria swimming near the capillary entrance move into the capillary toward higher concentrations of the chemical attractant. The relative response is measured by comparing the accumulation of bacteria within the capillary in the presence and in the absence of a chemical stimulus gradient. This simple approach is appropriate for qualitative comparison of relative responses to different chemicals or different concentrations of the same chemical, but it is not as useful for relating the population behavior to intrinsic cell parameters for several reasons. One reason is that bacterial accumulation within the capillary depends on the geometry of the assay and the particular experimental conditions under which it is performed. For example, the number of bacteria moving into the capillary will depend on the capillary diameter as well as the initial bacterial density, the stimulus concentration, and the length of time the experiment is run. Therefore it is difficult to compare results from experiments performed in different laboratory situations. Even more importantly, based on these measurements, it is impossible to make predictions regarding the population response in application systems outside the controlled conditions of the laboratory. Although a mathematical model could account for assay geometry and experimental variables, the capillary assay is difficult to characterize because the bacterial and stimulus concentrations at the capillary entrance are poorly defined. Vicker (1981) used dye experiments to study the concentration profiles of the chemical stimulus as it diffused out of the capillary into free solution. He concluded that the dye movements outside the capillary were entirely anomalous and virtually independent of diffusion kinetics. The effects of convection and turbulence on the concentration profile are especially difficult to predict under these circumstances. Futrelle and Berg (1972) reduced these complications by increasing the viscosity of the free solution with 0.18% methyl cellulose. The bacterial density near the capillary entrance is also difficult to characterize; Adler (1969) observed that a "cloud" of bacteria forms around the tip of the capillary as cells move toward attractant that has diffused into the chamber from the capillary. Mathematical models to describe this assay have either assumed that the bacterial suspension in the chamber is uniform, well-mixed by convection (Segel et al., 1977), or required a complex model to account for gradients near the capillary entrance (Futrelle and Berg, 1972; Rivero, 1986).

3.1.2. Laser Densitometry Assay. In an effort to develop an experiment more amenable to mathematical analysis, Dalquist et al. (1972) used a laser densitometer to monitor changes in the spatial distribution of an initially uniform bacterial suspension of *S. typhimurium* exposed to several different gradients of serine (an attractant). Step-change, linear, and exponential gradients in attractant concentration were created by using suitable mixing chambers prior to filling the observation chamber. A linear density gradient in glycerol (0.5–3%) helped to stabilize the gradients within the observation chamber. However, for initial step changes in concentration the gradients were not well defined. The bacterial density profile throughout the chamber was measured by detecting the light scattered at right angles from the beam of a HeNe laser directed upward through the bottom of the observation chamber. Moving the observation chamber by means of a screw drive allowed the bacterial density to be measured continuously at all positions within the cell. To scan the length of the observation chamber required from 1 to 3 min.

Because the chemotactic response is driven by the attractant concentration gradient, it is critical that an experimental assay be designed in such a way to generate and maintain well-defined gradients throughout the assay. Step-change gradients have the advantage of inducing large, easily measured responses. However, in both the capillary and the laser densitometry assays the step-change gradients are not easily characterized. The SFDC assay developed by Ford et al. (1991) to overcome the shortcomings of the capillary and laser densitometry assays generates a step-change gradient that is easy to characterize.

3.1.3. Stopped-Flow Diffusion Chamber Assay. The SFDC shown schematically in Fig. 7.3 was originally developed to observe the movement of latex particles in electrolyte gradients (Staffeld and Quinn, 1989). It is appropriate for the study of bacterial chemotaxis because it provides well-characterized chemical gradients. Operating on the same basic principle as the Mach–Zehnder interferometry cell, two suspensions differing only in attractant concentration are contacted by impinging flow (Caldwell et al., 1957; Schlichting, 1979). As long as the fluid is flowing no mixing occurs between the two suspensions and a step change in stimulus concentration is maintained.

For random motility studies, a dilute bacterial suspension is introduced into the lower half of the chamber and a buffer containing no chemical attractant is added to the upper half. During flow a sharp step change in bacterial density is maintained between the two halves of the chamber. Once flow is stopped, bacteria in the lower half of the chamber begin swimming into the upper half and the initial step change gradually decays with time.

For chemotaxis experiments, a dilute bacterial suspension is introduced into the upper half of the chamber while the lower half contains the same dilute bacterial suspension but with appropriate amounts of the chemical attractant added. Initially, while the solutions are flowing through the chamber, the bacterial density remains uniform throughout the chamber and a step change in the attractant con-

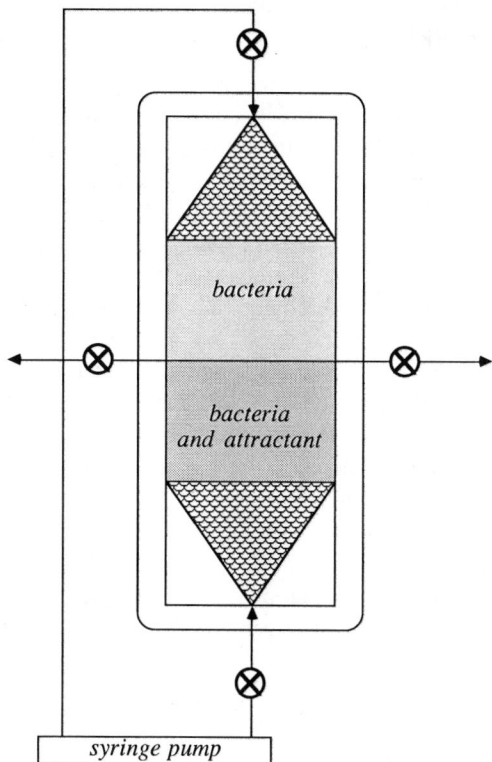

FIGURE 7.3. Diagram of the SFDC. Two glass microscope slides with a small gap (1.9 mm) between them form the walls of the chamber. Fluid enters the chamber through 0.6-mm slits in the top and bottom and exits through similar slits on either side. The position of these slits is critical in determining the flow pattern; the exit slits are directly across from each other and perpendicular to the centerline connecting the top and bottom slits. Three-way valves at each of the four openings are used to control flow to and from the chamber. The triangular-shaped packed beds help to dissipate the momentum of the incoming fluid and evenly distribute flow across the width of the chamber.

centration is maintained between the upper and lower halves. When flow is stopped attractant in the bottom half begins to diffuse into the upper half, creating a transient gradient in the attractant concentration. As bacteria sense and respond to this gradient, a band of high cell density forms immediately and moves downward toward higher attractant concentrations as time progresses and the attractant gradient decays. Light scattering is used to visualize and detect changes in bacterial density within the SFDC, which are then recorded on photographic negatives. The negatives are scanned and digitized using image analysis and calibrated to bacterial densities to provide experimental profiles for comparison to theoretical model predictions.

3.2. Modeling Equations

We model the experimental assays at the macroscopic level using the conservation equation, Eq. (1), combined with the flux expression of Keller and Segel, Eq. (2). The random motility and chemotactic velocity are taken from the RTBL model, which is based on the one-dimensional Segel equations [Eq. (10) and (11)] combined with the assumption that the probability of reversing directions is independent of the chemical gradient, as noted by Ford and Cummings (1992). Ford and Cummings also point out that the rigorous reduction of Alt's equations to a system with symmetry in all but one direction (the true geometry of the experimental assays) does not lead to the Segel equations. However, as recounted below, the expressions for random motility and chemotactic velocity from the one-dimensional RTBL model, with reduction of the net cell speed by the factor $\sqrt{3}$ (Macnab, 1980), have been demonstrated to account for experimental results in an attractant step-function concentration gradient quite satisfactorily for all three assays (Ford and Lauffenburger, 1991b, 1992; Ford et al., 1991). Hence, despite the approximate nature of these expressions for dimensions greater than one, we nevertheless consider them to be very useful.

We begin with conservation equations for the bacterial density c and the attractant concentration a in one spatial dimension, z:

$$\frac{\partial c}{\partial t} = -\frac{\partial}{\partial z}(J_{cz}), \tag{32}$$

$$\frac{\partial a}{\partial t} = -\frac{\partial}{\partial z}(J_{az}) - R_a. \tag{33}$$

The bacterial and attractant fluxes are J_{cz} and J_{az}, respectively. Since attractants such as amino acids and sugars are transported into the cell and metabolized, we include the consumption rate R_a to account for depletion of attractants, which are metabolizable. We may assume that bacterial growth is negligible since the experimental observation times are short compared to expected bacterial doubling times at the suboptimal growth conditions in the chemotaxis assays.

Substituting the Keller and Segel flux expression, Eq. (2), into the bacterial conservation equation gives

$$\frac{\partial c}{\partial t} = \mu \frac{\partial^2 c}{\partial z^2} - \frac{\partial}{\partial z}(V_c c). \tag{34}$$

The attractant flux is modeled according to Fickian diffusion with a diffusion coefficient D, and the consumption rate is assumed to follow Monod kinetics with specific uptake rate k_{max} and saturation constant K_s, which corresponds to the concentration at half the maximum consumption rate:

$$\frac{\partial a}{\partial t} = D \frac{\partial^2 a}{\partial z^2} - \left[\frac{k_{max} a}{K_s + a}\right] c. \tag{35}$$

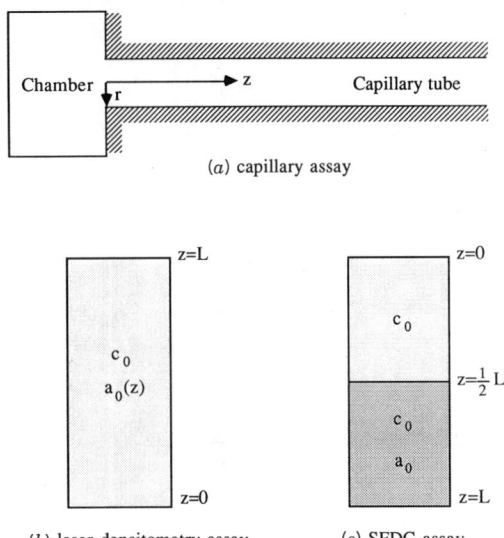

FIGURE 7.4. Geometry used for the experimental assays: (*a*) capillary, (*b*) laser densitometry, and (*c*) SFDC.

Initial and boundary conditions, as described in Section 3.3, depend on the particular experimental assay. The geometry for each assay is represented in Fig. 7.4.

3.3. Determination of Random Motility Coefficients from Population Assays

Random motility experiments are performed in the absence of chemical gradients, so only the first term describing diffusive motion is required in the Keller and Segel flux expression, leading to the following conservation equation for bacterial density:

$$\frac{\partial c}{\partial t} = \mu \frac{\partial^2 c}{\partial z^2}, \tag{36}$$

where the random motility coefficient is equal to the gradient-independent value, $\mu = \mu_0$. In Sections 3.3.1 and 3.3.2, we present the solution to Eq. (36) for the boundary and initial conditions corresponding to the capillary and SFDC assays. We have not analyzed the laser densitometry assays for random motility since none of these experiments are performed in the absence of a chemoattractant gradient. However, a random motility coefficient in the presence of a chemoattractant for the laser densitometry assay has been inferred by Segel and Jackson (1973) from the dispersion in the traveling bands and is reported in Table 7.1, but this type of analysis is beyond the scope of the present chapter.

TABLE 7.1. Values for Bacterial Transport Coefficients Determined from Capillary, Laser Densitometry, and SFDC Assays

Bacteria	Attractant	Chemotactic Sensitivity Coefficient	Random Motility Coefficient	Assay	References
P. fluorenscens	—	—	6×10^{-6} cm^2/s	Capillary	Segel et al. (1977)
E. coli B275	—	—	8×10^{-6} cm^2/s	Capillary	Segel et al. (1977), Adler (1969)
E. coli AW518	α-Methylasparate	7.5×10^{-4} cm^2/s	1.4×10^{-5} cm^2/s	Capillary	Rivero-Hudec and Lauffenburger (1986), Mesibov et al. (1973)
E. coli 1100	Galactose	7×10^{-4} cm^2/s	2.7×10^{-6} cm^2/s	Capillary	Ford and Lauffenburger (1992), Terracciano (1985)
S. typhimurium	L-Serine	7.5×10^{-4} cm^2/s	6×10^{-5} cm^2/s	Laser densitometry	Segel and Jackson (1973), Ford and Lauffenburger (1991b), Dalquist et al. (1972)
E. coli K12	Fucose	8.1×10^{-5} cm^2/s	1.1×10^{-5} cm^2/s	SFDC	Ford and Lauffenburger (1991a), Ford et al. (1991)

3.3.1. Capillary Assay.

Initial bacterial densities in the chamber ($z < 0$) and capillary tube are c_0 and zero, respectively. When the capillary is much longer than the distance bacteria can migrate due to random motility, we can represent the capillary as semi-infinite in one spatial dimension. This is a reasonable approximation for typical values of the random motility coefficient (1×10^{-5} cm^2/s) in a capillary 1 cm in length for an experiment lasting up to an hour. It implies that no bacteria reach the end of the capillary during the experiment. Throughout the experiment the bacterial density at the capillary entrance is assumed to be equal to the initial bacterial density in the chamber, c_0. This assumption is reasonable if the suspension in the chamber is subject to some convective mixing, as is usually the case (Vicker, 1981), and if the chamber volume is large compared to the capillary; that is, depletion of bacteria from the chamber as they swim into the capillary is negligible. The following boundary and initial conditions characterize the capillary assay:

$$c(z \to \infty, t) = 0 \tag{37a}$$

$$c(z, t) = c_0, \quad z \leq 0, \tag{37b}$$

$$c(z, 0) = 0, \quad z > 0, \tag{37c}$$

where $z \leq 0$ refers to the chamber and $z > 0$ refers to the capillary tube as illustrated in Fig. 7.4a. Under these conditions the solution to Eq. 36 is well known (Crank, 1975). We present the result for $z > 0$:

$$c(z, t) = c_0 \, \text{erfc}\left(\frac{z}{\sqrt{4\mu_0 t}}\right), \quad z > 0, \tag{38}$$

since we are interested in determining bacterial accumulation within the capillary. The net migration of bacteria from the chamber ($z < 0$) to the capillary ($z > 0$) is determined by

$$N_{\text{RM}}(t) = \pi r^2 \int_0^L c(z, t) \, dz \tag{39}$$

where $N_{\text{RM}}(t)$ is the number of bacteria accumulating in the capillary due to random motility, r is the capillary radius, and L is the length of the capillary. After substitution of Eq. (38) into Eq. (39) and integration, we obtain the following expression for the number of bacteria moving across the $z = 0$ plane as a function of the random motility coefficient, experimental time, and the chamber geometry:

$$N_{\text{RM}}(t) = \pi r^2 c_0 \left\{ L \, \text{erfc}\left[\frac{L}{\sqrt{4\mu_0 t}}\right] + \sqrt{\frac{4\mu_0 t}{\pi}} \left[1 - \exp\left(-\frac{L^2}{4\mu_0 t}\right)\right] \right\}. \tag{40}$$

For relatively short experimental times ($t \ll L^2/4\mu_0$), this reduces to

$$N_{RM}(t) = \pi r^2 c_0 \sqrt{\frac{4\mu_0 t}{\pi}}, \qquad (41)$$

which is the expression also developed by Segel et al. (1977). Based on this relationship, we expect the number of bacteria accumulating in the capillary to be proportional to $\sqrt{t}$. We can determine μ_0 from the slope of a plot of $N_{RM}(t)$ versus $\sqrt{t}$. The random motility coefficient was calculated from a rearrangement of Eq. (41) for bacterial accumulations reported by Terracciano (1985) of cells grown at different dilution rates in a chemostat. The values varied between 1.4 and 3.5×10^{-6} cm^2/s.

3.3.2. Stopped-Flow Diffusion Chamber Assay.

The boundary conditions for the SFDC differ from those for the capillary assay only in the bacterial density at the $z = 0$ position. For the SFDC, we replace Eq. (37b) by

$$c(z, t) = 0.5 c_0, \qquad z = 0. \qquad (42)$$

The equation for $N_{RM}(t)$ analogous to Eq. (41) then becomes

$$N(t) = A_x 0.5 c_0 \sqrt{\frac{4\mu_0 t}{\pi}}, \qquad (43)$$

where A_x represents the cross-sectional area of the SFDC. For the SFDC experiments, we can relate theoretical predictions to the image analysis of experimental profiles through the following relationship:

$$\frac{N(t)}{A_x} = 0.5 c_0 \sqrt{\frac{4\mu_0 t}{\pi}} = \Gamma \cdot \text{Area}, \qquad (44)$$

where the calibration constant Γ relates gray levels from image analysis to bacterial densities and the area is experimentally determined from bacterial density profiles [see Ford et al. (1991) for details of the analysis]. The resulting random motility coefficient measured in the SFDC for *E. coli* K12 was $1.1 \pm 0.4 \times 10^{-5}$ cm^2/s.

3.4. Determination of Chemotactic Sensitivity Coefficients from Population Assays

We consider the solution of Eq. (34) and (35) for the boundary and initial conditions specific to each assay in the following sections.

3.4.1. Capillary Assay. To facilitate their solution for the capillary assay, the conservation equations (34) and (35) have been nondimensionalized in terms of the dimensionless variables:

$$C = \frac{c}{c_0}, \tag{45a}$$

$$A = \frac{a}{K_d}, \tag{45b}$$

$$\xi_r = \frac{z}{r}, \tag{45c}$$

$$\tau_r = \frac{Dt}{r^2}, \tag{45d}$$

where r is the radius of the capillary. The resulting equations for the attractant concentration (nonmetabolizable) and bacterial density become

$$\frac{\partial A}{\partial \tau} = \frac{\partial^2 A}{\partial \xi_r^2} \tag{46}$$

and

$$\frac{\partial C}{\partial \tau_r} = \rho \frac{\partial^2 C}{\partial \xi_r^2} - \delta\rho \frac{\partial}{\partial \xi_r}(UC), \tag{47}$$

respectively, with U the dimensionless chemotactic velocity defined below. Two dimensionless groups result from the analysis:

$$\delta = \frac{\chi_0}{\mu_0} \tag{48a}$$

$$\rho = \frac{\mu_0}{D}. \tag{48b}$$

The dimensionless chemotactic sensitivity coefficient δ compares the directed migration of the bacterial population to its random dispersion. The dimensionless random motility coefficient ρ relates bacterial dispersion to attractant diffusivity.

Equation (25) combined with Eq. (30) yields a chemotactic velocity expression that is valid for experiments involving shallow attractant gradients, where the chemotactic response is expected to be linearly proportional to the attractant gradient. For steep gradients, the chemotactic velocity is given by Eq. (29) so that an infinitely steep gradient corresponds to a velocity no greater than the cell's swimming speed (i.e., totally directed motion toward increasing attractant concentrations). It has been demonstrated, though, that this refinement yields negligible advantage for the capillary assay (Rivero-Hudec and Lauffenburger, 1986); the steep gradients that are initially present quickly decay to shallow gradients due to diffusion and for typical experimental times of 1 h the effect on total bacterial accumulation is negligible. Therefore we use the following relationship for the chemotactic velocity when applied to the capillary assay:

$$U = \frac{V_c r}{\chi_0} = \frac{1}{(1+A)^2} \frac{\partial A}{\partial \xi_r}. \tag{49}$$

The dimensionless boundary conditions for the bacterial density follow from Eq. (37a)–(37c) and are given below:

$$C(\xi_r \to \infty, \tau_r) = 0, \tag{50a}$$

$$C(\xi_r, \tau_r) = 1, \quad \xi_r \leq 0, \tag{50b}$$

$$C(\xi_r, 0) = 0, \quad \xi_r > 0. \tag{50c}$$

For the attractant concentration we also approximate the capillary tube as semi-infinite in one spatial dimension, which is reasonable for typical values of the diffusion coefficients of attractants (1×10^{-5} cm^2/s). This implies that the attractant concentration at the closed end of the capillary does not change from its initial value as a result of diffusion. Initially, the capillary is filled with an attractant of dimensionless concentration A_∞ and placed in a suspension having dimensionless bacterial density C_c and dimensionless attractant concentration A_c. Because a true step change in attractant concentration is not physically attainable in an experimental assay, we approximate the attractant concentration at the capillary entrance as the average of the attractant concentration in the chamber and in the capillary, $A_0 = 0.5(A_\infty + A_c)$; this is the expected behavior for an initial step change in concentration that has decayed for a small time $\tau_r > 0$ and implies the chamber is well mixed except for a small boundary layer near the capillary entrance. When attractant consumption is negligible, the attractant concentration at the mouth of the capillary remains A_0 throughout the experiment. For cases with a significant consumption rate, an expression can be included to account for the depletion of attractant in the capillary (Ford and Lauffenburger, 1992). The dimensionless boundary and initial conditions for the attractant are

$$A(\xi_r \to \infty, \tau_r) = 0, \tag{51a}$$

$$A(\xi_r, 0) = A_\infty, \quad \xi_r > 0, \tag{51b}$$

$$A(0, \tau_r) = A_0. \tag{51c}$$

The attractant concentration in the capillary and the corresponding gradient from the solution of Eq. (46) with the boundary and initial conditions from Eq. (51) are given below:

$$A = A_0 + (A_\infty - A_0) \operatorname{erf}\left(\frac{\xi_r}{\sqrt{4\tau_r}}\right), \quad (52)$$

$$\frac{\partial A}{\partial \xi_r} = \frac{(A_\infty - A_0)}{\sqrt{\pi \tau_r}} \exp\left(\frac{-\xi_r^2}{4\tau_r}\right). \quad (53)$$

The equation for the bacterial density in the capillary has been solved using an integral balance method based on a truncated Taylor expansion of the model equation (Ford and Lauffenburger, 1992). The resulting migration behavior, including contributions due to both random motility and chemotaxis, is given by

$$\frac{N(\tau_r)}{c_0 \pi r^3} = \sqrt{\frac{4\rho \tau_r}{\pi}} \left(1 + \delta\sqrt{\rho}\, \frac{A_\infty - A_0}{(1 + A_0)^2}\right), \quad (54)$$

where N is the number of bacteria within the capillary at dimensionless time τ_r. The first term in Eq. (54) corresponds to random motility alone; the second term accounts for the chemotactic response. The result is an extremely simple expression for predicting bacterial accumulation in the capillary assay in terms of intrinsic cell parameters. Rewriting Eq. (54) yields a correspondingly simple expression for the value of the chemotactic sensitivity coefficient χ_0:

$$\chi_0 = \sqrt{\mu_0 D}\, \frac{(1 + A_0)^2}{A_\infty - A_0} \left(\frac{N(t)}{N_{RM}(t)} - 1\right). \quad (55)$$

We applied the model just described to data from Terracciano (1985) for *E. coli* 1100 responding to galactose in the capillary assay. In Fig. 7.5 the ratio of the bacterial accumulation due to a chemotactic response to the accumulation resulting from random motility alone, N/N_{RM}, is plotted as a function of the log of the initial concentration of attractant in the capillary. Superimposed on the data is the response predicted from the model equations, which accounts for metabolism of the attractant (see Ford and Lauffenburger, 1992 for details). A chemotactic sensitivity coefficient of 1.5×10^{-3} cm^2/s was determined based on the magnitude of the peak response.

We use slightly different reduced variables for distance and time to de-dimensionalize the conservation equations for application to the laser densitometry and SFDC assays:

$$\xi = \frac{zv}{D}, \quad (56a)$$

$$\tau = \frac{tv^2}{D}. \quad (56b)$$

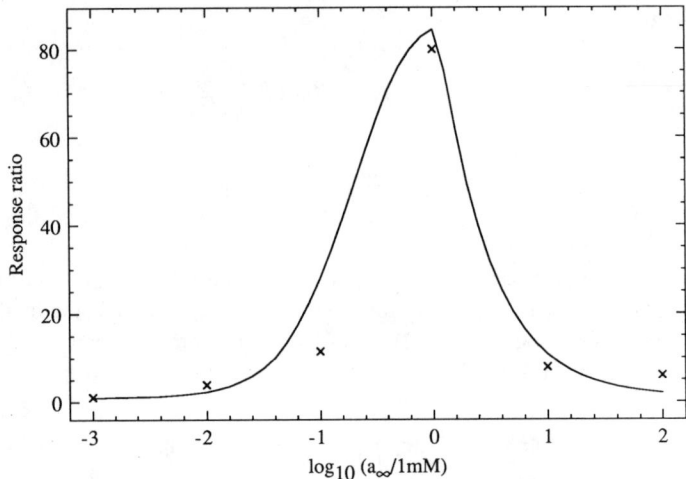

FIGURE 7.5. Comparison of the model predictions (——) with data from the capillary assay (×) reported by Terracciano (1985) for *E. coli* 1100 responding to galactose. For cells harvested from a chemostat at a dilution rate of 0.1 h^{-1}, a value of $\chi_0 = 1.5 \times 10^{-3}$ cm^2/s best characterized the data. Bacterial accumulation due to random motility was 2500 cells. Calculations accounted for the uptake of galactose by the cells.

The dimensionless conservation equations follow from Eq. (34) and (35):

$$\frac{\partial C}{\partial \tau} = \rho \frac{\partial^2 C}{\partial \xi^2} - \frac{\partial}{\partial \xi}(UC), \tag{57}$$

$$\frac{\partial A}{\partial \tau} = \frac{\partial^2 A}{\partial \xi^2} - \gamma \left(\frac{AC}{\kappa + A}\right). \tag{58}$$

The consumption term is included for generality, although for the application to the laser densitometry data we neglected consumption and for the SFDC we included it, but determined that it was inconsequential. The dimensionless random motility ρ is given by

$$\rho = \frac{\mu}{D} = \rho_0 \exp\left\{\delta\rho_0 \frac{1}{(1+A)^2} \frac{\partial A}{\partial \tau}\right\} \operatorname{sech}\left\{\delta\rho_0 \frac{1}{(1+A)^2} \frac{\partial A}{\partial \xi}\right\}, \tag{59}$$

although a gradient-independent random motility coefficient μ_0 has been shown to be satisfactory for the applications of interest in this chapter (see Ford and Lauffenburger, 1991a). The dimensionless chemotactic velocity is

$$U = \frac{V_c}{v} = \tanh\left(\delta\rho \frac{1}{(1+A)^2} \frac{\partial A}{\partial \xi}\right). \tag{60}$$

The following dimensionless groups have physical interpretations:

$$\rho_0 = \frac{\mu_0}{D}, \tag{61a}$$

$$\delta = \frac{\chi_0}{\mu_0}, \tag{61b}$$

$$\kappa = \frac{K_s}{K_d}, \tag{61c}$$

$$\gamma = \frac{k_{\max} c_0 D}{K_d v^2}. \tag{61d}$$

The ratio of bacterial dispersion in the absence of a chemical gradient to attractant diffusivity is characterized by ρ_0. The dimensionless chemotactic sensitivity coefficient δ, the ratio of the chemotactic sensitivity coefficient to the gradient-independent random motility coefficient, can be thought of as a Peclet number (Bird et al., 1960), a measure of the ratio of fluid convection to molecular diffusion. It essentially compares the directed motion of the bacterial population (chemotaxis) to its dispersion (random motility). The dimensionless group κ (with K_s the saturation constant for attractant uptake) is the affinity of the transport receptor for the attractant relative to the affinity of the chemotaxis receptor. This group gives an indication of the attractant concentrations for which consumption of the attractant will significantly affect the attractant gradient. The magnitude of the effect of consumption on the attractant gradient is given by γ. For values of γ greater than unity uptake controls the gradient and for values less than unity diffusion controls it. A study of the effects of these parameters on the bacterial density and attractant concentration profiles is reported by Ford and Lauffenburger (1991b).

Equations (57) and (58) with the boundary conditions for each assay as specified in the following sections were solved numerically using a Crank–Nicolson finite-differencing scheme and a predictor–corrector approach (Ford, 1989).

3.4.2. Laser Densitometry Assay. The experiments by Dalquist et al. (1972) are especially amenable to mathematical interpretation because of the effort made to establish and maintain well-defined attractant concentration gradients (except for step-change gradients) and the measurement of actual bacterial density profiles as opposed to total bacterial accumulation. In these respects it is a significant improvement over the capillary assay. Data from this assay for the response of *S. typhimurium* to serine in an exponential gradient are presented in Fig. 7.6. The chemotactic velocity was plotted as a function of the initial attractant concentration for an exponential gradient. These data have been analyzed by several other researchers and we begin by presenting their results.

Dalquist et al. (1972) neglected the contribution due to random motility in the flux expression from Eq. (2) and assumed the velocity depended on the attractant

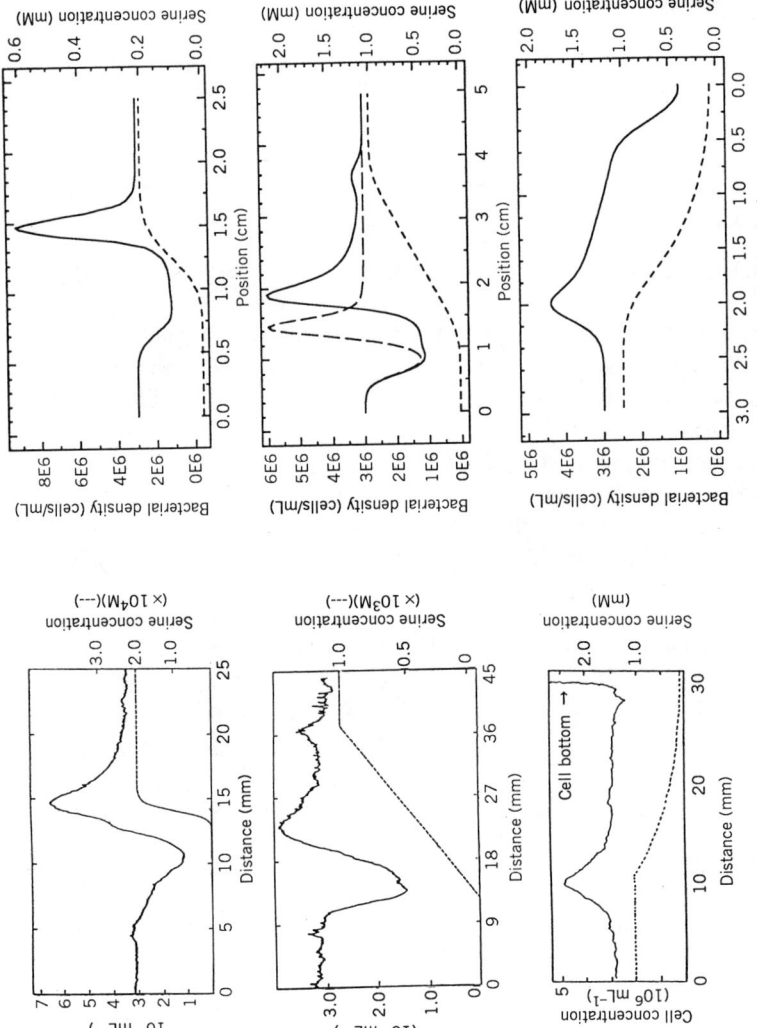

FIGURE 7.6. Experimental profiles of bacterial density and serine concentration as reported by Dalquist et al. (1972) for a step gradient at 11 min (top left), linear gradient at 18 min (middle left), and exponential gradient at 15 min (bottom left). The decay distance for the exponential gradient was 0.64 cm. (Reprinted with written permission from Macmillan Magazines Ltd.). On the right are the corresponding theoretical predictions for the bacterial density (——) and attractant concentration (– – –) calculated from the model equations. The long dashed line in the middle right figure shows the calculation excluding the low affinity receptor.

concentration according to Eqs. (25) and (26). Their model predicted a constant chemotactic velocity within an exponential gradient, but their experimental observations indicated that the velocity varied from 0.32 to 2.8 μm/s over a range of plateau concentrations from 10^{-6} to 10^{-2}M (see Fig. 7.6).

Segel and Jackson (1973) in their analysis of these experiments included random motility in the conservation equation but also assumed a constant average chemotactic velocity. They reported an average value of 2 cm^2/h for their proportionality factor δ, but note that it was indeed a function of the attractant concentration. Since the variation is fairly small, they proceeded to determine formulas for the random motility coefficient and average velocity based on the height and width of the profile peaks. The random motility coefficient was calculated to be 0.2 cm^2/h.

This same set of data was also analyzed by Segel (1977) using a model based on individual cell receptor-mediated processes. He also neglected random motility contributions but did include the dependence of the chemotactic velocity on the attractant gradient in the form given by Eq. (25) and (28). As stated previously in Section 2.4, the Hill formula (Rubinow, 1975) was included because it provided a better representation of the data. The best fit was for $n = \frac{1}{2}$, suggesting negative cooperativity in receptor binding. Negative cooperativity was one explanation offered by Segel (1977) to account for the sensitivity of the response over such a wide range of attractant concentrations. Another explanation suggested the existence of a heterogeneous population of receptors for serine chemotaxis (Segel, 1979). This hypothesis was supported by biochemical evidence for two receptor subpopulations reported by Clarke and Koshland (1979). They noted in binding assays that serine responses did not saturate but continued to increase at concentrations above 10 mM, which implied the existence of a second receptor with low affinity. Unsuccessful attempts to isolate serine receptor mutants in *E. coli* also pointed to the possibility of two different receptor sites since the probability of obtaining a mutant that lacked both receptors would be quite small (Dalquist et al., 1976).

We can include in our model the involvement of two receptor subpopulations, one with low affinity and a second one with high affinity. This requires a new scaling for the attractant concentration because we now have two dissociation constants, so we choose instead the initial attractant concentration giving

$$A' = \frac{a}{a_0}, \qquad (62a)$$

$$\kappa_i' = \frac{K_{d,i}}{a_0}. \qquad (62b)$$

The dimensionless chemotactic velocity then becomes

$$U' = \frac{V_c}{v} = \tanh\left(\sum_{i=1}^{2} \nu_i v^2 N_{T,i} \frac{\kappa_i'}{(\kappa_i' + A')^2} \frac{\partial A'}{\partial z}\right). \qquad (63)$$

Application of our model equations to the experimental set-up used by Dalquist et al. (1972) for the case with the exponential gradient requires the following boundary and initial conditions for the geometry pictured in Fig. 7.4b:

$$C(\xi, 0) = 1, \tag{64a}$$

$$\frac{\partial C}{\partial \xi}(0, \tau) = \frac{\partial C}{\partial \xi}(\xi_L, \tau) = 0, \tag{64b,c}$$

$$A' = \begin{cases} 1, & \xi \geq \frac{2}{3}\frac{vL}{D}, \\ \exp\frac{D(\xi + 2)}{\lambda v}, & 0 < \xi < \frac{2}{3}\frac{vL}{D}, \end{cases} \tag{65a}$$

$$\frac{\partial A}{\partial \xi}(0, \tau) = \frac{\partial A}{\partial \xi}(\xi_L, \tau) = 0, \tag{65b,c}$$

where λ is the distance over which the gradient changes by a factor of $1/e$. The model equations were solved numerically using the parameter values specified in Table III of Ford and Lauffenburger (1991b). Experimental times, distances, and initial concentrations were set to the reported experimental values. The dissociation constant for the high affinity receptor, $K_{d2} = 5$ μM, was the value measured by Clarke and Koshland (1979). The other dissociation constant, $K_{d1} = 0.2$ mM, was chosen to match the peak of the dose–response curve (Fig. 7.7) at the exper-

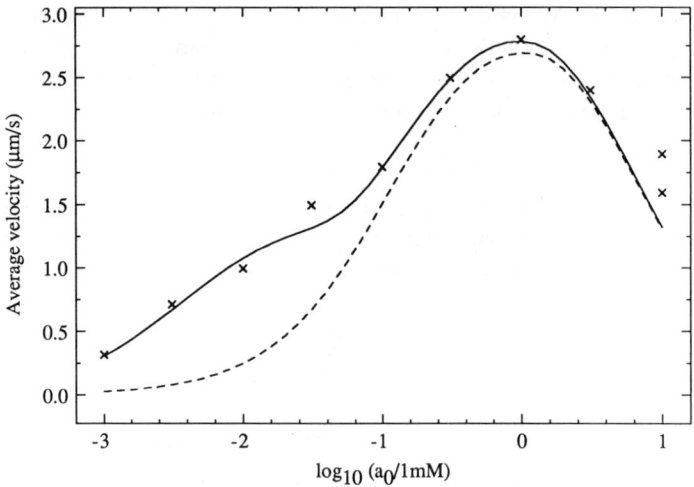

FIGURE 7.7. Comparison of model predictions with chemotactic velocities from the Dalquist et al. (1972) laser densitometry assay for an exponential gradient with decay distance λ of 0.64 cm. Data points are from Dalquist et al. (1972) and the model predictions are shown for a single low affinity receptor (---) with $K_d = 0.2$ mM and for both low and high ($K_d = 5$ μM) affinity receptors (——).

imentally measured attractant concentration. Values for the chemotactic sensitivity coefficients, $\chi_{01} = 7.5 \times 10^{-4}$ cm^2/s and $\chi_{02} = 2.5 \times 10^{-4}$ cm^2/s, were also determined by finding the value of χ_0 that yielded the best agreement between calculated and experimental cell density profiles.

With all model parameters specified from independent estimates and the data of Fig. 7.7, we compared *a priori* model predictions of the bacterial density profiles to the experimental profiles of Fig. 7.6. The model predictions, shown in Fig. 7.6, demonstrate excellent agreement with the qualitative features of the profiles. The most salient feature in these profiles is the second peak for the linear gradient. The dashed line is a model prediction with only the high affinity receptor (i.e., $N_{T,1} = 0$). This is certainly strong evidence for the existence of receptor subpopulations in serine taxis. Although the model comes close to quantitative accuracy, we really cannot expect complete agreement without a precise measurement of the random motility coefficient.

3.4.3. Stopped-Flow Diffusion Chamber Assay. The initial and boundary conditions for the SFDC geometry pictured in Fig. 7.4c follow:

$$C(\xi, 0) = 1, \tag{66a}$$

$$\frac{\partial C}{\partial \xi}(0, \tau) = \frac{\partial C}{\partial \xi}(\xi_L, \tau) = 0, \tag{66b,c}$$

$$A(\xi, 0) = \begin{cases} 0, & 0 < \xi < \tfrac{1}{2}\xi_L, \\ A_0, & \tfrac{1}{2}\xi \leq \xi \leq \xi_L, \end{cases} \tag{67a}$$

$$\frac{\partial A}{\partial \xi}(0, \tau) = \frac{\partial A}{\partial \xi}(\xi_L, \tau) = 0, \tag{67b,c}$$

where

$$\xi_L = \frac{Lv}{D}, \tag{68a}$$

$$A_0 = \frac{a_0}{K_d}. \tag{68b}$$

They describe a step change in the attractant concentration from 0 to a_0 initially imposed on a uniform density of bacteria, c_0, throughout the chamber of length L. The spatial derivatives in bacterial density and attractant concentration are zero at each end of the chamber, $x = 0$ and $x = L$.

Equations (57) and (58) were solved numerically with the above boundary conditions to generate bacterial density profiles for a range of values of the chemotactic sensitivity coefficient. A time sequence of these bacterial density profiles is shown by the dotted lines in Fig. 7.8. The area under the peak of the profile was calculated for each time and plotted versus the $\sqrt{t}$, which yielded a linear relation-

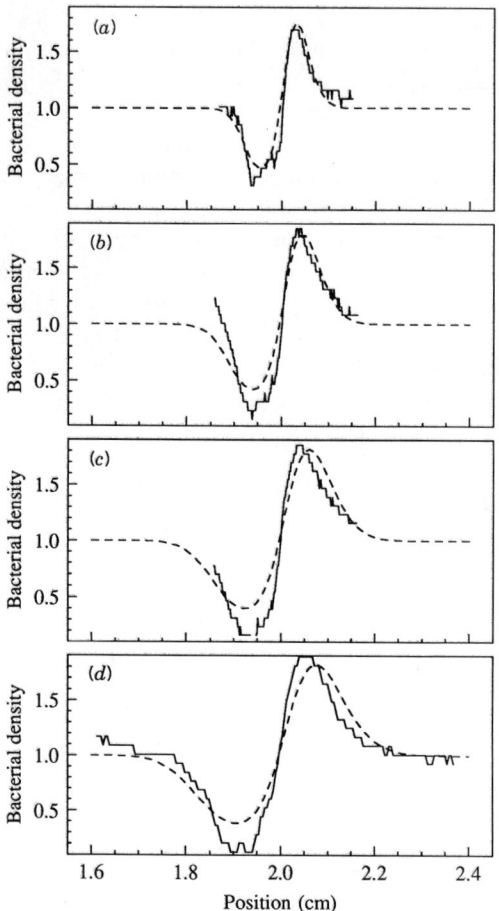

FIGURE 7.8. A time sequence of representative bacterial density profiles from the SFDC comparing experimental (—) and theoretical predictions (---) for (*a*) 1.25 min, (*b*) 2.5 min, (*c*) 4.25 min, and (*d*) 6.25 min. The experimental results are from an experiment with an initial attractant concentration of 0.2 mM and the theoretical predictions are for $\delta = 10$. The discrete step changes in the experimental profiles are due to the image analysis equipment, which uses a finite number of discrete gray levels for interpreting continuous changes in bacterial density.

ship. Experimental profiles were analyzed in the same way and also plotted as peak area versus $\sqrt{t}$. The value of δ for each experiment was determined by matching the slopes from experimental plots to the model results. This value was then used to compute bacterial density profiles for comparison to the experimentally measured profiles. Representative profiles of bacterial density for several experimental times were compared to corresponding model predictions in Fig. 7.8. Excellent agreement between the model predictions and experiments was achieved, providing strong support for the mathematical model.

3.5. Validation of Mathematical Model

To validate the functional form describing the dependence of the chemotactic velocity on the attractant concentration, we examined data from the laser densitometry and SFDC assays over a range of initial attractant concentrations. If Eq. (29) accurately describes this dependence, then a single value of the chemotactic sensitivity coefficient will be able to characterize the data over a wide range of concentrations. We also compare values of χ_0 determined from the population assays to values obtained using the relationship proposed by Rivero et al. in Eq. (31).

Values for the chemotactic sensitivity coefficient were determined in the SFDC over a large range of concentrations. In Fig. 7.9 we compared changes in the attractant concentration-dependent function $\chi(a)$ with changes in the attractant concentration-independent chemotactic sensitivity coefficient reported as χ_0/K_d, for attractant concentrations spanning two orders of magnitude. The variation of χ_0/K_d with concentration is small in comparison to the dependence of $\chi(a)$ on concentration. Clearly, we have successfully demonstrated that the dependence on attractant concentration is essentially captured by the receptor-mediated response mechanism included in the mathematical model from Eq. (30). The model accounts for all but about 5% of the variation in $\chi(a)$.

The chemotactic velocities measured by Dalquist et al. (1972) using the laser densitometry assay were shown in Fig. 7.7. The predicted velocities from our mathematical model are superimposed on the data and successfully characterize it

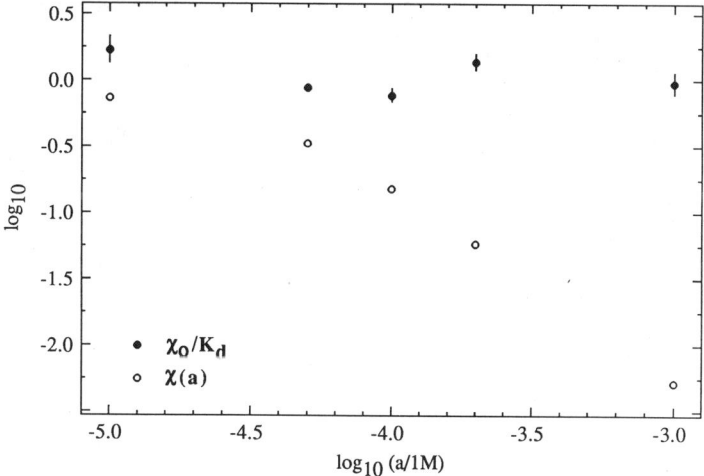

FIGURE 7.9. Comparison of the variation of $\chi(a)$ and χ_0/K_d over two orders of magnitude change in attractant concentration. Approximately 95% of the difference in $\chi(a)$ is accounted for by the expression for receptor-mediated signaling in Eq. (30). The closed circles represent experimental values of χ_0/K_d and the open circles represent the corresponding values of $\chi(a)$. The error bars were based on the percent standard deviation from a statistical analysis of the combined measurements.

over the entire range of concentrations. The dashed line shows a model solution including only the low affinity receptor and clearly illustrates the necessity of two receptor subpopulations to explain the dose–response curve for average velocity. Recall that this model was also required for the bacterial density profile in a linear attractant gradient as discussed in Section 3.4.2.

The relationships between the single-cell properties and population parameters derived by Rivero et al. (1989) enable us to make *a priori* estimates of the zero-gradient random motility coefficient and the chemotactic sensitivity coefficient. Evaluating Eq. (24) for a typical cell speed of 30 μm/s, tumbling probability of 1/s, and directional persistence of 0.3 yields an estimate for μ_0 of 4.3×10^{-6} cm^2/s. (Note that the three-dimensional swimming speed v must be reduced by a factor of $\sqrt{3}$ in these equations to account for projection of the three-dimensional bacterial motion to one dimension as discussed in Section 3.2.) The value measured in the SFDC assay, 1.1×10^{-5} cm^2/s, is reasonably close considering cell speeds for *E. coli* have been reported anywhere from 15 to 40 μm/s and μ is proportional to the square of the speed. The agreement gives credibility to the population parameters derived from single-cell properties in the RTBL model. The chemotactic sensitivity coefficient can also be estimated from single-cell properties by recalling that $\chi_0 = \nu v^2 N_T$ from Eq. (31). Berg and Brown (1972) measured the increase in the mean run times for *E. coli* in the presence of a serine gradient. Using their empirical relation based on Eqs. (21) and (22), we obtain a value for νN_T of 75 s. With a cell speed of 30 μm/s, we have $\chi_0 = 2.3 \times 10^{-4}$ cm^2/s. Our average value of χ_0 for *E. coli* K12 responding to fucose was 8.1×10^{-5} cm^2/s. The higher value for serine is not surprising; since it serves as both a carbon and nitrogen source for the bacterium, it may indeed elicit a stronger response than fucose.

For the Dalquist et al. data we have $\chi_{02} = 2.5 \times 10^{-4}$ cm^2/s, $N_{T,2} = 2500$, and a typical swimming speed of 30 μm/s, from which we estimate a value for the change in tumbling frequency per unit time change in bound receptors, $\nu = 0.033$ (receptor/s)$^{-1}$. This value compares well with a value of 0.030 (receptor/s)$^{-1}$ estimated from Eq. (31) for the Berg and Brown (1972) individual cell study of *E. coli* in the presence of a serine gradient.

Therefore we conclude that the model accurately reflects the population behavior and seems to support the proposed mechanisms of movement and signal transduction for individual cells.

4. CONCLUSIONS

In this chapter, we have reviewed the phenomenon of bacterial chemotaxis and its importance in microbial processes. We have surveyed the mathematical models that have been derived to describe this phenomenon, carefully distinguishing between population and subpopulation models and between balance equations and constitutive relations. We placed particular emphasis on the one-dimensional model of Rivero et al. (1989). The experimental assays used to measure population behavior were described and the role of the Rivero et al. model in analyzing the experimental results in terms of fundamental cell parameters (random motility and chemotactic sensitivity coefficients) was reviewed in detail.

The mathematical models we have described in this chapter play an important additional role in predicting the behavior of microbial population dynamics from which conclusions can be drawn about the role of such factors as motility, growth rate, and chemotaxis in the survival of competing bacterial species. For example, Lauffenburger and co-workers (Kelly et al., 1988; Lauffenburger, 1988; Lauffenburger and Calcagno, 1983; Lauffenburger et al., 1987) demonstrated that for certain ranges of the chemotactic sensitivity coefficient chemotaxis could provide a competitive advantage to a slower growing species allowing it to coexist along with and even outgrow a species having more favorable growth kinetics. Their approach illustrates the usefulness of mathematical modeling as a tool for predicting bacterial population behavior in natural systems. Note, however, that in order to apply these mathematical models to natural systems requires quantitative measurements of the random motility and chemotactic sensitivity coefficients and for the constitutive relations within the model to be confirmed experimentally. These last two aspects of mathematical modeling have been the focus of this chapter.

ACKNOWLEDGMENTS

The author is indebted to Professors Peter Cummings and Douglas Lauffenburger for helpful comments on an earlier version of this chapter. Partial financial support of this research by a grant (BCS-9009725) to the author from the National Science Foundation is gratefully acknowledged. Much of the research described in this chapter was performed by the author under the direction of Professor Lauffenburger at the University of Pennsylvania as part of her dissertation (Ford, 1989) and was supported by National Science Foundation Biotechnology Program Grant No. EET 86-12987.

REFERENCES

Adler, J. (1966), *Science*, **153**, 708–716.

Adler, J. (1969), *Science*, **166**, 1588–1597.

Adler, J. (1973), *J. Gen. Microbiol.*, **74**, 77–91.

Alt, W. (1980), *J. Math. Biol.*, **9**, 147–177.

Berg, H. C., and Brown, D. A. (1972), *Nature*, **239**, 500–504.

Bird, R. B., Stewart, W. E., and Lightfoot, E. N. (1960), *Transport Phenomena*, Wiley, New York.

Block, S. M., Segall, J. E., and Berg, H. C. (1983), *J. Bacteriol.*, **154**, 312–323.

Borkovich, K. A., and Simon, M. I. (1990), *Cell*, **63**, 1339–1348.

Caldwell, C. S., Hall, J. R., and Babb, A. L. (1957), *Rev. Sci. Instrum.*, **28**, 816.

Chapman, S., and Cowling, T. G. (1960), *The Mathematical Theory of Non-Uniform Gases*, Cambridge University Press, Cambridge.

Chet, I., and Mitchell, R. (1976), *Annu. Rev. Microbiol.*, **30**, 221-239.

Chet, I., Asketh, P., and Mitchell, R. (1975), *Appl. Microbiol.*, **30**, 1043-1045.

Clarke, S., and Koshland, D. E. Jr. (1979), *J. Biol. Chem.*, **254**, 9695-9702.

Crank, J. (1975), *The Mathematics of Diffusion*, 2nd ed., Clarendon Press, Oxford, pp. 11-27.

Dalquist, F. W., Lovely, P., and Koshland, D. E. Jr. (1972), *Nature (Lond.) New Biol.*, **236**, 120-123.

Dalquist, F. A., Elwell, R. A., and Lovely, P. S. (1976), *J. Supramolecular Structure*, **4**, 329-342.

Drake, D., and Montie, T. C. (1987), *Can. J. Microbiol.*, **33**, 755-763.

Drake, D., and Montie, T. C. (1988), *J. Gen Microbiol.*, **134**, 43-52.

Ford, R. M. (1989), Ph.D. Thesis, University of Pennsylvania, unpublished.

Ford, R. M., and Cummings, P. T. (1992), *SIAM J. Appl. Math.*, in press.

Ford, R. M., and Lauffenburger, D. A. (1991a), *Biotechnol. Bioeng.*, **37**, 661-672.

Ford, R. M., and Lauffenburger, D. A. (1991b), *Bull. Math. Biol.*, **53**:721-749.

Ford, R. M. and Lauffenburger, D. A. (1992), *Math. Biosci.*, **109**:127-149.

Ford, R. M., Phillips, B. R., Quinn, J. A., and Lauffenburger, D. A. (1991), *Biotechnol. Bioeng.*, **37**, 647-660.

Freter, R., O'Brien, P. C. M., and Macsai, M. S. (1979), *Am. J. Clin. Nutrition*, **32**, 128-132.

Futrelle, R. P., and Berg, H. C. (1972), *Nature*, **239**, 517-518.

Gristina, A. G. (1987), *Science*, **237**, 1588-1595.

Gulash, M., Ames, P., Larosiliere, R. C., and Bergman, K. (1984), *Appl. Env. Microbiol.*, **48**, 149-152.

Harwood, C. S., Parales, R. E., and Dispensa, M. (1990), *Appl. Env. Microbiol.*, **56**, 1501-1503.

Keller, E. F., and Segel, L. A. (1971a), *J. Theor. Biol.*, **30**, 225-234.

Keller, E. F., and Segel, L. A. (1971b), *J. Theor. Biol.*, **30**, 235-248.

Kelly, F. X., Dapsis, K., and Lauffenburger, D. (1988), *Microbial Ecol.*, **16**, 115-131.

Kennedy, M. J., and Lawless, J. G. (1985), *Appl. Environ. Microbiol.*, **49**, 109-114.

King, D. L., Wear, J. E., and Hazen, T. C. (1990), ASM National Meeting, Anaheim, CA, May 13-17, 1990.

Koshland, D. E. Jr. (1979), in *The Bacteria* (J. R. Sokatch and L. N. Ornston, Eds.), Vol. 7, Academic Press, New York, pp. 111-166.

Koshland, D. E. Jr. (1980), *Bacterial Chemotaxis as a Model Behavioral System*, Raven Press, New York.

Koshland, D. E. Jr. (1988), *Biochemistry*, **27**, 5829-5834.

Lapidus, I. R., and Schiller, R. (1976), *Biophys. J.*, **16**, 779-789.

Lauffenburger, D. A. (1988), in *CRC Handbook of Laboratory Model Systems for Microbial Ecosystems* (J. W. T. Wimpenny, Ed.), Vol. 2, CRC Press, Boca Raton, FL, pp. 141-176.

Lauffenburger, D. A., and Calcagno, P. B. (1983), *Biotechnol. Bioeng.*, **25**, 2103-2125.

Lauffenburger, D. A., Rivero, M., Kelly, F., Ford, R., and DiRienzo, J. (1987), *Ann. N.Y. Acad. Sci.*, **506**, 281-295.

López-de-Victoria, G. (1989), M.S. Dissertation, University of Puerto Rico, unpublished.

Macnab, R. M. (1980), in *Biological Regulation and Development* (R. F. Goldberger, Ed.), Vol. 2, Plenum, New York, pp. 377–412.

Macnab, R. M. (1987), in Escherichia coli and Salmonella typhimurium (F. C. Niedhardt, Ed.), Vol. 1, American Society for Microbiology, Washington, DC, pp. 732–759.

Macnab, R. M., and Koshland, D. E. Jr. (1972), *Proc. Natl. Acad. Sci. USA*, **69**, 2509–2512.

Mesibov, R., Ordal, G. W., and Adler, J. (1973), *J. Gen. Physiol.*, **62**, 203–223.

Nossal, R. (1979), in *Biological Growth and Spread: Mathematical Theories and Applications, Lecture Notes in Biomathematics* (S. Levin, Ed.), Vol. 38, Springer-Verlag, New York, pp. 410–439.

Nossal, R., and Chen, S. H. (1973), *Nature (Lond.) New Biol.*, **244**, 253–254.

Othmer, H., Dunbar, S., and Alt, W. (1988), *J. Math. Biol.*, **26**, 263–298.

Parkinson, J. S. (1988), *Cell*, **53**, 1–2.

Patlack, C. S. (1953), *Bull. Math. Biophys.*, **15**, 311–338.

Rivero, M. A. (1986), Ph.D. Thesis, University of Pennsylvania, unpublished.

Rivero-Hudec, M., and Lauffenburger, D. A. (1986), *Biotechnol. Bioeng.*, **28**, 1178–1190.

Rivero, M. A., Tranquillo, R. T., Buettner, H. M., and Lauffenburger, D. A. (1989), *Chem. Eng. Sci.*, **44**, 2881–2897.

Rubinow, S. I. (1975), *Introduction to Mathematical Biology*, Wiley, New York.

Schlichting, H. (1979), *Boundary-Layer Theory*, McGraw-Hill, New York.

Segall, J. E., Ishihara, A., and Berg, H. C. (1985), *J. Bacteriol.*, **161**, 51–59.

Segall, J. E., Block, S. M., and Berg, H. C. (1986), *Proc. Natl. Acad. Sci. USA*, **83**, 8987–8991.

Segel, L. A. (1977), *SIAM J. Appl. Math.*, **32**, 653–665.

Segel, L. A. (1979), in *Physical Chemical Aspects of Cell Surface Events in Cellular Regulation* (DeLisi and Blumenthal, Eds.), Elsevier, New York, pp. 293–302.

Segel, L. A., and Jackson, J. L. (1973), *J. Mechanochem. Cell Motil.*, **2**, 25–34.

Segel, L. A., Chet, I., and Henis, Y. (1977), *J. Gen. Microbiol.*, **98**, 329–337.

Spudich, J. L., and Koshland, D. E. Jr. (1975), *Proc. Natl. Acad. Sci. USA*, **72**, 710–713.

Staffeld, P. O., and Quinn, J. A. (1989), *J. Colloid. Interface Sci.*, **130**, 69–87.

Stewart, R. C., and Dalquist, F. W. (1987), *Chem. Rev.*, **87**, 997–1025.

Stock, J., and Stock, A. M. (1987), *Trends Biochem. Sci.*, **12**, 371–375.

Stock, J., Ninfa, A. J., and Stock, A. M. (1989), *Microbiol. Rev.*, **53**, 450–490.

Stroock, D. W. (1974), *Z. Wahrscheinlichkeitstheorie verw. Geb.*, **28**, 305–315.

Terracciano, J. S. (1985), Ph.D. Thesis, University of Massachusetts, unpublished.

Vicker, M. G. (1981), *Exp. Cell Res.*, **136**, 91–100.

Wolfe, A. J., and Berg, H. C. (1989), *Proc. Natl. Acad. Sci. USA*, **86**, 6973–6977.

Zigmond, S. H., Klausner, R. D., Tranquillo, R., and Lauffenburger, D. A. (1985), in *Membrane Receptors and Cellular Regulation* (M. P. Czech and C. R. Kahn, Eds.), Alan R. Liss, New York, pp. 347–356.

8 Network Analysis of Nitrogen Cycling in an Estuary

R. R. CHRISTIAN, J. N. BOYER, D. W. STANLEY, and W. M. RIZZO
Department of Biology and Institute for Coastal and Marine Resources, East Carolina University

1. Introduction 217
2. Background 218
   2.1. N cycling models of estuaries 218
   2.2. Network analysis 218
3. Neuse River Estuary 221
4. General model structure 223
5. Methods 228
6. Estimation of values 229
   6.1. Overview 229
   6.2. State variables 229
   6.3. Inputs 232
   6.4. Outputs 232
   6.5. Internal interactions 232
   6.6. Analyses of networks 235
7. Results and discussion 236
   7.1 Overview 236
   7.2. Mass balance of throughputs 236
   7.3. Structure analysis 237
   7.4. Biochemical cycle analysis 240
   7.5. Information indices and global attributes 242
   7.6. Comparison with other systems 243
   References 245

1. INTRODUCTION

The evaluation of metabolic and physiologic activities of microorganisms has historically been dominated by *in vitro* studies. Mathematical models have been developed to depict and explain these activities and predict kinetics within batch or continuous pure culture (Monod, 1949; Tempest, 1970). Models of microbial population interactions generally have come from simple and controlled experimental designs corresponding to laboratory microcosms (Fredrickson, 1977). Yet a major emphasis of microbial ecology is to understand the roles microbes play *in situ*. In the past 25 years increased effort has been placed on determining *in situ* microbial process rates. This effort has been associated with a dramatic improvement in techniques for such studies (Christian and Wetzel, 1991). For the most part, equations for mathematical models of these processes have been borrowed from the literature of *in vitro* activities and their models (Christian and Wetzel, 1991; Christian et al., 1986a). The application of modeling approaches developed from other disciplines has been less apparent. In this chapter we explore the use of network theory to evaluate the nitrogen cycle in the Neuse River Estuary, North Carolina. Through our analyses we attempt to describe this microbially dominated cycle in the context of ecosystem-level characteristics rather than as the summation of individual kinetic descriptions.

We contend that the description of ecosystem-level characteristics may well be served by an approach other than through the compilation of individual process-related equations into a mechanistic model. We present our logic as to the problems of integration in interpreting microbial process rate information to the ecosystem level. This section is used to focus the microbiologist on hierarchical interrelationships associated with nutrient cycling in ecosystems. The term integrate is used in its broadest sense as the act of forming, coordinating, or blending "into a functioning or unified whole" (Webster, 1985). We provide background to N cycling, to its modeling in estuaries, and to network theory and its applications in ecology. The results of a 4-year study of the N cycle within the Neuse River Estuary are the subject of our specific application of the approach. A general model is described, and each component variable and interaction is derived from the available data set. Analyses of two seasons, summer and winter, are compared to evaluate the value of network analyses.

2. BACKGROUND

2.1. N Cycling Models of Estuaries

Estuaries are often systems of high productivity and substantial nutrient recycling. All processes of the N cycle may be represented in an estuary, and microbial activities generally dominate biological cycling (Day et al., 1989). These ecosystems may be open to considerable exchange with neighboring fresh and marine waters, the atmosphere, as well as surrounding terrestrial ecosystems. The physical processes of exchange and mixing may rival the biological processes in importance. There is no "standard estuary," and the relative importances of biological, chemical, and physical processes vary considerably among estuaries and among times within an estuary. A major goal of estuarine ecosystem studies is to determine the conditions by which the various processes control nutrient cycling and energy flow. Ecosystem-level models of estuaries have been useful tools in attempting to meet this goal.

Models of estuarine ecosystems have taken a variety of forms. They include static or mass balance models (Baird and Ulanowicz, 1989; Davis et al., 1978), water quality models for management purposes (Thomann and Fitzpatrick, 1982; Wulff et al., 1990), and mechanistic simulation models (Baretta and Ruardij, 1987; Kremer and Nixon, 1978). It is not our intent to address all the various modeling approaches. We highlight two issues: N cycling models and the application of network theory to ecosystem models.

Wetzel and Wiegert (1983) reviewed models of N cycling in estuaries and other marine systems. They found very few estuarine simulation models that focused on N as the currency of flow. Most only included N as a variable that might limit primary productivity. The full cycling of the element was not incorporated. Since that time the progress in estuarine N cycle models has remained modest.

Kremer (1989) extended the simulations and analyses of the Narragansett Bay model previously described in 1978 (Kremer and Nixon, 1978). The model was constructed to reflect the trophic structure of the ecosystem and used C and N as currencies. Dissolved inorganic N (DIN) was represented as water column pools of NO_3^- and NH_4^+. Seasonal dynamics were simulated, and network analyses were conducted on results from each season. The results of these analyses are discussed later.

Billen and Lancelot (1988) described several idealized simulation models of N cycling in coastal ecosystems that may include estuaries. In two models of total system processing, they aggregated NH_4^+ and NO_3^- into a common DIN compartment. Dissolved organic N (DON) was represented in one of these models, but N_2 was not described as a state variable. They further developed a model of benthic cycling in which they separated NH_4^+ and NO_3^- as distinct state variables but treated N_2 production as an export.

Wulff et al. (1990) discussed two models involving N cycling in the Baltic Sea. One was highly aggregated and emphasized inputs, outputs, and physical factors, rather than internal cycling. A more spatially articulated model was used to evaluate N, P, and silicate budgets in the Baltic Sea. Even with this second model, internal cycling of N was not well represented. Thus neither model portrayed N cycling in detail.

2.2. Network Analysis

Simulation models have dominated the literature in systems ecology regarding nutrient cycling and energy flow (Dame, 1979; Mann et al., 1989). Significant gains in our understanding of estuaries have resulted from both the results of these simulations and the activities involved in developing the models. However, there remains concern as to the limitations of such models and their utility (Mann et al., 1989; Ulanowicz, 1988). Recently, network analysis has provided a different approach (Wulff et al., 1989). Network analysis, as it has been applied to ecological systems, encompasses and integrates three theoretical approaches: the analysis of flow structure through matrix manipulations, graph theory, and information theory. Pioneering work on the application of these approaches occurred in the early to mid-1970s (Finn, 1976; Hannon, 1973). Later, Ulanowicz (1980, 1986) developed his hypothesis concerning a system's attributes relating to ascendancy. Applications of network analysis and their comparison were the subject of the activities of the Scientific Committee on Oceanic Research (SCOR) Working Group 73 as described in 1989 in the volume edited by Wulff and co-workers. These analyses have been applied to static models and to discrete times for simulation models. More recently, Pahl-Wostl (1990) has proposed analyses to account for temporal organization. Rarely have complex or well-articulated models been evaluated. Perhaps the most ambitious use of network analysis published is that by Baird and Ulanowicz (1989) for a 36-compartment model of energy flow through Chesapeake Bay. In all, the use of network analyses is still in its developmental stages.

The following is a brief discussion of network analysis more fully developed by Ulanowicz (1986) and Kay et al. (1989). The order of presentation conforms to that found in NETWRK, a software package developed by Ulanowicz (1987). The set of approaches applied in this chapter involves analyses of flow structure, biogeochemical cycle distribution, and information indices. All begin with the construction of a network or multicompartmental model of a system depicting interactions as unidirectional, positive flows (fluxes) of material between compartments (state variables), and between compartments and the environment of the system (inputs and outputs). This network can be represented with boxes and arrows or, for calculations, in matrix form. A common currency of matter or energy is used for interactions. Compartments are represented as standing stocks and flows as process rates.

Analysis of flow structure involves manipulations of the matrix of within-system flows and of vectors of inputs and outputs. Where appropriate, outputs can be divided into a vector for exports and one for dissipative losses associated with respiration. From these matrices, a series of evaluations of flow structure are made. The total flows of material into and out of each compartment are calculated for evaluating mass balance and hence the predicted net increase or decrease of standing stock. Furthermore, both the direct and indirect relationships between compartments are determined. Direct relationships come from the "exchange matrix" of flows originally denoted by the investigator. Through matrix manipulation, another matrix, that of "total contribution coefficients," is calculated. In this, each coefficient (as a fraction of 1 or as a percentage) represents the fraction of total flow leaving a compartment and entering another independent of pathway taken. The coefficients along the major diagonal represent the portion of material leaving a compartment and then recycling into it (i.e., the degree of self-stimulation). Similarly, a "total dependency" matrix is constructed in which coefficients represent the fraction of input into a compartment coming directly or indirectly from each other compartment (i.e., its "extended diet"). Furthermore, calculations isolate the individual inputs to the system and determine their direct and indirect effects on each compartment.

The analysis of biogeochemical cycles is an application of graph theory. The network is divided into cycles in which a cycle represents the connection of individual flows (called arcs), which form a pathway that begins and ends with the same compartment. As flows are all positive values, each cycle is a positive feedback or autocatalytic loop. The cycles are grouped according to shared "weak arcs." A weak arc is the smallest flow within a cycle. Cycles grouped by a shared weak arc are called a nexus. As the weak arc may represent the controlling portion of a pathway, a nexus represents cycles potentially sharing a common process of control. The distribution of cycles and nexuses provides insight into the structure of the biogeochemical cycling of the system. Considerations include (1) the lengths of cycles and importance of length to throughput, (2) the distribution of cycles among compartments for recognition of subsystems, and (3) relative importance of cycling to inputs in total throughput (Finn, 1976, 1980). An algorithm is involved that systematically searches cycles and nexuses, and the latter consideration is evaluated by the Finn Cycling Index (Finn, 1980; Kay et al., 1989). This index describes the fraction of total throughput associated with recycling and hence a measure of efficiency of the system.

The last analyses considered here involve evaluating the overall system's ability for "growth and development" (Ulanowicz, 1980, 1986) by considering both throughput and organization. Organization is evaluated through the use of the Shannon-Wiener index of information content associated with flow structure. This index is scaled by the total throughput of the system. For a given throughput, there is a maximum "development capacity" based on maximizing the predictability of flows or interactions. This capacity is not achieved within ecological systems, but Ulanowicz (1986) postulated that systems develop or evolve to maximize their potential toward this capacity. The success of this development is called "ascendancy." Thus systems tend to maximize ascendancy within the constraints of capacity. Ascendancy can then be used to assess the degree of maturity of the system and the degree to which a system is stressed or disturbed. A system's ascendancy does not reach its capacity because of the "overhead" a system must pay to remain extant or to counter entropy. Overhead may take four forms associated with the uncertainties of imports, exports or tributes, dissipations or respiration, and redundancy or parallel pathways of flow within the system. Each of these may be considered as to its importance to the system's ability to develop. Furthermore, each may be considered in absolute terms (i.e., throughput × bits) or normalized as a fraction of capacity. The ecological significance of these concepts is thoroughly discussed by Ulanowicz (1986).

3. NEUSE RIVER ESTUARY

The Neuse River and its estuary in North Carolina have been the sites of a number of studies in the past two decades (Boyer et al., 1988; Christian et al., 1984, 1989, 1991; Hobbie and Smith, 1975; Paerl, 1987; Stanley, 1983, 1988). The studies have generally focused on two major issues: nutrient cycling and cyanobacterial blooms in the lower Neuse River. Concentrations of inorganic nutrients are high in the tributary rivers as a result of both agricultural and point sources (Christian et al., 1989; Paerl, 1987; Stanley, 1988). Concentrations decline within the estuary and the ratio of dissolved inorganic N to phosphate decreases to below 5 in the lower estuary, indicating the potential for N limitation of primary productivity (Christian et al., 1989, 1991). Recycling is a major source of N to phytoplankton within the estuary, and recycled N availability exceeds that of new or loaded N from the tributaries (Boyer et al., 1988; Stanley et al., unpublished data).

Cyanobacterial blooms dominated by *Microcystis aeruginosa* have been a recurring environmental problem in the Neuse River (Christian et al., 1986, 1988; Paerl, 1987). The blooms occur in summer under low flow conditions in the Neuse River (Christian et al., 1986). They are associated with large changes in the distribution of nutrients into and out of biomass (Christian et al., 1988; Heath, 1989) and disruption of the normal riverine food web (Fulton and Paerl, 1987a,b). These blooms reach considerable proportions in the river (Christian et al., 1988; Paerl, 1987) but as of yet have failed to make major intrusions into the estuary (Paerl, 1987).

The Neuse River basin is 16,000 km^2, and its estuary is 392 km^2. In Fig. 8.1 the symbols A through G represent the locations for sampling sites of the study presented here. Station A is near the estuary's mouth where water flows into Pamlico Sound. The Pamlico Sound is a major body of water separating river flow from the Atlantic Ocean. Because of the few narrow inlets among the barrier islands, the residence time of fresh water in the sound approaches 1 year and salinities are one-half to two-thirds seawater salinities (Giese et al., 1979). The lower estuary was further sampled at station B. Station C is at the prominent Wilkerson's Point of the estuary separating its middle region from the lower portion. Stations

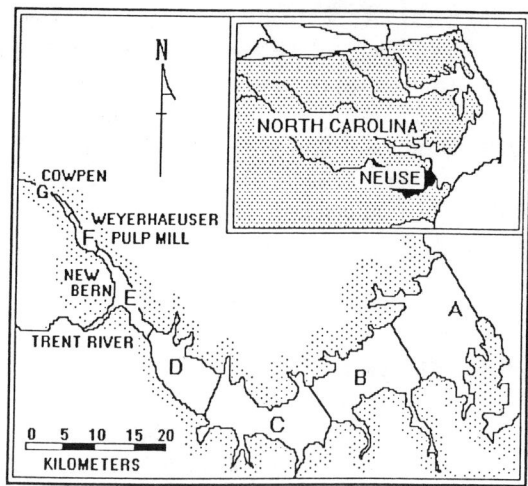

FIGURE 8.1. Map of the Neuse River Estuary, watershed, and sampling site locations.

TABLE 8.1. Physical Dimensions of Segments Within the Neuse River Estuary

Segment	Area (km²)	Average Depth (m)	Volume (m³)	Area (%)	Volume (%)
A	121.0	4.32	5.23×10^8	30.8	39.1
B	100.0	3.57	3.57×10^8	25.5	26.7
C	88.9	2.85	2.53×10^8	22.7	18.9
D	61.4	2.53	1.55×10^8	15.7	11.6
E	17.3	2.26	3.91×10^7	4.4	2.9
F	3.1	2.64	8.18×10^6	0.8	0.6
G	0.6	2.00	1.20×10^6	0.2	0.1
Total	392.3	3.41	13.36×10^8	100.0	100.0

D and E are in this middle region, and stations F and G are in the upper region. Measurable salinity rarely reaches station G (Christian et al., 1991). The major tributary is the Neuse River, but a second large tributary entering near station E is the Trent River.

The estuary was divided into representative segments as shown in Fig. 8.1. Each line separating segments was positioned equidistant between stations. The physical dimensions of each segment are shown in Table 8.1. Over two-thirds of the area and volume of the estuary are below Wilkerson's Point, and little of the study area is represented by stations F and G. Average depth per segment was calculated from soundings on nautical charts. No correction to depth has been made for fluctuations found with tides or seasons. Astronomical tides are insignificant, and wind tides and river discharges dominate short-term fluctuations of water level. The depth of the estuary generally increases from head to mouth. Both greater depth and area contribute to the relative importance of the lower estuary.

4. GENERAL MODEL STRUCTURE

The general model of N cycling within the Neuse River Estuary is structured to balance our general understanding of the N cycle with the information obtained from 9 years of study of the system including a 4-year intensive study of N cycling (Fig. 8.2). As such, we include some processes and compartments not directly measured, some are tailored to our measurements, and some are ignored or embodied in aggregated compartments or interactions. Here we provide an overview with more detailed explanations of values used. We discuss their derivation later. There are eight state variables (compartments) considered within the estuary (Table 8.2). DON, NO_x, and NH_4 (definitions given in Table 8.2) were directly measured during the study. PN-phyto and PN-abiotic were considered fractions of the measured total particulate N(PN) along with a portion of PN-hetero. PN-phyto values were derived from chlorophyll *a* concentrations. PN-hetero theoretically

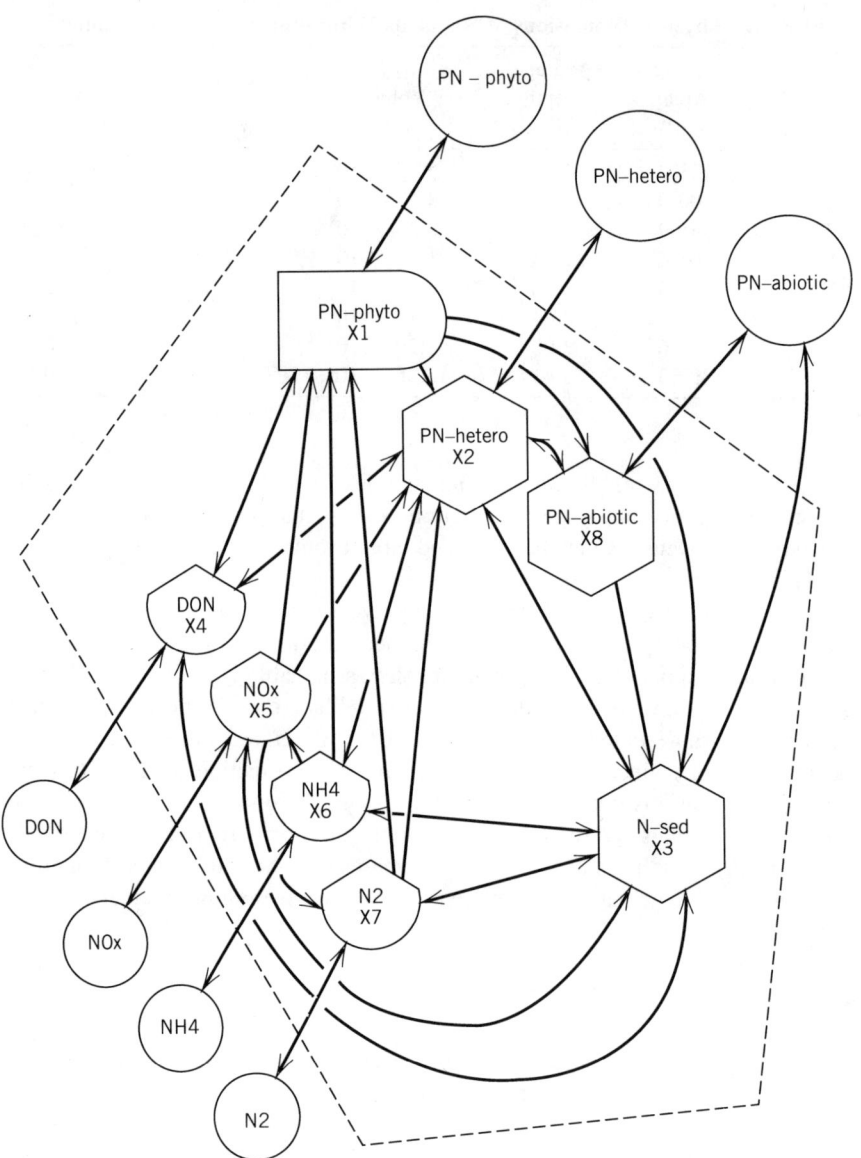

FIGURE 8.2. Diagram of N cycle in the Neuse River Estuary. Definitions of state variables are given in Table 8.2 and those of interactions in Table 8.3. Area inside the dashed lines represents the estuarine system under consideration. Symbol format from Odum (1983).

TABLE 8.2. State Variables[a] of the N Cycling Model for the Neuse River Estuary

Compartment Number	State Variable	Abbreviation	Comment
1	Phytoplankton N	PN-phyto	Estimated from chlorophyll *a* concentration
2	N in heterotrophs	PN-hetero	Highly aggregated
3	Sediment N	N-sed	Total particulate N in sediment, highly aggregated
4	Dissolved organic N	DON	Includes biologically available and refractory compounds
5	Nitrate plus nitrite	NO_x	Dominated by nitrate
6	Ammonia	NH_4	Independent of ionic form
7	Diazo N	N_2	Inferred from solubility characteristics
8	Nonliving seston N	PN-abiotic	Calculated by difference

[a] All are in units of mmol N $\times$ m^{-2} and values are area weighted.

includes N in bacteria, microplankton, zooplankton, and nekton. Only part of these were properly sampled for PN analyses. None was directly analyzed during the 4-year study and therefore literature estimates were used. N-sed is a highly aggregated, but measured, compartment of both living and nonliving parts within the surface 1 cm of sediment. N_2 was not directly measured and was calculated from solubility characteristics.

The interactions between state variables are flows of N with the units mmol N $\times$ m^{-2} $\times$ season^{-1}. They include the 44 potential flows that may or may not be present under all conditions or may or may not be quantitatively significant. These are defined in Table 8.3. Seven are input flows. Most material entering the estuary is from riverine loading. Precipitation adds DON, NO_x, and NH_4. A minor input may be contributed with the intrusion of sound waters but this is ignored. Atmospheric inputs of N_2 are possible in response to N_2 fixation. Seven output flows largely involve discharge to the sound. Furthermore, N_2 from denitrification may be released to the atmosphere, and emigration of heterotrophs to the sound or river is possible. The remaining output (F_{30}) represents sediment burial.

There are 29 internal flows depicted in the model. Generally, they represent well-known processes of the N cycle, such as inorganic N uptake, ammonification, nitrification, N_2 fixation, and denitrification. Explicit trophic transfers are restricted to planktonic herbivory and detritivory and to benthic/water column transfers, which include components of suspension feeding and bottom feeding. Uptakes of NO_x, NH_4, and urea were measured under a variety of environmental conditions. Separations of heterotrophic from autotrophic uptake rates were based on the differences between total integrated uptake and rates derived from information on primary productivity. Nitrification, N_2 fixation, excretion, and denitri-

TABLE 8.3. Interactions[a] **of the N Cycling Model for the Neuse River Estuary**

Type	Designation	Definition
Inputs	F_{01}	Loading from surface water of PN-phyto
	F_{02}	Loading from surface water and immigration of PN-hetero
	F_{04}	Loading from surface water and precipitation of DON
	F_{05}	Loading from surface water and precipitation of NO_x
	F_{06}	Loading from surface water and precipitation of NH_4
	F_{07}	Loading from surface water and atmosphere of N_2
	F_{08}	Loading from surface water of PN-abiotic
Outputs	F_{10}	Discharge to sound of PN-phyto
	F_{20}	Discharge to sound and emigration of PN-hetero
	F_{30}	Burial of N-sed
	F_{40}	Discharge to sound of DON
	F_{50}	Discharge to sound of NO_x
	F_{60}	Discharge to sound of NH_4
	F_{70}	Discharge to sound and release to atmosphere of N_2
	F_{80}	Discharge to sound of PN-abiotic
Internal interactions	F_{41}	Light uptake of urea with DON
	F_{51}	Light uptake of NO_x
	F_{61}	Light uptake of NH_4
	F_{71}	N_2 fixation by cyanobacteria
	F_{12}	Feeding on phytoplankton
	F_{32}	Feeding on benthos
	F_{42}	Dark uptake of urea and other available DON
	F_{52}	Dark uptake of NO_x
	F_{62}	Dark uptake of NH_4
	F_{72}	N_2 fixation by heterotrophic bacteria
	F_{82}	Feeding on detrital substrate
	F_{13}	Sedimentation of phytoplankton and suspension feeding
	F_{23}	Sedimentation of heterotrophs and suspension feeding
	F_{43}	Benthic uptake of DON including urea
	F_{53}	Benthic uptake of NO_x
	F_{63}	Benthic uptake of NH_4
	F_{73}	Benthic N_2 fixation
	F_{83}	Sedimentation of detritus and suspension feeding

TABLE 8.3. (*Continued*)

Type	Designation	Definition
	F_{14}	Photosynthetic release of DON
	F_{24}	Excretion of DON including urea
	F_{34}	Release of DON from sediments
	F_{35}	Release of NO_x from sediments
	F_{65}	Nitrification
	F_{26}	Ammonification and excretion of NH_4
	F_{36}	Release of NH_4 from sediments
	F_{37}	Denitrification in sediments
	F_{57}	Denitrification in the water column
	F_{18}	Nonpredatory mortality of phytoplankton
	F_{28}	Nonpredatory mortality of heterotrophs and egestion

[a] All interactions represent flows of N (mmol N $\times$ m^{-2} $\times$ season^{-1}).

fication were not measured directly in our study. These rates were inferred from literature values and indirect estimates.

Matrix representation of the interactions of the general model is shown in Table 8.4. The total connectivity of the general model was 51.8% (29 of a possible 56 interactions). The numbers of internal flows into abiotic N state variables were small; NO_x, NH_4, N_2, and PN-abiotic each received two flows, and DON received three flows. PN-phyto received four flows supplying N requirements. PN-hetero, by including bacteria that can use inorganic N as well as organic N-requiring populations, received seven flows. N-sed contained both autotrophic and heterotrophic components as well as an abiotic component. It also received the maximum seven flows. Thus aggregated, biotic compartments were recipients to a greater diversity of flows than abiotic compartments.

TABLE 8.4. Matrix of Interactions of N Cycling Model of the Neuse River Estuary

Variable[a]	1	2	3	4	5	6	7	8
1	—	F_{12}	F_{13}	F_{14}				F_{18}
2		—	F_{23}	F_{24}		F_{26}		F_{28}
3		F_{32}	—	F_{34}	F_{35}	F_{36}	F_{37}	
4	F_{41}	F_{42}	F_{43}	—				
5	F_{51}	F_{52}	F_{53}		—		F_{57}	
6	F_{61}	F_{62}	F_{63}		F_{65}	—		
7	F_{71}	F_{72}	F_{73}				—	
8		F_{82}	F_{83}					—

[a] The variable numbers shown in this table correspond to the compartment numbers listed in Table 8.2. Interaction designations correspond to those presented in Table 8.3. The columns in this table represent the recipience of flows. The rows in this table represent the donation of flows.

This pattern of relative diversity of recipient flows was less evident for donor flows. The compartment that donated the greatest number of flows (five) is still N-sed. However, NO_x, NH_4, PN-phyto, and PN-hetero each donated four flows. DON and N_2 donated three flows and PN-abiotic is a source for only two flows. Thus PN-abiotic was least connected to other compartments. Inorganic N pools tended to be more limited in their sources than in the diversity of their sinks. PN-hetero and N-sed were potentially important processors for a diversity of flows. PN-phyto played a somewhat more restricted role.

5. METHODS

Details of methodologies regarding data acquisition have largely been described elsewhere (Boyer et al., 1988; Christian et al., 1991; Rizzo et al., unpublished data; Stanley et al., unpublished data). Here we give an overview of sampling, variables analyzed, and general model analysis procedures. Descriptions of integration methods are given in the following section.

Data were collected for up to 4 years (February 1985 to February 1989) for hydrographic information, light availability, nutrient and chlorophyll a concentrations, and process rates. The most complete data set comes from the final year. During that time rates of sediment/water exchange were estimated in support of the more comprehensive study of the water column. For this chapter we use data from the summer of 1988 and winter of 1988–1989 for model construction. This permits a comparison of conditions of high and low biological activity.

Loading to the estuary from riverine sources was calculated from data resulting from weekly samples at station G for the Neuse River, the New Bern sewage treatment plant (STP), and the Trent River at Pollocksville. Measured, relevant concentrations included NO_x, NH_4, PN, dissolved Kjeldahl N (DKN), urea, and chlorophyll a. For loading estimates, concentrations were multiplied by appropriate river flow or STP discharge. River flow was calculated from daily discharge values at the nearest gauging stations (Kinston for the Neuse and Trenton for the Trent) corrected for basin surface area below the stations. Precipitation was estimated from NOAA climatological data for New Bern, NC, and concentrations of nutrients in rainwater were multiplied by precipitation volume to estimate loading from the atmosphere directly to estuarine surface waters.

To estimate estuarine concentrations and process rates in the water column, samples were collected at the seven stations, shown in Fig. 8.1, on a 2–3-week schedule. Concentrations of the variables listed above were determined at each station. Extinction coefficients of photosynthetically active radiation (PAR), salinity, temperature, and dissolved oxygen concentrations were also measured. Water samples were returned to the laboratory to estimate process rates under various light conditions at near ambient temperatures. Process rates estimated were primary productivity using $^{14}CO_2$; and NH_4, NO_x, and urea uptake and ammonification using ^{15}N methodology. On a 6-week schedule, sediment cores were taken at three stations (A, D, and G) for sediment/water exchange rate estimates. Cores were incubated under various light conditions at ambient temperatures, and changes in concentrations of NH_4, NO_x, and dissolved oxygen in the overlying water were measured. Rates were calculated from these changes.

6. ESTIMATION OF VALUES

6.1. Overview

State variable (mmol N $\times$ m^{-2}) and flux (mmol N $\times$ m^{-2} $\times$ season^{-1}) values were estimated for both the summer 1988 and winter 1988–1989 (Table 8.5). Our approaches of estimation varied among values from interpolations of data taken during the season within the Neuse River Estuary, to extrapolations from neighboring estuaries, to extrapolations from the literature on N cycling in estuaries, and finally to mass balances of individual compartments. Some values were derived from more than one of these approaches. The reliabilities of estimations were thus varied. In Table 8.6, we summarize our confidence in values relative to their mode of estimation. In general, values considered highly reliable were those from Neuse data requiring little extrapolation. Those considered moderately reliable were from the Neuse or neighboring estuaries and requiring some extrapolation. Those considered as having low reliability were derived from the general literature or mass balance calculations. In all, 52 pairs of values were estimated, and the distribution of reliability was nearly even among the three categories. State variable and input values were generally well characterized. Planktonic uptake and sediment/water exchange of DIN were reliably characterized. The least confidence was placed on values associated with the food web, N_2, and DON. Information on estimation procedures is presented below. We have tried to provide insight to our procedures without detailing every nuance.

6.2. State Variables

Volume-weighted mean concentrations of NO_x, NH_4, and DON for the two seasons were presented in Christian et al. (1991), and these values were multiplied by average depth (3.41 m) to obtain required values directly. Volume-weighted PN and chlorophyll a (chl a) concentrations were also given and used to compute PN-phyto, PN-hetero, and PN-abiotic. Chl a concentrations were converted to N using conversions of 67 g ash-free dry mass (AFDM): 1 g chl a (APHA 1989), 2 g AFDM: 1 g C, and 106 atoms C: 16 atoms N (Redfield, 1958). The difference of PN-phyto subtracted from PN provided the PN available to PN-hetero (minus fish) and PN-abiotic. We assumed that PN-hetero consisted of bacteria, μ-flagellates, microzooplankton, zooplankton, and fish in ratios similar to those described by Baird and Ulanowicz (1989) for the Chesapeake Bay. Bacterial and μ-flagellate N values were derived from Neuse River Estuary data presented in Christian et al. (1984) and Heath (1989). Using ratios of trophic groupings from Baird and Ulan-

TABLE 8.5. N Cycling in the Neuse River Estuary as Represented by the Following Values of Each State Variable and Flux for the Two Seasons: Summer 1988 and Winter 1988–1989

State Variables	Summer	Winter
	(mmol N × m^{-2})	
X_1	24	20
X_2	22	13
X_3	1300	1300
X_4	78	47
X_5	7	4
X_6	5	2
X_7	1600	2400
X_8	24	9

Inputs	Summer	Winter	Outputs	Summer	Winter
	(mmol N × m^{-2} × season^{-1})			(mmol N × m^{-2} × season^{-1})	
F_{01}	7	7	F_{10}	5	2
F_{02}	3	5	F_{20}	11	6
F_{04}	27	94	F_{30}	19	85
F_{05}	64	159	F_{40}	25	20
F_{06}	23	28	F_{50}	1	1
F_{07}	453	1009	F_{60}	2	4
F_{08}	1	7	F_{70}	478	1030
			F_{80}	2	1

Internal Interactions	Summer	Winter		Summer	Winter
	(mmol N × m^{-2} × season^{-1})			(mmol N × m^{-2} × season^{-1})	
F_{41}	450	190	F_{14}	670	220
F_{51}	450	190	F_{24}	450	190
F_{61}	2440	680	F_{34}	0	0
F_{71}	0	0	F_{35}	16	25
F_{12}	2415	821	F_{65}	560	200
F_{32}	2	1			
F_{42}	672	294	F_{26}	4424	1371
F_{52}	290	140	F_{36}	220	29
F_{62}	1560	490			
F_{72}	0	0	F_{37}	34	25
F_{82}	993	203	F_{57}	0	0
F_{13}	204	7	F_{18}	15	12
F_{23}	2	1	F_{28}	981	333
F_{43}	0	0			
F_{53}	0	0			
F_{63}	38	5			
F_{73}	9	4			
F_{83}	2	148			

[a]All values were area weighted. Designations for state variables shown in this table correspond to the compartment numbers listed in Table 8.2. Designation of flux for inputs, outputs, and internal interactions correspond to those presented in Table 8.3.

TABLE 8.6. Degree of Reliability Attributed to Each State Variable or Flux Value

High	Degree of Reliability Moderate	Low
State Variable or Flux		
X_1	X_2	X_7
X_4	X_3	F_{07}
X_5	X_8	F_{71}
X_6	F_{02}	F_{12}
F_{01}	F_{08}	F_{32}
F_{04}	F_{10}	F_{72}
F_{05}	F_{20}	F_{82}
F_{06}	F_{30}	F_{13}
F_{51}	F_{40}	F_{23}
F_{61}	F_{50}	F_{43}
F_{52}	F_{60}	F_{83}
F_{62}	F_{70}	F_{14}
F_{53}	F_{80}	F_{24}
F_{63}	F_{41}	F_{34}
F_{35}	F_{42}	F_{65}
F_{36}	F_{73}	F_{57}
	F_{26}	F_{18}
	F_{37}	F_{28}

"Summer and winter values were computed similarly and given the same degree of reliability. The variable numbers shown in this table correspond to the compartment numbers listed in Table 8.2. Interaction designations correspond to those presented in Table 8.3.

owicz (1989), PN-hetero values without fish were estimated as 21.5 and 12.6 mmol $N \times m^{-2}$ for summer and winter, respectively. These values were subtracted from the PN values remaining after PN-phyto subtraction to obtain PN-abiotic. For the contribution of fish N to PN-hetero, we used trawl information from the Neuse River Estuary for seasonal dynamics and from the Albemarle Sound for biomass (Hester, 1975). The resultant fish N contributed only 0.6 and 0.3 mmol $N \times m^{-2}$ for summer and winter, respectively.

Other standing stocks came from literature values. The value for N-sed was considered constant throughout the year, corresponded to the surface 1 cm as the active zone of sediment/water exchange, and was obtained from measurements in the Neuse River Estuary (Matson et al., 1983). Standing stock values for N_2 were assumed to be at 100% saturation at all times. Concentrations were estimated for average seasonal water temperatures using *Lange's Handbook of Chemistry* (Dean, 1985).

6.3. Inputs

During the time under consideration, we directly measured the concentrations of N species in waters representing the major sources of loading to the estuary: Neuse River, Trent River, a paper-pulp mill, a sewage treatment plant, and precipitation (Stanley et al., unpublished data). Concentrations times flow (or precipitation) provided direct input estimates for DON, NO_x, and NH_4. PN loading was subdivided into the separate components based on the measured chl a at station G and on the projected heterotrophic contribution based on information from station G when possible. Generally, procedures were as described for standing stocks. Values for N_2 again assumed 100% saturation of water, but input included a flux from the atmosphere equal to N_2 fixation.

6.4. Outputs

Discharge of water to the Pamlico Sound was considered to equal the freshwater inflow to the Neuse River Estuary as averaged by season for flushing times given in Christian et al. (1991). No effort was made to account for either inputs or outputs of saline Pamlico Sound water. Nutrient concentrations were normally low near the mouth of the Neuse River Estuary (Christian et al., 1989), and inputs and outputs of sound water were considered essentially equal.

Measured concentrations of N species from station A were averaged and multiplied by freshwater flow to estimate appropriate outputs. Again, PN was subdivided based on chl a concentrations and estimates of heterotroph N. As the standing crop of nekton was a minor component of PN-hetero, we ignored outputs due to emigration. N_2 output combined flux to the sound at 100% saturation and release to the atmosphere equal to denitrification. Sediment N buried below the active zone was calculated by combining information on long-term sedimentation rate for the Neuse River Estuary from Benninger and Martens (1983) with net differences between seasonal inputs and outputs to the estuary. Long-term burial was calculated as 52 mmol N $\times$ m^{-2} $\times$ season^{-1}. This rate was apportioned to reflect a larger influx of materials to the estuary in winter than in summer.

6.5. Internal Interactions

The strength of our data set is the information on planktonic uptake of DIN and primary productivity. We measured process rates under controlled temperature and light conditions within the laboratory. Rates from bottle experiments were then integrated over depth and time to provide areal daily rates for each station. To integrate over the estuary for the day of sampling, the influence of each station's rate corresponded to percent area as shown in Table 1. Rates for each day between the sampling dates were estimated by linear interpolation. Details of this integration scheme are in Boyer et al. (1988) and Stanley et al. (unpublished data).

Total N uptake by phytoplankton was calculated from integrated primary productivity assuming an atomic ratio of C:N of 6.625:1 (Redfield, 1958). Total primary productivity was assumed to be 1.25 times the particulate primary productivity measured by ^{14}C-CO_2 incorporation. The correction accounts for a 20% release of photosynthate N as DON. The actual percentage is unknown but similar to the 25% DOC release used by Baird and Ulanowicz (1989). We assumed that released dissolved organic matter would be slightly nitrogen-poor relative to cellular needs. The total phytoplankton N requirements were apportioned among NH_4, NO_x, and DON. DON uptake was considered to be essentially urea uptake, and urea uptake was considered to be equal to NO_x uptake (Paasche, 1988; Christian et al., unpublished data). The relative proportions of measured uptake rates of NH_4 and NO_x and the assumed DON uptake rate were used to subdivide total phytoplankton N uptake rates (F_{41}, F_{51}, F_{61}). Planktonic N_2 fixation rates (F_{71}) have been considered negligible based on attempts at measurement by Paerl (1987).

Seven potential fluxes into heterotrophs were considered. One, N_2 fixation (F_{72}), was set to zero for the reason described above. Uptake rates of NH_4 (F_{62}) and NO_x (F_{52}) by the bacterial component were calculated as the differences between measured integrated uptakes and those ascribed to phytoplankton uptake. DON uptake (F_{42}) was indirectly estimated by assuming that the DON pool remained in steady state. After accounting for other inputs and outputs to the DON pool, the difference was considered taken up by heterotrophs. Similarly, feeding on phytoplankton (F_{12}) was assumed to be equal to that necessary to maintain the steady state after accounting for other inputs and outputs. Detrital feeding (F_{12}) was also calculated by mass balance. Flux from the benthos (F_{32}) was considered to reflect feeding by the minor nekton component. The values were the lowest capable of being chosen that would reflect seasonal differences (i.e., 1 and 2 for winter and summer, respectively). This small exchange is similar to that described by Baird and Ulanowicz (1919). As can be seen, the fluxes into PN-hetero are rather poorly understood with the exception of DIN uptake. The compartment is highly aggregated, which might be perceived to complicate estimations. However, disaggregation would only increase the number of required values, and few would have been measured for the Neuse.

The benthic system potentially received N from all other compartments. Two of these potential fluxes were set to zero: benthic uptakes of DON (F_{43}) and NO_x (F_{53}). The first was set to zero because no data existed to portray a positive value. Little has been done on this subject in the Neuse River or other estuary. The fluxes of O_2, NO_x, and NH_4 between sediments and water were measured in the Neuse River Estuary during the two seasons under consideration. Cores from stations A, D, and G were incubated in the laboratory under a variety of light availabilities. Integrations of fluxes from these measured rates involved calculations of *in situ* light availability and depth distribution of the sediment surfaces. Photosynthetically active radiation (PAR) at the sediment surface was calculated from surface light, measured extinction coefficients, and hours of daylight. DIN requirements for photosynthesis were computed from O_2 exchange using a photosynthetic quotient of 1.0 and Redfield ratio (Redfield, 1958). These values were compared to measured NH_4 and NO_x exchanges in the light. NO_x exchange in the light integrated over the estuary was small and out of the sediments (2 and 4 mmol $N \times m^{-2} \times season^{-1}$ for summer and winter, respectively). Thus the fluxes of NO_x to the integrated sediments (F_{53}) were set to zero. We note, however, that within the upper portion of the estuary where water-column NO_x concentrations were large (Christian et al., 1989), NO_x was found to enter sediments. The fluxes in the light for NH_4 were into the sediments (F_{63}), but interestingly they were sufficient to meet only 29 and 6% of the sediment photosynthetic N requirements in summer and winter, respectively. Thus recycling of N within the sediments appears to be important in this ecosystem.

Fluxes of phytoplankton (F_{13}) into the sediments were based on information from Baird and Ulanowicz (1989), reflecting the lack of information on this process for the Neuse. We compared their phytoplankton C production to uptake by suspension feeders and the bivalve *Mya*. In the Chesapeake Bay 4.5 and 0.4% of available phytoplankton C entered these compartments in summer and winter, respectively. These percentages were doubled to reflect the fact that the average depth of the Neuse is approximately half that of Chesapeake Bay, and the doubled percentages of PN-phyto production were used for the fluxes into the sediment. For PN-hetero flux into the sediments (F_{23}) we assumed that relatively little occurred. Heterotrophic N covered a large size range dominated by bacteria or small protozoans, which would be colloidal. We assumed that only 10% of PN-hetero biomass would be consumed by benthos in a season. Both F_{13} and F_{23} are based on little information concerning the Neuse, but we assume a similar pattern to that described for Chesapeake Bay (Baird and Ulanowicz, 1989).

The remaining fluxes into the sediments were from N_2 fixation (F_{73}) and sedimentation and suspension feeding of detritus (F_{83}). Various sediment processes including N_2 fixation rates were measured in the neighboring Pamlico River Estuary during the early 1980s by Kuenzler et al. (1984). We used their rates for appropriate seasons. The remaining detrital input flux was calculated by mass balancing sediment inputs and outputs by season.

Two potential fluxes into the DON compartment were quantified, and one (F_{34}) was assumed to be zero for the lack of other information. Photosynthetic release of DON (F_{14}) was calculated as a portion of total primary production as described. Dissolved organic matter in estuaries is operationally defined by the methods employed and represents a large diversity of molecules. The bulk of DON standing stock acts as if it is largely refractory, but a minor component is readily available for biological activity (Christian et al., 1991). We consider that DON was in steady state, and release of DON through excretion was allowed to balance the uptake of DON by phytoplankton. In this way, we assumed that the major source of DON was from grazers or their consumers.

Nitrate plus nitrite was resupplied from benthic and water-column nitrification. Benthic flux (F_{35}) was directly measured as described. However, no measurements of water-column nitrification (F_{65}) were available for the Neuse River Estuary. Therefore we extrapolated rates from literature values. McCarthy et al. (1984) presented summer volumetric nitrification rates for the York River and Chesapeake Bay. We averaged these rates after conversion to areal and seasonal scales and used the average for the summer flux. Winter flux was considered to be less than half the summer rate because of the lower temperatures (20°C lower) and lack of hypoxia/reaeration cycles in winter. The resulting fluxes were higher than the limited estimation by Wanat (1983) for the Pamlico River Estuary as cited by Kuenzler et al. (1984) (560 and 200 mmol N $\times$ m^{-2} $\times$ season^{-1} versus 20 mmol N $\times$ m^{-2} $\times$ season^{-1}).

Ammonification (F_{26}) and release of NH_4 from sediments (F_{36}) recycled N to the DIN (i.e., NH_4) pool. Benthic release of NH_4 was measured as described and is the dominant efflux of N from sediments. Ammonification represented the major flux overall and was calculated as the average of two separate mass balance estimates: one for the NH_4 pool itself and one for the PN-hetero compartment. These two estimates differed from their mean by 1.5% in summer and 3.5% in winter. Also, the rates of ammonification are in keeping with measured rates in the Neuse River Estuary (Stanley et al., unpublished data). Mean values were used, preventing an actual mass balancing of the two compartments involved (X_2 and X_6).

Denitrification replenishes N_2, but again no direct measurements of this process were made in the Neuse River Estuary. Because denitrification is an anaerobic process, its occurrence in the water column would be restricted largely to conditions of hypoxia and anoxia. Hypoxic conditions associated with stratification occurred in summer especially in the upper estuary. However, the volume of impacted water is small relative to the Neuse's total volume. We assumed that the integrated flux would be small and set F_{57} to zero. Benthic denitrification (F_{37}) was assayed in the Pamlico River Estuary by Kuenzler et al. (1984) who found undetectable rates (from the acetylene block technique) if NO_3 was not added to the assay chamber. The lack of measurable rates may have resulted from close coupling of benthic nitrification and denitrification and the inhibition of nitrification by acetylene. We therefore assumed that denitrification equaled their sediment nitrification rates minus that NO_x we measured as released from sediments. These resultant fluxes were generally less than those summarized by Koike and Sørensen (1988) for other estuarine sediments.

The last two fluxes requiring estimation were for nonpredatory mortality of phytoplankton (F_{18}) and comparable mortality of heterotrophs plus egestion (F_{28}). The former was considered to be small relative to primary productivity and was set at 1% $\times$ day^{-1} of phytoplankton standing stock. The calculation for F_{28} was more elaborate. We constructed a food web based on our measured information on input fluxes and standing stocks and fractional transfers used by Baird and Ulanowicz (1989). From this food web we calculated rates of egestion and considered these rates to represent F_{28}. Nonpredatory mortality of heterotrophs was assumed to be insignificant.

6.6. Analyses of Networks

All analyses of our network models were conducted through *NETWRK4: A Package of Computer Algorithms to Analyze Ecological Flow Networks* (Ulanowicz, 1987). This programming package was supplied by R. E. Ulanowicz, University of Maryland, Chesapeake Biological Laboratory, Solomons, MD 20688. Two computer runs were made, one for summer and one for winter values. Data as shown in Table 8.5 were entered using DATAIN, available with the package. Subsequent analyses were conducted with the NETWRK program. The program directed all matrix calculations, cycle sorting, and index calculations described in Section 2.2. Details of the specific mathematics may be found in Ulanowicz (1986) and Wulff et al. (1989). Documentation of package functions was incorporated with the package.

7. RESULTS AND DISCUSSION

7.1. Overview

The static models, as diagrammed in Fig. 8.2 with quantities given in Table 8.5, were analyzed relative to their networks. Because six of the potential fluxes were assumed to be 0 mmol N $\times$ m^{-2} $\times$ season^{-1}, the realized connectivity was 41.1% as opposed to the potential connectivity of 51.8%. Connectivities of nutrient cycling models tend to be greater than energy flow models where dissipative flows of energy, or its surrogate carbon, are not involved in many feedback loops. As we compare our findings with those of others and especially with those of Baird and Ulanowicz (1989), the importance of this increased connectivity will be evident. We discuss our results in the following order: (1) the mass balance of throughputs, (2) structure analysis, (3) biogeochemical cycle analysis, and (4) information indices.

7.2. Mass Balance of Throughputs

Although several fluxes were estimated by assuming steady-state conditions for individual state variables, no attempt was made to balance the system as a whole. Thus four state variables were balanced, but four were not (Table 8.7). Overall, the system was out of balance by 0.2 and 1.9% of total throughput in summer and winter, respectively. Given the degree of uncertainty associated with numerous fluxes (Table 8.6), we do not wish to ascribe much significance to the imbalances. For example, the estuary is shown to be a source for heterotroph N and NH_4 during both seasons. If, however, the inputs to PN-hetero were reduced by 1.1% in summer and 2.7% in winter with concomitant decrease in ammonification, these two state variables would be balanced. Also, although the estuary is a sink for NO_x in summer and a source in winter, the estimates used for nitrification were derived from few measured values from other estuaries (McCarthy et al., 1984). The resultant fluxes were large relative to the other source of NO_x, measured fluxes from sediments. Therefore the major inputs to NO_x were values for which confidence is low. To improve our understanding of N cycling, it would be important to measure nitrification directly.

TABLE 8.7. Differences Between Inputs and Outputs of Each State Variable for Summer and Winter[a]

State Variable	Summer	Winter
	(mmol N $\times$ m^{-2} $\times$ season^{-1})	
X_1, PN-phyto	2	5
X_2, PN-hetero	67	53
X_3, N-sed	0	0
X_4, DON	0	0
X_5, NO$_x$	-101	53
X_6, NH$_4$	67	49
X_7, N$_2$	0	0
X_8, PN-abiotic	0	0
Total difference	35	160
Total throughput	18,054	8,037

[a]Positive values indicate inputs > outputs. Negative values indicate inputs < outputs.

7.3. Structure Analysis

Various matrix manipulations allow analyses of the network structure. The total contribution matrices in Table 8.8 describe the percentage of N leaving a compartment, X_i, that eventually enters X_j by either direct or indirect flows. In summer considerable recycling is evident in several ways. Greater than 93% of N from PN-phyto (X_1), PN-hetero (X_2), DON (X_4), NO$_x$ (X_5), NH$_4$ (X_6), and PN-abiotic (X_8) reentered the food web at X_1, X_2, or X_6. In contrast, less than 20% of N$_2$ entered any single state variable. Although the largest percentage contributions were between compartments within the water column, more than 50% of all outputs from water-column compartments (excepting N$_2$) passed through sediments.

During winter, the percent contributions were lower than in summer. The only paths greater than 90% were into heterotrophs (from X_1, X_4, and X_6). Thus recycling was reduced. Recycling, as indexed through percent contributions, was quite insignificant for N$_2$ outputs. Also, the pattern of contributions from compartments to sediments changed. The only contribution greater than 50% was from the sedimentation of detritus (X_8). The planktonic food web's contribution to sediments decreased and presumably so did the biological activity in the sediments dependent on feeding on water-column organisms.

The diagonal coefficients (where $X_i = X_j$) represent indices of "self-stimulation" or the percentage of a compartment's output that reenters the compartment. Again, recycling is shown to be important to the aquatic food web. In summer, phytoplankton and heterotrophs had greater than 90% of the outputs reenter the respective compartments. In winter, the percentages decreased to 77.4 and 86.3%, respectively, but these remained large values. During both seasons, NH$_4$ was found to be involved in strong feedback; 95.2% (summer) and 82.9% (winter) of NH$_4$

TABLE 8.8. Total Contribution Matrices of N Cycling Networks for Each Season[a]

Summer Donor (X_i)	X_1	X_2	X_3	Recipient (X_j) X_4	X_5	X_6	X_7	X_8
X_1	93.2	97.7	62.3	85.5	70.7	96.0	14.4	79.7
X_2	94.5	96.2	59.6	83.5	72.0	97.9	13.8	81.5
X_3	79.0	79.6	49.8	68.4	61.7	81.8	23.2	64.9
X_4	94.6	96.9	59.3	82.5	70.0	95.0	13.8	79.0
X_5	97.7	98.5	61.2	84.6	71.1	96.6	14.2	80.3
X_6	96.3	97.0	60.6	83.4	73.6	95.2	14.1	79.1
X_7	1.5	1.5	1.9	1.3	1.1	1.5	0.4	1.2
X_8	94.3	99.8	59.5	83.3	71.9	97.7	13.8	81.3
Winter Donor (X_i)	X_1	X_2	X_3	Recipient (X_j) X_4	X_5	X_6	X_7	X_8
X_1	77.4	97.5	49.4	66.0	47.3	87.2	8.8	66.4
X_2	77.4	86.3	49.6	58.2	48.3	89.3	8.8	67.4
X_3	27.2	29.8	15.0	18.9	24.8	29.6	17.8	20.2
X_4	82.8	95.1	47.5	58.9	46.0	85.0	8.5	64.4
X_5	77.7	84.7	42.5	53.9	41.0	75.7	7.6	57.4
X_6	85.1	92.7	46.8	59.0	53.2	82.9	8.4	62.9
X_7	0.1	0.1	0.4	0.1	0.1	0.1	0.1	0.1
X_8	56.1	70.2	70.6	41.5	38.3	64.0	12.6	47.4

[a]Values are percentages of flow. Designations of state variables (X's) correspond to the compartment numbers listed in Table 8.2.

outputs returned to this compartment. The importance of NH_4 is reiterated throughout our discussions.

Another view of the relationships between compartments through both direct and indirect flows is given through total dependency matrices (Table 8.9). These characterize the percentage of input to a recipient compartment (X_j) that passed through a direct or indirect donor (X_i). The coefficients of the dependency matrix provide information on the "extended diet" of a population in a food web or extended source of a nutrient pool. As an example, in summer, 96.6% of the inputs to sediment N (X_3) passed through phytoplankton; 93.2% passed through heterotrophs, but only 6.1% had once been associated with N_2. In fact, for both seasons, N_2 was the least important source of N to all compartments. This results from the insignificance of N_2 fixation in both the water column and sediments.

The high percentage values throughout the summer's matrix again indicate considerable recycling. For all state variables, except NO_x and N_2, more than 90% of the N extended source came through each compartment of the aquatic food web (X_1 and X_2) and DIN (X_5 or X_6). Between 60 and 90% of the N for all but N_2 passed through detritus in each of its forms, dissolved (X_4) and particulate (X_8).

TABLE 8.9. Total Dependency Matrices of N Cycling Networks for Each Season[a]

Summer Donor (X_i)	Recipient (X_j)							
	X_1	X_2	X_3	X_4	X_5	X_6	X_7	X_8
X_1	93.2	96.8	96.6	96.4	74.9	96.3	6.7	96.7
X_2	95.4	96.2	93.2	95.0	77.0	99.2	6.5	99.8
X_3	50.9	50.9	49.8	49.7	42.2	52.9	7.0	50.8
X_4	84.0	85.2	81.7	82.5	65.7	84.6	5.7	85.1
X_5	92.3	92.2	89.6	90.1	71.1	91.6	6.3	92.1
X_6	95.9	95.7	93.7	93.6	77.6	95.2	6.5	95.6
X_7	3.1	3.1	6.1	3.1	2.6	3.3	0.4	3.1
X_8	77.7	81.4	76.1	77.3	62.7	80.8	5.3	81.3
Winter Donor (X_i)	Recipient (X_j)							
	X_1	X_2	X_3	X_4	X_5	X_6	X_7	X_8
X_1	77.4	87.7	84.9	76.7	50.3	86.0	2.1	86.4
X_2	86.0	86.3	94.8	75.2	57.2	97.9	2.3	97.5
X_3	15.8	15.6	15.0	12.8	15.4	17.0	2.4	15.3
X_4	71.3	73.6	70.3	58.9	42.1	72.1	1.7	72.1
X_5	73.0	71.6	68.7	58.9	41.0	70.2	1.7	70.2
X_6	86.3	84.6	81.7	69.6	57.4	82.9	2.0	83.0
X_7	0.5	0.4	2.9	0.4	0.4	0.5	0.1	0.4
X_8	43.1	48.5	93.3	37.1	31.3	48.5	2.3	47.4

[a] Values are percentages of flow. Designations of state variables (X's) correspond to the compartment numbers listed in Table 8.2.

The dependency of compartments (except N_2, X_7) on sediment N was less than on aquatic N. Approximately 40–50% of N needs of all compartments but X_7 were associated with sediments at some point in the cycle.

Again, in winter most percentages decreased. Only four percentages were greater than 90%. Three were associated with heterotrophs as an extended source. The fourth percentage paired particulate detritus (X_8) as a source of N to sediments. This was associated with increased sedimentation in winter (Table 5). Although two of the four coefficients that were greater than 90% entered the sediments, the sediments were a minor source of N for other compartments. Thus the sediments in winter are a less significant contributor to the continued cycling of N than in summer.

An extension of input–output analysis involves decomposing the graph relative to individual external sources (F_{oj}). The effects of each system's source of N to various compartments is computed independent of other sources. The resultant matrices (not shown) indicate the number of N atoms entering a compartment from one atom entering the estuary from a source. Considered in the context of recycling, the value also refers to the number of passes into a compartment relative to

TABLE 8.10. Contribution of N Sources to Primary Production[a]

Source	Summer	Winter
NO_x	14.5	3.1
NH_4	14.2	3.8
DON	14.0	3.7
N_2	0.2	<0.1

[a] Units equal the number of N atoms entering primary producers from 1 atom imported from source.

an atom's original input. We consider four sources of N available directly to primary producers (NO_x, NH_4, DON, and N_2) and their contribution to primary production (Table 8.10). During each season, the contributions per molecule of NO_x, NH_4, and DON were similar and contributions exceeded 1 atom into primary producers per atom entering the estuary. Imported N_2 provided little N to primary producers. In summer, each source cycled at least 14 times into primary producers. In winter recycling decreased to 3–4 times. From these results we infer that, as a result of recycling within the estuary, the species of DIN (i.e., NO_x or NH_4) is immaterial to controlling primary production. Efforts to manage N loading to favor NO_x or NH_4 would not be particularly valuable. Although we found similar results for DON, caution must be given to broad interpretation. The bulk of DON is probably refractory, while its importance to biological activity is through urea and similar available molecules. Loading of solely the refractory material may reduce DON's contribution to primary production.

7.4. Biochemical Cycle Analysis

The next analyses arose from the distributions of feedback loops. Loops are defined as cycles, each representing a unique path by which material flows out of a compartment to at least one other and returns to the original. A nexus is one or more cycles sharing a common weak arc. A weak arc is an interaction between compartments, which represents the smallest flux within a cycle (Ulanowicz, 1986). Analyses of cycle distributions and characteristics can give insight to the patterns of flow and the potential for homeostatic control.

Forty-two cycles were found during each season (Table 8.11). The cycles were distributed among 14 nexuses each season, but the distribution of cycles and weak arcs differed between seasons. For both seasons most nexuses possessed only one cycle. In summer the nexus with a weak arc from detritus to sediments had 16 cycles, whereas weak arcs from sediments to NH_4 and to NO_x combined for 10 cycles. Similarly, in winter sediments were the recipient with weak arcs in three nexuses totaling 13 cycles and were the donor in two nexuses totaling nine cycles. Cycles involving weak arcs associated with fluxes between benthic and aquatic subsystems numbered 28 and 22 for summer and winter, respectively. The re-

TABLE 8.11. Distribution of Biogeochemical Cycles and Respective Weak Arcs[a]

Summer		Winter	
Weak Arc of Nexus	Number of Cycles	Weak Arc of Nexus	Number of Cycles
8, 3	16	7, 3	1
7, 3	1	6, 3	5
1, 8	4	1, 3	7
3, 5	6	1, 8	10
6, 3	1	3, 5	3
3, 6	4	3, 6	6
5, 2	1	5, 2	1
2, 4	1	2, 4	1
4, 1	2	4, 1	2
5, 1	2	5, 1	2
1, 4	1	8, 2	1
2, 8	1	1, 4	1
6, 2	1	6, 2	1
4, 2	1	6, 1	1

[a] Weak arc is smallest common flux. Information given in order of ascending weak arc flux. Designation numbers assigned to weak arcs correspond to the compartment numbers listed in Table 8.2.

maining cycles involved weak arcs of only aquatic compartments. Thus most potential control over cycles involves benthic/aquatic coupling; however, the quantity of flow from this coupling is decidedly less than that from aquatic cycling. The weak arcs of benthic/aquatic coupled nexuses did not exceed 220 or 29 mmol $N \times m^{-2} \times season^{-1}$ in summer and winter, respectively. Weak arcs for aquatic cycles during these seasons were generally larger (up to 2415 and 680 mmol $N \times m^{-2} \times season^{-1}$ for the respective seasons). Therefore sediments have many routes of recycling interaction with the aquatic subsystem, but the majority of recycling actually occurs within the water column.

For our model of N cycling in the Neuse River Estuary, cycles ranged from two to seven lengths. The distribution of cycle lengths can be normalized as a percentage of total throughput associated with each length. Most flow was associated with cycles of two through four lengths (Table 8.12). The distribution of flow favored shorter cycles in summer than in winter. We postulate that the increased biological activity in summer favors short cycles of N passing rapidly between NH_4 and biotic compartments and within the aquatic food web.

The sum of values by season from Table 8.12 is called the "Finn Cycling Index" (Finn, 1980; Kay et al., 1989), a calculation of the percentage of total throughput associated with recycling as opposed to flows into and out of the system. The Finn Cycling Indices for N in the Neuse were very high: 91.5 and 62.7% for summer and winter, respectively. The high values confirm our previous find-

TABLE 8.12. Normalized Distributions of Cycle Lengths

Length	Percentage of Total Throughput	
	Summer	Winter
2	35.2	23.3
3	42.1	28.9
4	13.0	8.3
5	1.2	1.9
6	<0.1	0.3
7	<0.1	<0.1
8	0	0
Total	91.5	62.7

ings as to the amplification of effects of source inputs. Just as the effects of sources were more amplified in summer than winter, the Cycling Index in summer was much greater.

7.5. Information Indices and Global Attributes

The last series of analyses involve the calculations of several global or system-wide characteristics derived from merging input–output analysis with information theory (Ulanowicz, 1986). With information theory a network of interactions is analyzed relative to the predictability of flow within the network. A network that is "maximally articulated" or has a perfectly predictable flow pattern is said to have a maximum information content. Less predictability and hence greater diversity of potential flows reduces the information index based on the Shannon–Wiener formula. Several information indices of system organization are scaled by total throughput to characterize both size and structure of the system. Details of these analyses were given in Ulanowicz (1986) and Kay et al. (1989).

The "development capacity" of a system is considered to reflect the system's potential of maximum articulation or predictability of the network for a given total throughput. Ulanowicz (1986) postulated that systems succeed to or evolve toward an organizational structure that approaches the development capacity. The success of the approach is measured as "ascendancy." The deterrents to ascendancy-equaling capacity are embodied in "overhead" terms. These refer to decreased predictability relative to imports, exports, dissipations, and redundancy. Imports and exports are self-evident. Dissipations are associated with respiratory losses, which are applicable to energy flow but not necessarily to nutrient cycling. Redundancy refers to the duplication of flow pathways. All are scaled by total throughput; however, it is common to normalize indices as fractions of development capacity. Both approaches are given for our N cycle model in Table 8.13.

The development capacity in summer was over twice that in winter, owing

7. RESULTS AND DISCUSSION

TABLE 8.13. Information Indices and Global Attributes of N Cycling in the Neuse River Estuary

	Summer[a]	Winter[a]
Total system throughput (mmol N $\times$ m^{-2} $\times$ season^{-1})	18,504	8,037
Development capacity ((mmol N $\times$ m^{-2} $\times$ season^{-1}) $\times$ bits)	65,659	31,415
Ascendancy ((mmol N $\times$ m^{-2} $\times$ season^{-1}) $\times$ bits)	25,946 (0.395)	14,833 (0.472)
Overhead		
Import ((mmol N $\times$ m^{-2} $\times$ season^{-1}) $\times$ bits)	1,353 (0.021)	2,279 (0.073)
Export ((mmol N $\times$ m^{-2} $\times$ season^{-1}) $\times$ bits)	850 (0.013)	1,014 (0.032)
Dissipation ((mmol N $\times$ m^{-2} $\times$ season^{-1}) $\times$ bits)	0	0
Redundancy ((mmol N $\times$ m^{-2} $\times$ season^{-1}) $\times$ bits)	37,511 (0.571)	13,290 (0.423)

[a] Values in parentheses are as fraction of capacity.

largely to the larger throughput in summer. Ascendancy in winter was also less than that in summer. But the differences in ascendancy were proportionally less, as the normalized ascendancy in winter was greater than that in summer. Ascendancy was 39.5 and 47.2% of capacity for summer and winter, respectively. Thus less than half of the potential organizational articulation was achieved. The major overhead costs were associated with redundancy, an indication of ambiguity of flow associated with the considerable recycling activity of the system. Little overhead was associated with imports or exports, and, as previously noted, none was associated with dissipation. The overheads involving exchange between the estuary and its environment were greater in winter than summer; and redundancy decreased in winter along with its relative importance. Again, the internal biological activity within the estuary decreased in winter as loading and removal increased. This decrease was actually associated with an increase in relative ascendancy, indicating that tight biogeochemical cycling is not necessarily a requisite to maximizing relative ascendancy.

7.6. Comparison with Other Systems

Network analyses as done here have not been conducted for models of many real systems. Ulanowicz (1986) provided a limited number of examples in his book on the subject. Recently, two sources have given more complete information. Baird and Ulanowicz (1989) provided a detailed analysis of energy flow (using C) in the Chesapeake Bay. *Network Analysis in Marine Ecology*, edited by Wulff et al. (1989), includes papers involving analyses of several coastal and marine ecosystems. The models described in the book were compared by Mann et al. (1989). The papers that described networks of N flow were by Kremer (1989) for Narragansett Bay, by Field et al. (1989b) for the southern Benguela upwelling system, and by Ducklow et al. (1989) for oceanic warm core rings. We extend the comparison to our model of N cycling in the Neuse River Estuary.

Comparisons are difficult in that analyses are sensitive to model structure. This includes sensitivity to the number of compartments, their designation, currency, and hence postulated interactions (Mann et al., 1989). Furthermore, the variety of network analyses described in papers differs among the limited number of publications available. For example, of the papers on N cycling cited in the previous paragraph, only one presented the Finn Cycling Index (Field et al., 1989), and two presented normalized information indices (i.e., ascendancy/capacity) (Field et al., 1989; Kremer, 1989). Our N cycling model had lower normalized ascendancies than any given in these papers. Furthermore, they were lower than those presented for energy flow in the Chesapeake Bay (Baird and Ulanowicz, 1989) or others cited by Mann et al. (1989). Conversely, our normalized redundancy values are high relative to literature values (Baird and Ulanowicz, 1989; Kremer, 1989). We propose that this high redundancy (and hence low ascendancy) is associated with the considerable potential for alternate paths of recycling within the Neuse. The large amount of recycling has been indicated by several of our analyses, and our model holds more true to a generalized N cycle than others cited. The generalized cycle provides numerous transformations not directly associated with conventional trophic flows of C or energy. In addition, no dissipation was represented in our model. These factors would lead to multiple directions of flow and to increased redundancy and lower ascendancy. Until comparable models are analyzed, intersystem comparisons are suspect.

The results of our analyses, which indicate the vast importance of recycling in the Neuse River Estuary, are in keeping with more conventional analyses for the Neuse (Boyer et al., 1988) and for the neighboring estuaries of the Pamlico River (Kuenzler et al., 1979) and the Chowan River (Stanley and Hobbie, 1981). These are river-dominated estuaries, but they have extended residence times for water, which promotes the development of a community capable of processing considerable quantities of N through both organic and inorganic phases. This development, however, did not lead to particularly high ascendancies, as might be inferred from Ulanowicz (1986). Much of this processing occurs in the water column, and in warmer months the processing is more rapid than in winter months. Interactions with the benthic subsystem are numerous but of limited importance quantitatively. Most of the biological production is dissipated within the estuaries, releasing energy but remineralizing nutrients for further support of production. The availability of secondary production to humans through harvest appears to be a minor component of total throughput.

ACKNOWLEDGMENTS

Support for this work was provided by the National Oceanic and Atmospheric Administration Office of Sea Grants NA85AA-D-SG022 and NA86AA-D-SG046 and the State of

North Carolina. The grant was administered by the University of North Carolina Sea Grant College Program. Also, support was provided by the U.S. Environmental Protection Agency under grant agreement R-812475-01-0. Further support was from Texasgulf Chemicals, Inc. We thank R. E. Ulanowicz for the use of NETWRK4. We thank the following individuals for their assistance in field, laboratory, and clerical activities: A. Anderson, D. Daniel, K. Evans, L. Harper, M. Jones, B. King, G. Lackey, and R. Willis.

REFERENCES

APHA, AWWA, and WPCF (1989), *Standard Methods for the Examination of Water and Wastewater*, 17th ed., American Public Health Association, New York, pp. 10-39, 10-40.

Baird, D., and Ulanowicz, R. E. (1989), *Ecol. Monogr.*, **59**, 329-364.

Baretta, J. W., and Ruardij, P. (1987), *Continental Shelf Res.*, **7**, 1471-1476.

Benninger, L. K., and Martens, C. S. (1983), *Sources and Fates of Sedimentary Organic Matter in the White Oak and Neuse River Estuaries*, Rept. No. 194, Water Resources Research Institute of the University of North Carolina, Raleigh, 60pp.

Billen, C., and Lancelot, C. (1988), in *Nitrogen Cycling in Coastal Marine Environments* (T. H. Blackburn and J. Sorensen, Eds.), Wiley, Chichester, pp. 341-378.

Boyer, J. N., Stanley, D. W., Christian, R. R., and Rizzo, W. M. (1988), Proceedings North Carolina American Water Research Association, Symposium on Coastal Water Resources, TPS 88-1, AWRA, Bethesda, MD, pp. 165-176.

Christian, R. R., and Wetzel, R. L. (1991), *Microbial Ecol.*, **22**, 111-125.

Christian, R. R., Stanley, D. W., and Daniel, D. A. (1984), in *The Estuary as a Filter* (V. S. Kennedy, Ed.), Academic Press, New York, pp. 349-366.

Christian, R. R., Wetzel, R. L., Harland, S. M., and Stanley, D. W. (1986a), in *Prospectives in Microbial Ecology* (F. Megusar and M. Gantar, Eds.), Slovene Society for Microbiology, Ljubljana, Yugoslavia, pp. 38-45.

Christian, R. R., Bryant, W. L., Jr., and Stanley, D. W. (1986b), *The Relationship Between River Flow and* Microcystis aeruginosa *Blooms in the Neuse River, North Carolina*, Rept. No. 223, Water Resources Research Institute of the University of North Carolina, Raleigh, 100pp.

Christian, R. R., Stanley, D. W., and Daniel, D. A. (1988), *Characteristics of a Blue-Green Algal Bloom in the Neuse River, North Carolina*, Working Paper No. 87-2, University of North Carolina Sea Grant College Program, Raleigh, 71pp.

Christian, R. R., Rizzo, W. M., and Stanley, D. W. (1989), National Undersea Research Program, Res. Rept. 89-2, Department of Commerce, NOAA, pp. 19-40.

Christian, R. R., Boyer, J. N., and Stanley, D. W. (1991), *Mar. Ecol. Prog. Ser.*, **71**, 259-274.

Dame, R. F. (Ed.) (1979), *Marsh-Estuarine Systems Simulation*, University of South Carolina Press, Columbia, 260pp.

Davis, G. J., Brinson, M. M., and Burke, W. A. (1978), *Organic Carbon and Deoxygenation in the Pamlico River Estuary*, Rept. No. 131, Water Resources Research Institute of the University of North Carolina, Raleigh, 123pp.

Day, J. W. Jr., Hall, C. A. S., Kemp, W. M., and Yanez-Arancibia, A. (1989), *Estuarine Ecology*, Wiley, New York, pp. 257–308.

Dean, J. A. (Ed.) (1985), *Lange's Handbook of Chemistry*, 13th ed., McGraw-Hill, New York, p. 10-5.

Ducklow, H. W., Fasham, M. J. R., and Vezina, A. F. (1989), in *Network Analysis in Marine Ecology: Methods and Applications* (F. Wulff, J. G. Field, and K. H. Mann, Eds.), Springer-Verlag, Berlin, pp. 159–205.

Field, J. G., Wulff, F., and Mann, J. H. (1989a), in *Network Analysis in Marine Ecology: Methods and Applications* (F. Wulff, J. G. Field, and K. H. Mann, Eds.), Springer-Verlag, Berlin, pp. 3–14.

Field, J. G., Moloney, C. L., and Attwood, C. G. (1989b), in *Network Analysis in Marine Ecology: Methods and Applications* (F. Wulff, J. G. Field, and K. H. Mann, Eds.), Springer-Verlag, Berlin, pp. 132–158.

Finn, J. T. (1976), *J. Theor. Biol.*, **56**, 363–380.

Finn, J. T. (1980), *Ecology*, **61**, 562–571.

Fredrickson, A. G. (1977), *Annu. Rev. Microbiol.*, **31**, 63–89.

Fulton, R. S., III, and Paerl, H. W. (1987a), *Limnol. Oceanogr.*, **32**, 634–644.

Fulton, R. S. III, and Paerl, H. W. (1987b), *J. Plankton Res.*, **9**, 837–855.

Giese, G. L., Wilder, H. B., and Parker, G. G. (1979), *Hydrology of Major Estuaries and Sounds of North Carolina*, Rept. No. 79-46, U.S. Geological Survey, Water Resources Investigations, Raleigh, 175pp.

Hannon, B. (1973), *J. Theor. Biol.*, **41**, 535–546.

Heath, J. K. (1989), *The Decomposition of* Microcystis aeruginosa *in Estuarine Microcosms*, M.S. Thesis, East Carolina University, Greenville, NC, 115pp.

Hester, J. M. Jr. (1975), *Nekton Population Dynamics in the Albemarle Sound and Neuse River Estuaries*, M.S. Thesis, North Carolina State University, Raleigh.

Hobbie, J. E., and Smith, N. W. (1975), *Nutrients in the Neuse River Estuary*, Rept. No. UNC-SG 75-21, University of North Carolina Sea Grant Program, Raleigh, 183pp.

Kay, J. J., Graham, L. A., and Ulanowicz, R. E. (1989), in *Network Analysis in Marine Ecology: Methods and Applications* (F. Wulff, J. G. Field, and K. H. Mann, Eds.), Springer-Verlag, Berlin, pp. 15–61.

Koike, I., and Sørensen, J. (1988), in *Nitrogen Cycling in Coastal Marine Environments* (T. H. Blackburn and J. Sorensen, Eds.), Wiley, Chichester, pp. 251–274.

Kremer, J. N. (1989), in *Network Analysis in Marine Ecology: Methods and Applications* (F. Wulff, J. G. Field, and K. H. Mann, Eds.), Springer-Verlag, Berlin, pp. 119–131.

Kremer, J. N., and Nixon, S. W. (1978), *A Coastal Marine Ecosystem—Simulation and Analysis*, Springer-Verlag, New York, 217pp.

Kuenzler, E. J., Stanley, D. W., and Koenings, J. P. (1979), *Nutrient Kinetics of Phytoplankton in the Pamlico River, North Carolina*, Rept. No. 139, Water Resources Research Institute of the University of North Carolina, Raleigh, 163pp.

Kuenzler, E. J., Albert, D. B., Allgood, G. S., Cabaniss, S. E., and Wanat, C. G. (1984), *Benthic Nutrient Cycling in the Pamlico River*, Rept. No. 215, Water Resources Research Institute of the University of North Carolina, Raleigh, 148pp.

Mann, K. H., Field, J. G., and Wulff, F. (1989), in *Network Analysis in Marine Ecology: Methods and Applications* (F. Wulff, J. G. Field, and K. H. Mann, Eds.), Springer-Verlag, Berlin, pp. 259–282.

Matson, E. A., Brinson, M. M., Cahoon, D. D., and Davis, G. J. (1983), *Biogeochemistry of the Sediments of the Pamlico and Neuse River Estuaries, North Carolina*, Rept. No. 191, Water Resources Research Institute of the University of North Carolina, Raleigh, 103pp.

McCarthy, J. J., Kaplan, W. A., and Nevins, J. L. (1984), *Limnol. Oceanogr.*, **29**, 84–98.

Monod, J. (1949), *Annu. Rev. Microbiol.*, **3**, 371–394.

Odum, H. T. (1983), *Systems Ecology: An Introduction*, Wiley, New York, p. 8.

Paasche, E. (1988), in *Nitrogen Cycling in Coastal Marine Environments* (T. H. Blackburn and J. Sorensen, Eds.), Wiley, Chichester, pp. 33–58.

Paerl, H. W. (1987), *Dynamics of Blue-Green Algal* (Microcystis aeruginosa) *Blooms in the Lower Neuse River, North Carolina: Causative Factors and Potential Controls*, Rept. No. 229, Water Resources Research Institute of the University of North Carolina, Raleigh, 164pp.

Pahl-Wostl, C. (1990), *Oikos*, **58**, 293–305.

Redfield, A. C. (1958), *Am. Sci.*, **46**, 205–221.

Smith, S. V., and Hollibaugh, J. T. (1989), *Mar. Ecol. Prog. Ser.*, **52**, 103–109.

Stanley, D. W. (1983), *Nitrogen Cycling and Phytoplankton Growth in the Neuse River, North Carolina*, Rept. No. 204, Water Resources Research Institute of the University of North Carolina, Raleigh, 85pp.

Stanley, D. W. (1988), Proceedings North Carolina American Water Research Association, Symposium on Coastal Water Resources, TPS 88-1, AWRA, Bethesda, MD, pp. 155–164.

Stanley, D. W., and Hobbie, J. E. (1981), *Limnol. Oceanogr.*, **26**, 30–42.

Tempest, D. W. (1970), in *Methods in Microbiology* (J. R. Norris and D. W. Ribbons, Eds.), Vol. 2, Academic Press, London, pp. 259–276.

Thomann, R. W., and Fitzpatrick, J. J. (1982), *Calibration and Verification of a Mathematical Model of the Eutrophication of the Potomac Estuary*, Hydro Qual, Inc., Mahwah, NJ, 500pp.

Ulanowicz, R. E. (1980), *J. Theor. Biol.*, **85**, 223–245.

Ulanowicz, R. E. (1986), *Growth and Development: Ecosystems Phenomenology*, Springer-Verlag, New York, 203pp.

Ulanowicz, R.W. (1987), *NETWRK4: A Package of Computer Algorithms to Analyze Ecological Flow Networks*, University of Maryland, Chesapeake Biological Laboratory, Solomons.

Ulanowicz, R. W. (1988), *Ecol. Modelling*, **43**, 45–56.

Wanat, C. G. (1983), *Ammonium Production and Nitrification in the Sediments of the Pamlico River, North Carolina*, M.S.P.H. Technical Rept., Department of Environmental Sciences and Engineering, University of North Carolina, Chapel Hill, 90pp.

Webster, Merriam (1985), *Webster's Ninth New Collegiate Dictionary*, Merriam-Webster Inc., Springfield, MA, p. 628.

Wetzel, R. L., and Wiegert, R. G. (1983), in *Nitrogen in the Marine Environment* (E. J. Carpenter and D. G. Capone, Eds.), Academic Press, New York, pp. 869–892.

Wulff, F., Field, J. G., and Mann, K. H. (Eds.) (1989), *Network Analysis in Marine Ecology: Methods and Applications*, Springer-Verlag, Berlin, 284pp.

Wulff, F., Stigebrandt, A., and Rahm, L. (1990), *Ambio*, **19**, 126–133.

9 Modeling Carbon Utilization by Bacteria in Natural Water Systems

JOHN P. CONNOLLY
Environmental Engineering & Science Program, Manhattan College

RICHARD B. COFFIN
U.S. Environmental Protection Agency, Environmental Research Laboratory, Gulf Breeze, FL

ROBIN E. LANDECK
Environmental Engineering & Science Program, Manhattan College

1. Introduction 249
2. Equations describing bacterial growth and death 250
3. Analysis of bacteria–substrate interactions 253
   3.1. Laboratory substrate additions 253
       3.1.1. Yield coefficients 253
       3.1.2. Substrate utilization 257
   3.2. *In situ* natural substrates 258
       3.2.1. Operational definitions of substrate 258
       3.2.2. Phytoplankton-generated substrate 265
4. Application of the model to a microcosm study 269
5. Conclusions 272
   References 273

1. INTRODUCTION

A significant focus of aquatic microbial ecology has been the development of a quantitative understanding of bacterial dynamics in natural water systems. The impetus for much of the research into bacterial dynamics has been the finding that the microbial loop is an important component of aquatic food webs. The results of this research have provided a basis for addressing questions regarding the role of bacteria in biotechnology risk assessment and global climate.

With regard to biotechnology, the ability of a genetically engineered microorganism (GEM) released to the environment to survive and compete in a natural setting and possibly transfer the engineered trait to other organisms has significant implications with regard to the risk of adverse ecological or public health impacts. Survival and growth of a GEM is dependent on its response to various environmental factors, its ability to compete with the indigenous community for resources, and its vulnerability to predation. Thus an assessment of risk is predicated on a quantitative description of the dynamics of both the GEM and the indigenous community.

The role that bacteria play in oceanic carbon cycling may be a significant factor in the response of the global environment to increases in atmospheric carbon dioxide concentrations. Bacteria and the microheterotrophs that graze on them are believed to be critical components of carbon cycling in natural water systems. Together they process much of the carbon entering through primary production. The ultimate fate of that carbon depends on the rates at which it is metabolized at each trophic level and transferred between levels. Experimental studies have suggested that bacterioplankton return to the food chain carbon lost as dissolved organic carbon (Azam, et al., 1983; Williams, 1984), but also that the microbial food chain metabolizes a significant fraction of the carbon passing through it and thus also may be a carbon sink (Caron et al., 1985; Ducklow et al., 1986).

Both of the above areas require a means of predicting bacterial dynamics. It is thus necessary to develop mathematical models of the microbial food web. Simple models, ranging from empirical relationships between bacteria and phytoplankton (Bird and Kalf, 1984; Cole et al., 1988) to whole system mass balances of carbon flux (Coffin and Sharp, 1987) to steady-state analyses of the relationships between substrate, bacteria, and predators (Billen et al., 1980; Taylor and Joint, 1990; Wright et al., 1987), have been developed for various natural systems. However, while such models provide a basis for understanding relationships between the components of the microbial food web, they are not capable of addressing the major questions put forth by those concerned with biotechnology and global climate. As examples: What is the potential for a specific GEM to survive in a natural system? What effect would an increase in primary productivity have on the flux of carbon to the deep ocean? Such questions require a dynamic modeling framework capable of predicting the response of the microbial food web to changes in such variables as substrate quality and quantity and substrate and nutrient recycle rates. Development of such a framework has been hampered by an inability to define what constitutes substrate given the myriad of organic compounds present in a system. The purpose of this chapter is to present a review of bacterial growth kinetics in laboratory systems, to propose a methodology for defining bacterial substrate in a natural system, and to present the application of this methodology to predicting the dynamics of the microbial food web that have been observed in a microcosm study of a lake.

2. EQUATIONS DESCRIBING BACTERIAL GROWTH AND DEATH

Growth of bacteria and their uptake of substrate are classically described using the Monod equation. Death is determined by a nonpredatory loss rate (i.e., respiration

2. EQUATIONS DESCRIBING BACTERIAL GROWTH AND DEATH

and loss of viability) and a predation (and/or parasitism) rate. The equation describing these mechanisms is

$$\frac{dB}{dt} = \mu_m f(S, N, O_2)B - (k_d + FZ)B, \qquad (1)$$

where B = bacterial biomass concentration (μg C/L)
μ_m = maximum growth rate (day^{-1})
f = limitation factor
S = substrate carbon concentration (μg C/L)
N = concentration of additional, potentially limiting, nutrients (μg/L)
O_2 = oxygen concentration (mg O_2/L)
Z = predator biomass concentration (μg C/L)
k_d = nonpredatory loss rate (day^{-1})
F = filtration or clearance rate of the predator (L/μg C/day)

The limitation factor f accounts for the reduction in growth due to substrate carbon, nutrient, or oxygen limitation. For any one of these effects the limitation factor is given by the Monod expression. For substrate the equation is

$$f = \frac{S}{K_M + S}, \qquad (2)$$

where K_M = Michaelis half-saturation constant (μg/L).

Monod kinetics have been applied successfully in the analysis of bacterial productivity in aquatic ecosystems. For example, Billen et al. (1980) successfully predicted steady-state concentrations of several organic substrates in three sites representative of estuarine, coastal, and open ocean conditions by using Monod kinetics defined by measured yield coefficients and literature estimates of a constant maximum substrate utilization rate and half-saturation constants for each substrate.

The interaction of factors limiting growth has been the subject of considerable discussion. Models have been proposed in which growth can simultaneously be affected by several factors. In most cases the total limitation is assumed to be the product of the individual limitation factors given by Eq. (2) (Bader, 1978; McGee et al., 1972; Thingstad and Pengerud, 1985). Such multiplicative limitation has been demonstrated for carbon source and dissolved oxygen limitation of *Sphaerotilus natans* in continuous culture (Lau et al., 1984). An alternate formulation assumes that growth is controlled solely by the most limiting factor (Ryder and Sinclair, 1972; Sykes, 1972). This noninteractive limitation has been demonstrated for nutrient control of algal growth (Ahlgren, 1980; Rhee, 1978). A theoretical analysis using a two-substrate enzyme model suggests that both interactive and noninteractive limitation are reasonable approaches depending on the substrates involved (Bader, 1978). In the analyses presented here, substrate carbon concentration is assumed to be the sole factor affecting bacterial growth.

The concentrations of carbon and other nutrients are controlled by the rate at which they enter the system from external sources, the rate at which they are produced in the system through algal excretion, grazing, and solubilization of particulate organic material (i.e., internal sources), the rate at which they are transported through the system by advection and dispersion, and the rate at which they are taken up by the bacteria. Uptake by the bacteria is defined by the growth rate and a carbon or nutrient yield coefficient (μg biomass carbon or nutrient produced per μg carbon or nutrient taken up):

$$\frac{dS}{dt} = -\frac{\mu}{Y} B, \qquad (3)$$

where μ = net growth rate = $\mu_m f(S, N, O_2)$
Y = yield coefficient

The yield coefficient defines the ratio of growth to substrate uptake. Its value is dependent on the substrate source and, to some extent, on the cell stoichiometry and energetics of the bacteria; varying as the nutritional and physiological status of the cell changes.

The predation term included in Eq. (1) defines the loss of bacterial carbon to the next trophic level. Models of zooplankton predation have generally described grazing using a Monod formulation (Canale et al., 1973; Fenchel, 1980; Thingstad and Pengerud, 1985; Williams, 1980; Wright, 1988). The equation defining the grazer biomass (Z) is

$$\frac{dZ}{dt} = \alpha G_m \frac{B}{K_z + B} Z - (R + k_{dz})Z, \qquad (4)$$

where α = assimilation efficiency of the predator
G_m = maximum grazing rate (μg C consumed/μg C of grazer body mass/day)
K_z = half-saturation constant for grazing rate limitation (μg C/L)
R = respiration rate (μg C respired/μg C of grazer body mass/day)
k_{dz} = death rate of the grazers (day^{-1})

The assimilation efficiency defines the fraction of grazed bacteria that is converted to biomass. The remaining fraction is excreted as particulate or dissolved organic matter. The filtration or clearance rate term of Eq. (1) is defined from Eq. (4) as

$$F = \frac{G_m}{K_z + B}. \qquad (5)$$

3. ANALYSIS OF BACTERIA-SUBSTRATE INTERACTIONS

3.1. Laboratory Substrate Additions

3.1.1. Yield Coefficients. The rates of carbon, nitrogen, and phosphorus utilization by bacteria for biosynthesis are defined by the rate of bacterial biomass production and the stoichiometric balance set by the physiological state of the cells. Because carbon may be used catabolically, growth rate and cell stoichiometry alone are not sufficient to define the rate of carbon uptake. An empirical constant, termed the growth yield coefficient [presented as yield coefficient Y in Eq. (3)], is used to relate growth and carbon utilization. In contrast, the net rates of uptake of nitrogen and phosphorus may be defined from the rate of increase of biomass carbon and stoichiometric ratios between carbon and nitrogen and phosphorus. The absolute rates of uptake of nitrogen and phosphorus may be greater than defined by cell stoichiometry, depending on the ratios of carbon to nitrogen and carbon to phosphorus in the substrate. However, excess nitrogen and phosphorus in the substrate are released back to the environment as ammonium or phosphate (Billen, 1984; Goldman et al., 1987) so that net uptake is defined by stoichiometry.

We have compiled several hundred carbon growth yield coefficient values encompassing more than 70 compounds. Analysis of the data has indicated that values are relatively constant for any one substrate. Significant differences exist among substrates, although substrates of similar structure appear to have similar yields. The highest yields (~ 0.8) are observed for the amino acids. Alcohols, alkanes, and sugars all have similar yields (~ 0.6). Yields on dissolved organic carbon (DOC) and particulate organic carbon (POC) from plant materials are generally much lower (Fig. 9.1). Consistent with the work of Linton and Stephenson (1978) we found that growth yields are linearly related to the heat of combustion for substrates with values less than about 11 kcal/g C (Fig. 9.2). Above this value the yields are approximately constant. This relationship between yield coefficient and heat of combustion is related to the energy content of the substrate. Gommers et al. (1988) have shown, theoretically, that at high heats of combustion the substrates are completely assimilated and growth yield is defined by obligatory CO_2 production within the metabolic pathways. At lower energy contents a portion of the substrate must be shunted to catabolic pathways to satisfy energy needs.

Recent studies have indicated that the stoichiometric ratio that defines nitrogen utilization is approximately constant. Goldman et al. (1987) found that the elemental composition of natural marine bacteria assemblages cultured on combinations of C and N sources spanning a range of substrate C:N ratios from 1.5:1 to 10:1 was relatively invariant at a C:N:P ratio of about 45:9:1. In these experiments phosphorus was never limiting, so that the constancy of the phosphorus cell stoichiometry was not fully tested. Billen's (1984) analysis of ammonia regeneration data from studies using natural seawater supplemented with various organic substrates indicated that the ratio of biomass nitrogen produced to carbon used was nearly constant at about 0.1 g N/g C. This result suggests that either yield and stoichiometry covaried identically in these studies or, more likely, that the cell stoichiometry was constant.

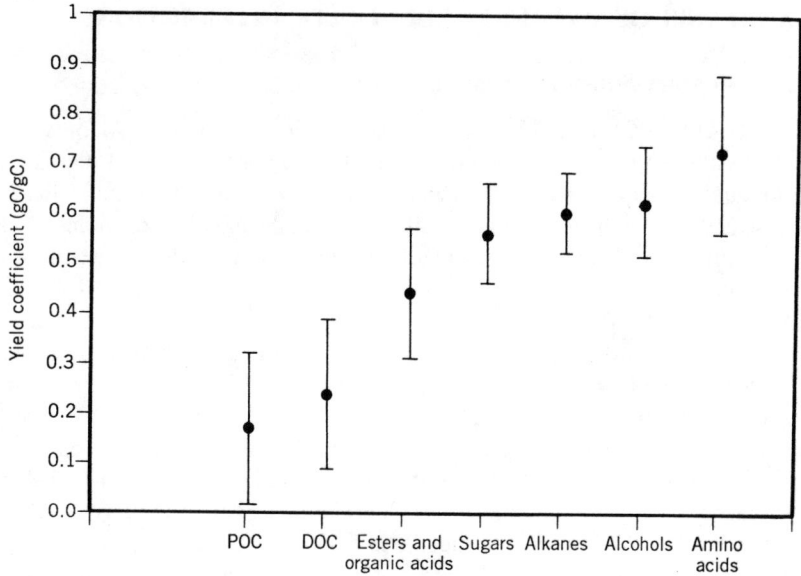

FIGURE 9.1. Mean ± standard deviation of yield coefficients for bacterial growth on various classes of substrate.

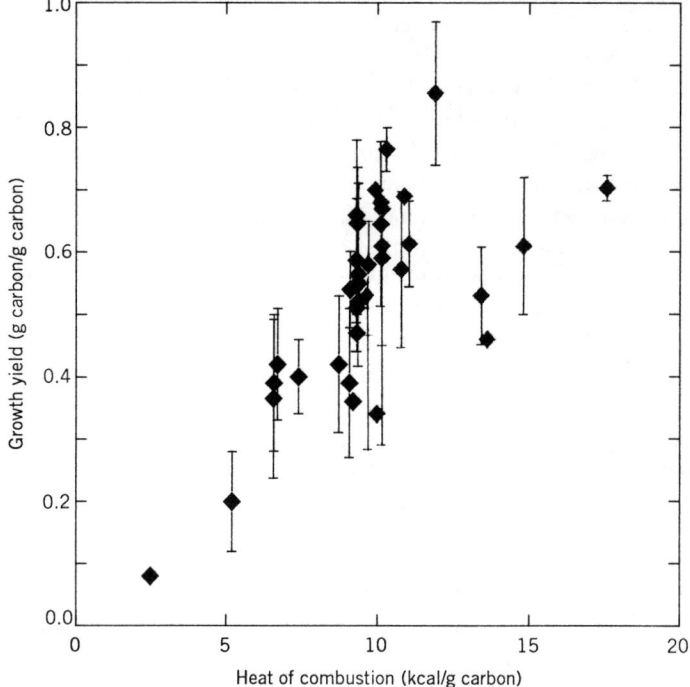

FIGURE 9.2. Mean ± standard deviation of yield coefficients for bacterial growth on individual substrates in relation to the heat of combustion of the substrate.

In an analysis of chemostat data published by Pengerud et al. (1987) we found some variation in cellular phosphorus content. In this study of bacterial growth, eight parallel chemostats were arranged along a gradient of glucose concentration that forced orthophosphate limitation at the higher glucose levels and carbon limitation at the lower glucose levels. The authors reported evidence of a changing biomass composition and/or a changing respiration coefficient. Our analysis of their data indicates that the cellular phosphorus content was approximately constant at 4.5×10^{-10} μM-P/cell in the region of glucose limitation (substrate C:P < 140), decreased in a range where both glucose and phosphorus are completely consumed, and was constant in the region of phosphorus limitation (substrate C:P > 250) at about 2.4×10^{-10} μM-P/cell. Carbon growth yield increased from 1.1×10^7 cells/μM-C in the region of phosphorus limitation to 1.6×10^7 cells/μM-C in the region of carbon limitation, indicating a decline in yield with increasing substrate C:P. This decline is consistent with the growth yield–substrate C:N relationship reported by Goldman et al. (1987) in which yield coefficient declined from about 0.8–0.9 at substrate C:N of 1.5:1 to about 0.4–0.5 at substrate C:N > 6:1. Goldman and co-workers attributed this phenomenon to an increase in conservation of carbon in proportion to the degree of carbon limitation. Where carbon is not the limiting constituent, a threshold growth yield is reached.

Changes in cell stoichiometry can be accommodated by considering intracellular carbon and nutrient pools whose concentrations are dependent on the growth rate of the bacteria and uptake from the environment. The inclusion of intracellular pools allows the computation of a variable cell stoichiometry and thus a variable cell yield. Such a formulation has been proposed for algal growth (Droop, 1974) and for bacterial growth (Thingstad and Pengerud, 1985). In addition to computing variable stoichiometry, this formulation computes growth limitation using the intracellular concentration of carbon or nutrient. Nyholm (1976) presents data indicating that in phosphate-magnesium-, and potassium-limited chemostats, the specific growth rate of bacteria is approximately linearly related to the intracellular concentration of the limiting substrate.

The necessity to consider variable stoichiometry and intracellular substrate concentration in the control of growth depends on their ecological significance. In addition, the ability to apply this more complicated model formulation must be considered. Modeling an intracellular pool requires the determination of two additional coefficients relative to the Monod type formulation: a minimum cell quota of the constituent and a maximum specific uptake rate of the constituent. These coefficients are difficult to quantify and thus add two additional degrees of freedom that may add uncertainty to the computation.

Di Toro (1980) has shown that, for algae, nutrient uptake kinetics are much faster than the population growth kinetics and the intracellular pool is essentially in equilibrium with the extracellular pool. This justifies, at least for algae, the use of extracellular concentration in computing growth limitation. The issue of cell stoichiometry may be of more importance since it affects the computation of the maximum biomass that can be supported by a given substrate input. However, in this regard it may be sufficient to assume a minimum stoichiometry based on the yield coefficient at limitation. As an example, we computed the phosphorus, glucose, and bacteria concentrations for the chemostat experiments of Pengerud et al. (1987) mentioned above using the Monod equations presented above and the minimum phosphorus cell stoichiometry. The results are shown in Fig. 9.3. The model computes dissolved phosphorus concentrations that are higher than observed when glucose limitation is apparent. This is a consequence of the use of the minimum

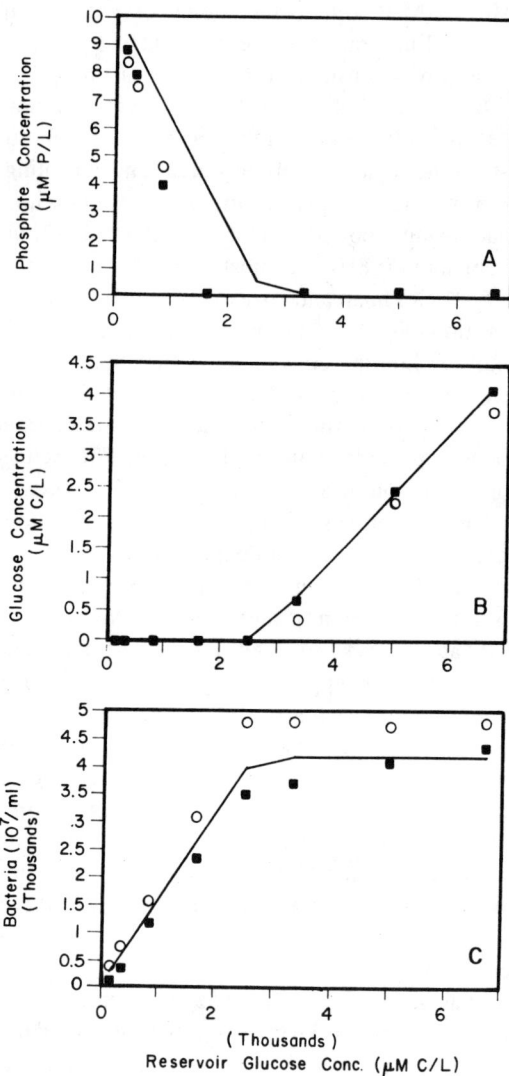

FIGURE 9.3. Comparison of phosphate, glucose, and bacterial abundance measured in a chemostat study (Pengerud et al., 1987) with values computed using the Monod model. Open and closed symbols designate measurements from replicate chemostats. The model prediction is shown by the line.

phosphorus cell stoichiometry rather than the higher value appropriate to this region. The calculation does, however, accurately predict the bacteria concentration in this region because growth is controlled by glucose, whose yield coefficient is correctly represented. In the region of phosphorus limitation the model also accurately predicts the bacteria since the correct phosphorus stoichiometry is used.

3.1.2. Substrate Utilization. Substrate utilization is controlled by the half-saturation constant defining growth limitation of the bacteria. Published half-saturation constants for individual substrates provide a guide for the establishment of half-saturation constants for empirically defined natural substrates and a substantial database exists in the literature. However, only a fraction of these data are relevant to natural communities. Much of the published data are for pure cultures or for mixed cultures and natural populations studied in chemostats. Pure culture data are difficult to interpret because studies have shown that the growth characteristics of a population change as a function of the enrichment technique used to isolate and maintain the organism (Jannasch, 1968; Powell, 1967). This generalization is supported by the observation that, for a given species, maximum growth rate and half-saturation constant are positively correlated (Jannasch, 1965). Mixed populations in chemostats undergo selective processes and do not reflect the original population. In addition, growth characteristics of pure cultures in chemostats appear to be dependent on the maximum growth rate of the organism and the dilution rate of the chemostat (Jannasch, 1967).

Batch incubations with natural water samples appear to provide the most ecologically relevant estimates of maximum growth rate and half-saturation constant. Simkins and Alexander (1985) have demonstrated that the batch experiment provides a robust estimate of these parameters, which is independent of initial substrate concentration and population density.

A further complication in using published kinetic parameters to define appropriate ranges is the observation that half-saturation values appear to be dependent on environment type. Vaccaro and Jannasch (1966) found that glucose half-saturation values determined in Atlantic Ocean water off South America decreased from roughly 100 μg C/L near the coast to 10 μg/L in the open ocean. Billen (1984) compiled published acetate half-saturation constants and natural utilization rates and showed a positive relationship between these parameters that correlated with environment type. Marine system half-saturation constant values range from 25 to 120 μg C/L in open ocean water and 120 to 240 μg C/L in coastal waters; thus they are similar to the glucose values reported by Vaccaro and Jannasch. Higher values are seen for inland waters, sediments, and activated sludge with values increasing in relation to the expected carbon concentrations in these systems.

Half-saturation constants reported for various amino acids may also be environment specific. Although Billen (1984) did not find a strong relationship for alanine, half-saturation constants plus *in situ* amino acid concentrations [i.e., $K_M + S$ in Eq. (2)] range from about 0.1 μg C/L in eastern Long Island Sound and New York Bight (Fuhrman and Ferguson, 1986), where DOC concentrations are about

1–2 mg/L to about 1–10 μg/L in several lakes and coastal marine sites with DOC concentrations of 4–7 mg/L (Jorgensen and Sondergaard, 1984). These values are about a factor of 100 lower than those for glucose and acetate.

3.2. *In Situ* Natural Substrates

Natural assemblages of bacteria are exposed to a myriad of organic carbon compounds that may serve as substrate for growth. The specific compounds used by bacteria will depend on their rate of production and the efficiency with which the bacteria can metabolize them. Obviously, compounds for which constitutive enzyme systems exist would most likely serve as substrate. Based on the laboratory experiments with single substrates discussed above, amino acids may be classified as "optimum" substrate because of their high yield and low half-saturation constants. Conversely, substrates that support low bacterial growth yields, and for which half-saturation constants are high, are less satisfactory. The relative importance of varying quality substrates depends on their availability.

3.2.1. Operational Definitions of Substrate.
To quantify substrate and its utilization in a natural water system, we have assumed that DOC can be separated into three components: two classifications of labile substrate and a refractory or slowly degraded component. The first classification of labile substrate, L1, is presumed to be representative of simple compounds that are readily used by bacteria. The second classification of labile substrate, L2, presumably represents more complex compounds that are also readily used but that are less optimal for bacterial growth. Finally, refractory substrate R represents compounds that are not normally degraded by bacterioplankton in aquatic ecosystems. Under such a scheme all inputs of DOC are subdivided into these three components. Presumably, fractionation of any DOC source into the components could be accomplished using values defined by experiment. Growth kinetics for the two labile components are then required. To formulate these kinetics we have further assumed that growth and substrate utilization can be described using the Monod model and that L2 substrate is only used after the L1 substrate has been exhausted (i.e., not until growth limitation using L1 substrate exceeds growth limitation using L2 substrate). A similar approach has successfully been used to describe carbon mineralization in sediments by dividing carbon into "G" classes based on lability (Berner, 1980).

Evidence of a split between substrate components may be seen in 20°C biochemical oxygen demand (BOD) assays of natural water samples. Figure 9.4 shows the cumulative oxygen utilization (BOD) in filtered and unfiltered water samples from Santa Rosa Sound, Florida. Data are presented for samples collected on 2/7/89 (Fig. 9.4a) and 4/10/89 (Fig. 9.4b). The sample sets have similar long-term or ultimate BOD values, suggesting similar levels of degradable substrate. However, significant differences are evident in the pattern of oxygen consumption over time. In the February sample a rapid oxygen utilization occurs over the first few days and the ultimate BOD is reached by about 15–20 days. In the April

3. ANALYSIS OF BACTERIA–SUBSTRATE INTERACTIONS 259

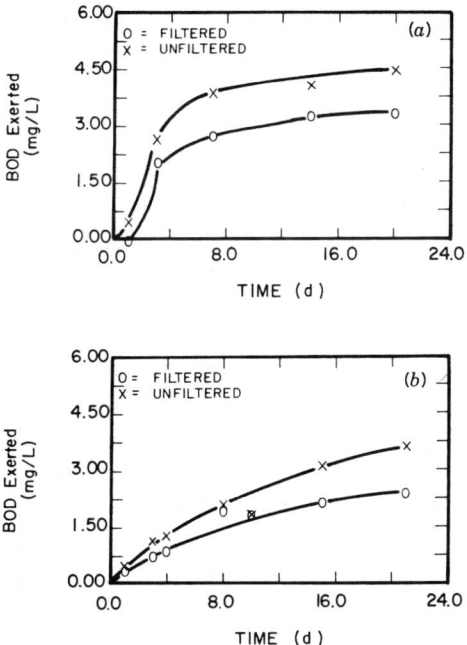

FIGURE 9.4. BOD observed in water samples taken from Santa Rosa Sound on (*a*) February 7, 1989 and (*b*) April 10, 1989.

sample the oxygen utilization rate is much slower, with little evidence of an early rapid utilization and no attainment of the ultimate BOD during the experiment. These data suggest that significant differences existed in the quality of the substrate present on the two sampling dates. Apparently, much more of the substrate in the February sample was readily available for bacterial growth.

Data from a BOD assay of water taken from an enclosed salt marsh on Santa Rosa Sound (Range Point) are presented in Fig. 9.5. In this study data are available for both BOD and bacterial abundance (Acridine Orange Direct Counts). The BOD data indicate an initial rapid oxygen consumption that coincides with an increase in bacterial numbers. Note that the bacterial abundance in the filtered sample (which was inoculated with 1% unfiltered water) does not approach that of the unfiltered sample. After the first day the bacterial numbers begin to decrease and the oxygen utilization rate decreases. These data suggest that the bacteria were growing on a very labile substrate that was depleted within the first day of the experiment.

To estimate parameter values of a Monod model considering two substrate components, we have analyzed published studies of bacterial utilization of substrates that may be classified as either L1 or L2. A batch reactor study of glucose utilization by aerobic heterotrophs (Seto and Alexander, 1985) provides an example of L1 substrate utilization. In their study the uptake of radiolabeled glucose was monitored by measuring particulate and dissolved radiolabeled carbon at various

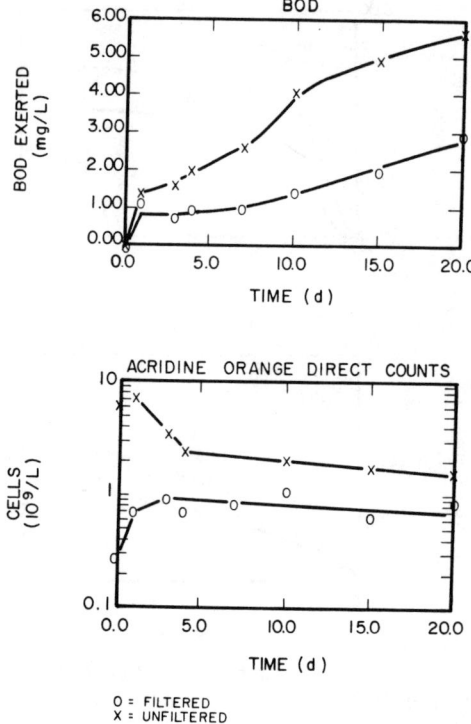

FIGURE 9.5. BOD and bacterial abundance in a water sample taken from Range Point salt marsh.

times. Figure 9.6 presents the Seto–Alexander data and computed glucose and bacterial carbon concentrations. Similarly, to quantify uptake on a less labile (i.e., L2) carbon, we analyzed a study of bacterial degradation of secondary treatment plant effluent (Stamer et al., 1979). In this study the bacterial biomass was not measured but was inferred from measurements of dissolved (DOC) and total organic carbon (TOC). Cumulative oxygen utilization (BOD) was used in conjunction with the organic carbon measurements to determine the growth and substrate kinetics defined by Eqs. (1)–(3). Figure 9.7 presents observed and computed BOD, DOC, and TOC concentrations. The computed TOC is the sum of computed DOC and bacterial carbon. Note that TOC concentrations follow the data trend but are consistently lower than the data, suggesting the presence of a relatively nondegradable particulate organic carbon component that is not included in the computed TOC.

The parameter values that produced the computed values shown in Figs. 9.6 and 9.7 are presented in Table 9.1. Significant differences in bacterial utilization of the two carbon sources are evident. As expected, maximum growth rate and

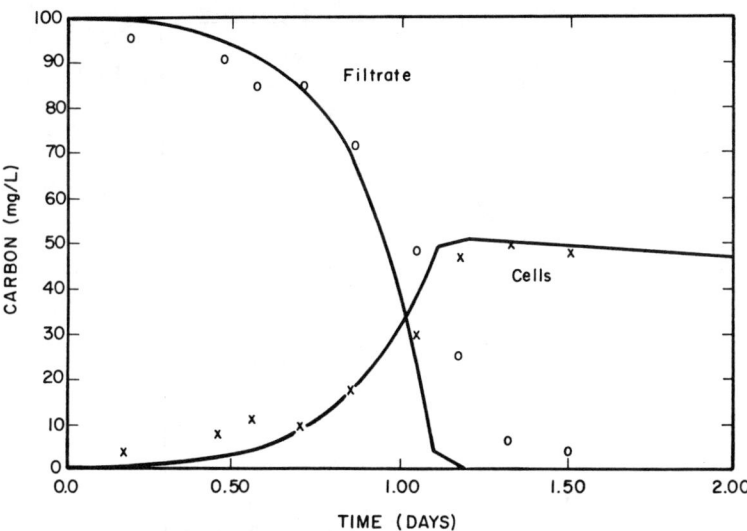

FIGURE 9.6. Comparison of particulate (cells) and dissolved (filtrate) carbon measured in a batch [^{14}C]-glucose uptake study (Seto and Alexander, 1985) with values computed using the Monod model. The symbols designate measured data and the line designates the model results.

growth yield are much higher for glucose than for secondary effluent. The glucose growth yield of 0.52 is consistent with the values reported in Section 3.1.1. The half-saturation constants are also much different, with the glucose value being about two orders of magnitude less and about equal to values reported for coastal waters. Respiration rates are about the same, reflecting the bacterial rate of endogenous metabolism. These values provide some guidance for the application of this model to carbon sources that contain L1 and L2 components.

We have applied the model to the April 1989 Santa Rosa Sound BOD data presented in Fig. 9.4b by dividing the total degradable substrate calculated from the ultimate BOD into L1 and L2 components. The division between components may be set by visually estimating the BOD at the break in oxygen utilization presumed representative of depletion of the most labile substrate. For this experiment, the BOD at day 1 was used as the break point. Both the filtered and unfiltered data were fit with the same parameter values. The computed substrate carbon and the comparison of the observed and computed BOD and bacterial carbon are presented in Fig. 9.8. Bacterial carbon was computed from bacterial abundance using a conversion factor of 10^{-10} mg C/cell. Measurements of bacterial volumetric carbon contents and bacterial biovolume yield conversion factors that range from about 10^{-11} to 2×10^{-10} mg C/cell. This factor is essentially an additional model coefficient that affects the shape of the computed BOD profile by increasing or decreasing the rates of bacterial production and carbon mineralization necessary

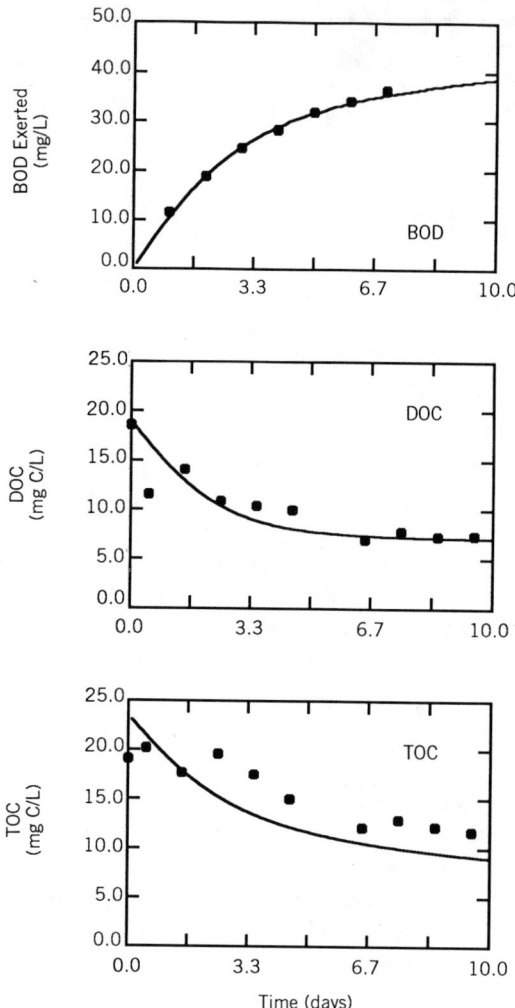

FIGURE 9.7. Comparison of BOD, DOC, and TOC measured in a BOD assay study of municipal secondary effluent (Stamer et al., 1979) with values computed using the Monod model. The symbols designate measured data and the line designates the model results.

to fit the observed data. The value used was determined such that the coefficient set resulted in a computation that was consistent with the observed BOD data and with the observed range of values for each coefficient.

The coefficient values used in the Santa Rosa Sound model are presented in Table 1. The maximum growth rate of 4/day and half-saturation constant of 300 μg C/L for L1 carbon are similar to the values obtained for glucose. However, the growth yield is 0.3 compared to a yield on glucose of 0.52. This lower

3. ANALYSIS OF BACTERIA–SUBSTRATE INTERACTIONS

TABLE 9.1. Bacterial Growth and Death Parameter Values Used in the Models

Parameter	Glucose	Secondary Effluent	Santa Rosa Sound BOD Assay		Lake Bagsvaard Microcosm Study	
			L1 Carbon	L2 Carbon	L1 Carbon	L2 Carbon
Maximum growth rate (day^{-1})	4.5	0.65	4.0	0.8	4.0	2.0
Growth yield	0.52	0.25	0.30	0.20	0.50	0.20
Half-saturation constant (mg C/L)	0.20	18.0	0.30	5.0	0.20	2.0
Respiration rate (day^{-1})	0.10	0.15		0.07		0.10

yield is consistent with the general belief that bacterioplankton in marine waters are nitrogen limited and with the observation that the carbon yield coefficient is low under such conditions (Goldman et al., 1987; Hopkinson et al., 1989). It also may suggest that detrital material constituted the primary carbon source in this system since lower growth yields have been observed on this type of substrate (Linley et al., 1983). The L2 carbon maximum growth rate of 0.8/day, growth yield of 0.2, and half-saturation constant of 5 mg C/L are close to the values obtained for secondary effluent.

To further test the model we used the same coefficient values and computed oxygen consumption and bacterial growth for the February 1989 experiment. We assumed that all the oxidizable carbon in this experiment was L1. The comparison of observed and computed BOD is presented in Fig. 9.9. The model fits the data fairly well.

The data and analyses presented here suggest that defining oxidizable substrate as L1 and L2 components provides a basis for quantifying bacterial growth and substrate utilization. They further suggest that oxygen utilization may be used as a measure of these substrate components. Finally, they confirm the utility of Monod kinetics in the simulation of bacteria and substrate dynamics.

Our preliminary work suggests that the labile substrate pool is small and exhausted rapidly by bacteria (see Fig. 9.5). Therefore to examine the dynamics of the L1 pool and bacterial growth in response to organic matter, measurements of BOD and bacterial abundance must be on the scale of hours or even minutes rather than days. Figure 9.10a presents a 24-h experiment for which BOD and bacterial abundance are measured on August 7, 1990 from the dock at Gulf Breeze Environmental Research Laboratory (GBERL) over hourly intervals. For this and similar experiments the results suggest that the most rapid changes in bacterial abundance and oxygen concentrations are observed within 6 h. Figure 9.10b presents a similar experiment from November 15, 1990 with water from Range Point, in which bacteria and oxygen were monitored for 5 h. For this experiment bacteria

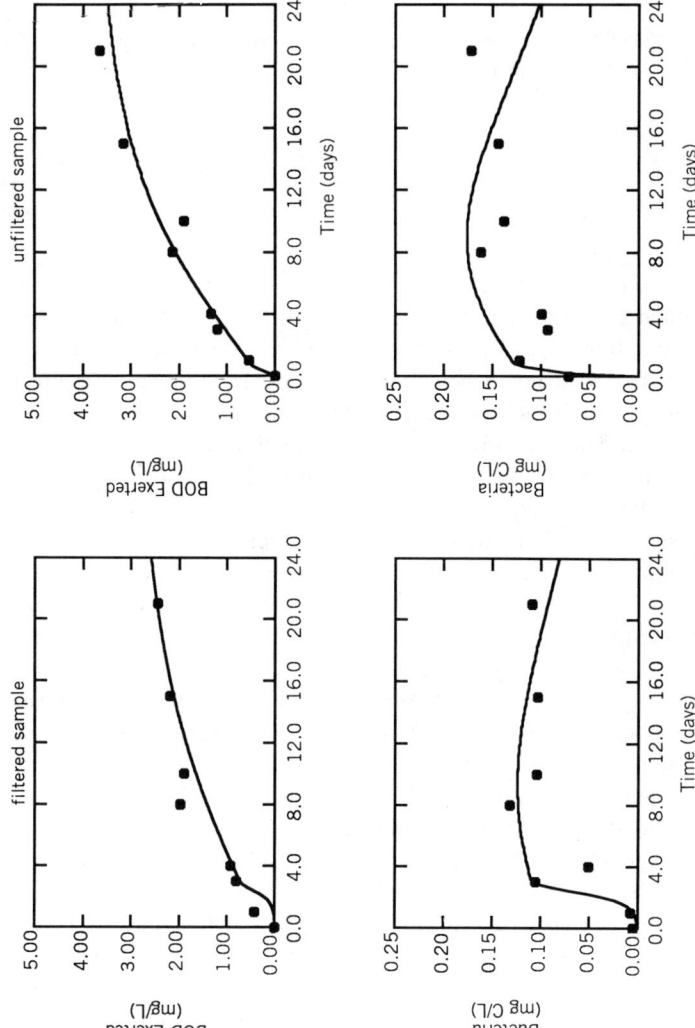

FIGURE 9.8. Comparison of BOD and bacterial abundance measured in BOD assays of filtered and unfiltered water samples taken from Santa Rosa Sound on April 10, 1989 with values computed from the L1–L2 substrate model. The symbols designate measured data and the line designates the model results.

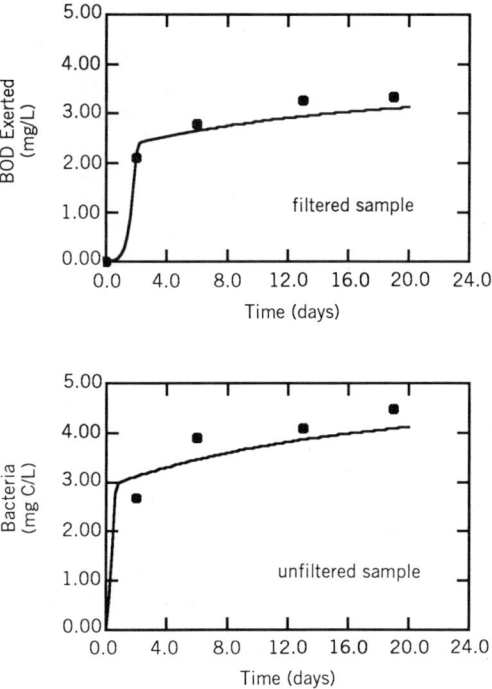

FIGURE 9.9. Comparison of BODs measured in BOD assays of filtered and unfiltered water taken from Santa Rosa Sound on April 10, 1989 with values computed from the L1–L2 substrate model. The symbols designate measured data and the line designates the model results.

and oxygen vary significantly over this short time scale. We have compiled data from similar experiments on waters from GBERL and Range Point from March 1990 to January 1991. Bacterial growth and oxygen depletion rates were measured from these experiments by taking the linear segment of the change in bacteria or oxygen through time. For each experiment, three to eight points were analyzed over 2–6 h. From this analysis bacterial growth covaries significantly with the BOD that is measured over the same time scale (Fig. 9.11). The slope of the line predicts that 0.2 mg of O_2 are required for every 1×10^9 bacteria that are produced. These data further support the use of oxygen as a tool for understanding microbial interaction with substrate and suggest that oxygen utilization may be used to study bacterial production.

3.2.2. Phytoplankton-Generated Substrate. In estuarine, coastal, and open ocean systems and in many lakes, phytoplankton are the dominant source of bacterial substrate. Phytoplankton organic matter becomes available to bacteria through extracellular release from healthy cells, cell senescence and lysis, and "sloppy"

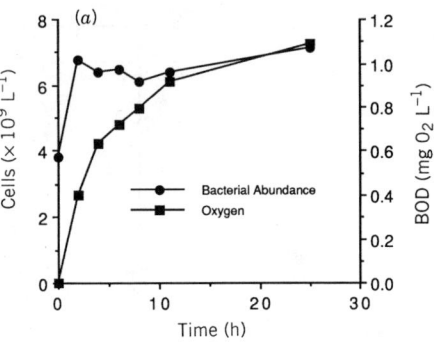
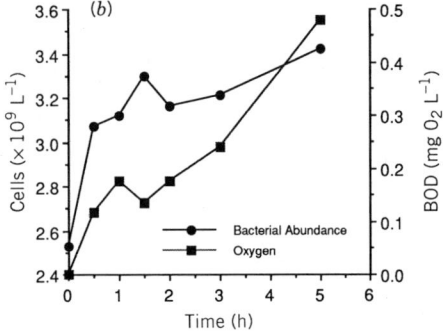

FIGURE 9.10. Bacterial abundance and BOD measured in BOD assays. (*a*) BOD bottle study of an August 7, 1990 water sample from the dock at GBERL. Each data point is an average of triplicate bottles sacrificed at the appropriate time. (*b*) The 3-L intravenous bag study of a November 15, 1990 water sample from Range Point. Each data point is an average of samples drawn from the three bags. In both studies bacteria were measured by AODC and oxygen concentrations by micro-Winkler titrations.

feeding and excretion by zooplankton. Each of these processes will contribute differently to the various substrate pools.

Organic material from extracellular release by phytoplankton (generally referred to as extracellular organic carbon or EOC) is primarily L1 substrate. A number of studies have indicated that EOC is a major carbon source for planktonic bacteria (Larsson and Hagstrom, 1982; Riemann and Sondergaard, 1984; Robarts and Sephton, 1989). Experiments have indicated that bacteria readily utilize 70–100% of this material (Chrzanowski and Hubbard, 1989; Iturriaga and Hoppe, 1977; Wiebe and Smith, 1977; Wolter, 1982). Uptake rates as high as 0.25 per hour have been measured for particular size fractions of this material (Lancelot, 1984). Bacterial growth yields on EOC range from about 0.45 to 0.70 (Bell and Sakshaug, 1980; Chrost and Faust, 1983; Jensen, 1985). These values are higher than those reported for growth on natural water DOC, which average about 0.20 (Fig. 1) (Bell and Kuparinen, 1984; Bjornsen, 1986). This difference suggests the importance of other carbon sources that would likely be categorized as L2 substrate.

The rate of production of EOC is a much disputed issue. Expressed as a percentage of primary production, reported EOC rates vary from a few percent to as high as 70% (Table 9.2). This variability may be related to various factors associated with the water system, the phytoplankton community composition, and the

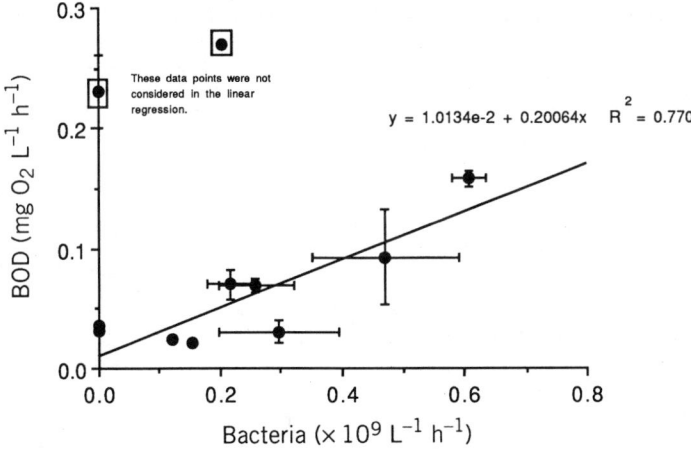

FIGURE 9.11. Linear regression of bacterial production on BOD from water samples that were incubated in BOD bottles or intravenous bags from GBERL or Range Point, FL between March 7, 1990 and January 15, 1991. Error bars represent the standard deviations of triplicate samples.

assumptions underlying the experimental design (Chrzanowski and Hubbard, 1989). Specifically, the phase of the algal bloom, stresses imposed on the algal populations, the time course of the kinetics, population density, nutrient deficiencies, and temperature may affect the amount of exudate released (Bell and Sakshaug, 1980; Fogg, 1983).

Phytoplankton death and subsequent cell lysis have been found to result in a rapid release of DOC (Cole et al., 1984; Hansen et al., 1986). Hansen et al. (1986) found that within the first 24 h after death from 11 to 43% of the ^{14}C-labeled cell content was released as dissolved organic carbon. The highest percentages were for lake water samples dominated by cyanobacteria. These studies found that the initial lysis products were rapidly taken up by bacteria at a growth efficiency of about 60%. Longer term incubations indicate that the remaining phytoplankton carbon is released much more slowly at a rate that is temperature dependent (Cole et al., 1984). This carbon also appears to be rapidly taken up by bacteria. DOC data from long-term (8-day) incubations of the diatom *Phaeodactylum tricornutum* in the dark (Caron et al., 1985) indicate that bacteria take up the carbon at about the rate it is released.

Grazing may also contribute to substrate production through damage to prey cells ("sloppy feeding") and leaching of DOC from fecal material. For example, Riemann et al. (1986) found that dissolved free amino acids were released during herbivorous grazing at rates as high as 6–12% of calculated ingestion rates. In experiments with *Daphnia pulex*, Lambert (1978) measured DOC release rates of 10–17% of the ingested carbon from large diatoms and 4% from a small green

TABLE 9.2. Literature Survey of Algal Exudation as Percentage of Primary Production

Reference	Percentage
Hellebust (1965) as cited in Watson (1978)	4–38
Fogg (1966) as cited in Sharp (1977)	5–35
Parsons and Seiki (1970) as cited in Watson (1978)	15
Ignatiades (1973) as cited in Larsson and Hagstrom (1982)	<5
Bell, Lang, and Mitchell (1974)	7
Wetzel (1975) as cited in Lampert (1978)	>10
Al-Hasan and Coughlan (1976) as cited in Fogg (1983)	34
Iturriaga and Hoppe (1977) as cited in Larsson and Hagstrom (1979)	12–31
Sharp (1977)	0–5
Lee and Nalewajko (1978) as cited in Larsson and Hagstrom (1982)	<5
Lancelot (1979) as cited in Fogg (1983)	20–60
Larsson and Hagstrom (1979)	45
Bell (1980)	7–10 and 10
Mague et al. (1980)	5–10
Joiris et al. (1982) as cited in Azam et al. (1983)	20–70
Larsson and Hagstrom (1982)	12–16
Wolter (1982)	5.1–12.5, 10–40, 20, and 27.6
Fogg (1983)	5–40
Shailaja and Pant (1986)	44
Robarts (1988)	3.6–7.3
Chrzanowski and Hubbard (1989)	30.8
Pett (1989)	22
Robarts and Sephton (1989)	8.1
Zlotnik and Dubinsky (1989)	1–55

algae that was swallowed whole. Using incubations of diatom cells in the dark, Caron et al. (1985) found that about 10% of the POC ingested by microflagellates was released as DOC and another 10% was released as egested POC. Egested POC or fecal material will release DOC due to diffusion of included DOC and solubilization of the particulate material, either through the activity of attached bacteria or further ingestion and digestion by zooplankton. Diffusion of any included DOC will occur on the order of minutes (Jumars et al., 1989), whereas solubilization will occur more slowly, although still relatively rapidly. Observations of sinking particles in the ocean indicate that most particulate organic material is decomposed within the upper few hundred meters of the water column (Cho and Azam, 1988; Karl et al., 1988) at rates as high as 0.2–0.3/d (Banse, 1990). A significant fraction of this POC appears to be mineralized through zooplankton respiration (Banse, 1990; Caron et al., 1985); however, the available data are insufficient to provide a clear picture of the fate of this carbon.

4. APPLICATION OF THE MODEL TO A MICROCOSM STUDY

The bacterial growth and death equations presented in Section 2 coupled with the L1-L2 substrate categorization have been applied to a microcosm study of nutrient and carbon cycling in eutrophic Lake Bagsvaard, Denmark (Landeck, 1990). This 24-day study (Kroer et al., 1991) included measurements made in triplicate of the abundance of heterotrophic bacteria, phytoplankton, and bacterial grazers; primary and bacterial production; and ammonia nitrogen and orthophosphate phosphorus concentrations. Additionally, light intensity, light-dark cycle times, and temperature were recorded.

The equations presented in Section 2 were combined with equations describing phytoplankton, nutrients, and zooplankton (see Landeck, 1990). The kinetics of phytoplankton production and interaction with nutrients were based on the formulations used in eutrophication models of the Great Lakes (Di Toro and Connolly, 1980). The bacteria and their interactions with carbon substrates and higher trophic levels were coupled to the phytoplankton kinetics through substrate production by the phytoplankton. Because the phytoplankton population of Lake Bagsvaard during the period of the microcosm study (August 9 to September 1, 1989) was dominated by cyanobacteria, we assumed that all phytoplankton nitrogen requirements were supplied by fixation of N_2. The consequence of this assumption is that phytoplankton growth was not nitrogen limited. Note that this modeling approach agrees with formulations used previously to simulate nitrogen fixation by blue-green algae (Bierman et al., 1980; Canale et al., 1976; Scavia et al., 1976; Tetra Tech, Inc., 1979).

The model did not include bacterial utilization of nitrogen and phosphorus. Nutrient recycling was modeled as in previous eutrophication models, that is, a simple temperature-dependent rate constant proportional to the phytoplankton biomass. We recognize that bacteria control mineralization and in some environments may compete with the phytoplankton for nutrients. However, nutrients were never limiting in this microcosm experiment and this refinement of the model was not necessary.

Primary production was assumed to be the only source of substrate carbon for the bacteria. Substrate carbon was generated through exudation, senescence, and lysis of the phototrophic assemblage, grazed but unassimilated carbon, and solubilization of particulate organic carbon. Exudation was fixed at 10% of primary production, with 80% of this carbon routed to the L1 pool and 20% to the L2 pool. Carbon from senescence and lysis of the algae and grazing was routed to the DOC and POC pools in proportion to a presumed carbon composition of the biota and the assimilation efficiency of the grazer (see Landeck, 1990). The values of the parameters defining bacterial growth and utilization of the carbon were based on the BOD assays discussed in Section 3.2.1 and the growth yield and substrate utilization studies discussed in Sections 3.1.1 and 3.1.2 (Table 9.1). An L1 carbon growth yield of 0.5 was used based on the studies of bacterial utilization of phytoplankton EOC and initial lysis products (see Section 3.2.2). The L2 growth yield was specified as 0.2.

Three grazer trophic levels were included, representing the nano-, micro-, and mesozooplankton. The nanozooplankton were presumed to graze only on the bacteria. The microzooplankton grazed on the phytoplankton biomass and 60% of the nanozooplankton biomass. The mesozooplankton grazed on 50% of the nanozooplankton biomass and all the microzooplankton biomass and had a fixed death rate representing grazing loss to higher trophic levels. The maximum grazing rate and the respiration rate in Eq. (4) were determined from allometric relationships between consumption rate and body weight (Moloney and Field, 1989) and respiration rate and body weight (Fenchel and Finlay, 1983, for protozoa; Ikeda, 1985, for metazoa). Respiration of the protozoa (i.e., the nano- and microzooplankton levels) was composed of a routine component defined as 5% of the maximum respiration rates defined by the Fenchel and Finlay relationship and a growth-related component defined as 40% of the growth rate (Fenchel and Finlay, 1983). Assuming weights of the three zooplankton groupings equal to 5, 100, and 10,000 pg C, G_m values at 20°C of 40, 20, and 6.3 µg C/µg C/day and maximum respiration rates at 20°C of 3.3, 1.6, and 0.5 µg C/µg C/day were obtained from the allometric relationships. The grazing rate half-saturation constant was 3 mg C/L for the nanozooplankton, 10 mg C/L for the herbivorous microzooplankton, and 0.5 mg C/L for the mesozooplankton. A schematic representation of the systems included in the model and the interactions between these systems is presented in Fig. 9.12.

The microcosm water column was modeled as a single, completely mixed segment. Microcosm inflow and outflow were simulated in the model by a periodic

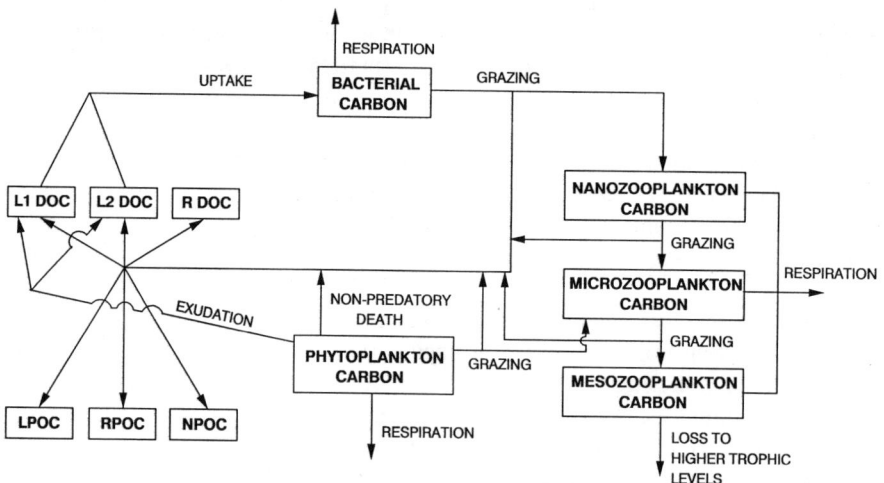

FIGURE 9.12. A schematic diagram of the components of the model of the Lake Bagsvaard microcosm.

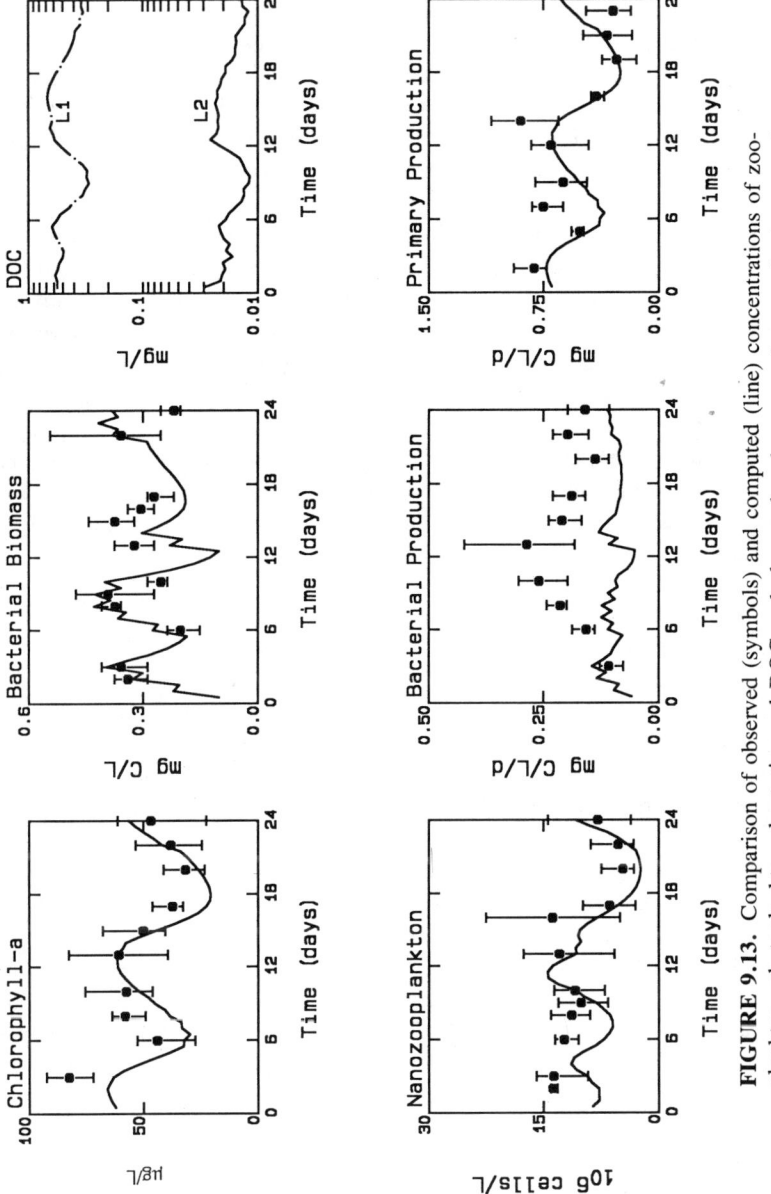

FIGURE 9.13. Comparison of observed (symbols) and computed (line) concentrations of zooplankton, phytoplankton, bacteria, and DOC and observed and computed rates of primary and secondary production in the Lake Bagsvaerd microcosms. Data are the mean ± standard deviation of triplicate microcosms.

function, which approximated the daily except Saturday and Sunday replacement of 10% of the microcosm water column with lake water. Data collected in Lake Bagsvaard itself provided the time variable boundary conditions for the inflow as well as the microcosm initial conditions.

To compare the model results with the measurements from the microcosm, AODC values were converted to bacterial biomass assuming 0.02 pg C/cell, nanoflagellate counts were converted to biomass assuming 5 pg C/cell, and [^{3}H]thymidine incorporation rates were converted to secondary production rates assuming 10^{10} μg C/nmol thymidine (see Landeck, 1990, for supporting references).

The model successfully reproduced both the magnitudes of the populations and production rates and the major features of the dynamics occurring in the microcosm (Fig. 9.13). The model does underpredict the secondary production indicated by the thymidine data. However, these data are subject to considerable uncertainty because of uncertainty regarding the conversion between thymidine uptake and carbon production. The model indicates that the dynamics in the microcosm are largely the result of predator–prey interactions. The bacteria respond rapidly to substrate production resulting mainly from grazing on the primary producers. The bacterial biomass oscillations computed by the model and indicated by the data are the result of variations in grazing pressure as the nanozooplankton are grazed by the upper trophic levels.

5. CONCLUSIONS

The utilization of DOC by pelagic bacteria is dependent on both the compounds comprising the DOC and environment-specific growth characteristics of the bacteria. Laboratory studies have shown that yield coefficients are related both to the energy content of the compound and the availability of other required nutrients such as nitrogen. Half-saturation constants describing substrate or nutrient limitation of growth rate appear to be a function of environment, higher values being associated with higher nominal substrate levels.

To make a modeling analysis of bacterial growth tractable, it is necessary to define a scheme for classifying DOC according to its ability to be used as substrate. We have proposed three categories of DOC: L1, L2, and R carbon. The fractionation of carbon into these categories may be accomplished through a BOD assay in which breaks in the oxygen utilization curve are used to signify the exhaustion of first the L1 and then the L2 component.

Our preliminary laboratory assays suggest that the L1 substrate pool is small. The bacteria are able to readily use this substrate at maximum growth rates calculated to be about 4/d. Because of the small size of this substrate pool and its rapid use, it is exhausted in the assays within a time scale of hours or even minutes.

A review of experimental studies of the fate of carbon produced through primary production indicates that a significant fraction of the DOC released from phytoplankton through exudation, cell lysis, and sloppy feeding by grazers would be classified as L1 substrate. Furthermore, a portion of the detrital POC generated through grazing appears to be rapidly solubilized to L1 DOC.

Use of the L1–L2 substrate categorization, routing phytoplankton carbon to these classes in accordance with published experimental results, and specification of bacteria and grazer dynamics according to a Monod formulation [Eqs. (1)–(5)] successfully modeled the microbial dynamics of a microcosm study. Additional applications of this modeling framework are necessary to further validate this approach.

ACKNOWLEDGMENTS

This investigation was supported by a cooperative agreement (CR-815310) with the USEPA Environmental Research Laboratory, Gulf Breeze, Florida. The support of the Project Officer, Parmely H. Pritchard, is gratefully acknowledged.

We would also like to acknowledge the participation of Peggy Harris of the Gulf Breeze Lab who conducted the BOD assays.

REFERENCES

Ahlgren, G. (1980), *Arch. Hydrobiol.*, **89**, 43–53.

Azam, F., Fenchel, T., Field, J. G., Gray, J. S., Meyer-Reil, L. A., and Thingstad, F. (1983), *Mar. Ecol. Prog. Ser.*, **10**, 257–263.

Bader, F. G. (1978), *Biotechnol. Bioeng.*, **20**, 183–202.

Banse, K. (1990), *Deep-Sea Res.*, **37**, 1177–1195.

Bell, R. T., and Kuparinen, J. (1984), *Appl. Environ. Microbiol.*, **48**, 1221–1230.

Bell, W. H. (1980), *Limnol. Oceanogr.*, **25**(6), 1007–1020.

Bell, W. H., and Sakshaug, E. (1980), *Limnol. Oceanogr.*, **25**, 1021–1033.

Bell, W. H., Lang, J. M., and Mitchell, R. (1974), *Limnol. Oceanogr.*, **19**(5), 833–839.

Berner, R. A. (1980), *Early Diagenesis. A Theoretical Approach*, Princeton University Press, Princeton, NJ.

Bierman, X., et al. (1980), *The Development and Calibration of a Multi-Class Phytoplankton Model for Saginaw Bay, Lake Huron. Great Lakes Environmental Planning Study*, Contribution No. 33. Great Lakes Basin Commission, Ann Arbor, MI.

Billen, G. (1984), in *Heterotrophic Activity in the Sea* (J. E. Hobbie and P. J. LeB. Williams, Eds.), Plenum Press, New York, p. 313–355.

Billen, G., Joiris, C., Wijnant, J., and Gillain, G. (1980), *Est. Coast. Mar. Sci.*, **11**, 279–294.

Bird, D. F., and Kalf, J. (1984), *Can. J. Fish. Aquat. Sci.*, **41**(7), 1015–1023.

Bjornsen, P. K. (1986), *Mar. Ecol. Prog. Ser.*, **30**, 191–196.

Canale, R. P., Lustig, T. D., Kehrberger, P. M., and Salo, J. E. (1973), *Biotechnol. Bioeng.*, **15**, 707–728.

Canale, R. P., Depalma, L. M., and Vogel, A. H. (1976), in *Modeling Biochemical Processes in Aquatic Ecosystems* (R. P. Canale, Ed.), Ann Arbor Science Publishers, Ann Arbor, MI, pp. 33-74.

Caron, D. A., Goldman, J. C., Andersen, O. K., and Dennett, M. R. (1985), *Mar. Ecol. Prog. Ser.*, **24**, 243-254.

Cho, B. C., and Azam, F. (1988), *Nature*, **332**, 441-443.

Chrost, R. J., and Faust, M. A. (1983), *J. Plankton Res.*, **5**(4), 477-493.

Chrzanowski, T. H., and Hubbard, J. G. (1989), *Hydrobiologia*, **179**, 61-71.

Coffin, R. B., and Sharp, J. H. (1987), *Mar. Ecol. Prog. Ser.*, **41**, 253-266.

Cole, J. J., Likens, G. E., and Hobbie, J. E. (1984), *Oikos*, **42**, 257-266.

Cole, J. J., Findley, S., and Pace, M. L. (1988), *Mar. Ecol. Prog. Ser.*, **43**, 1-10.

Di Toro, D. M. (1980), *Ecol. Modelling*, **8**, 201-218.

Di Toro, D. M., and Connolly, J. P. (1980), USEPA EPA-600/3-80-065, Duluth, MN.

Droop, M. R. (1974), *J. Mar. Biol. Assoc. U.K.*, **54**, 825-855.

Ducklow, H. W., Purdie, D. A., Williams, P. J. LeB., and Davies, J. M. (1986), *Science*, **232**, 865-867.

Fenchel, T. (1980), *Limnol. Oceanogr.*, **25**(4), 733-738.

Fenchel, T., and Finlay, B. J. (1983), *Microb. Ecol.*, **9**, 99-122.

Fogg, G. E. (1983), *Botanica Marina*, **26**, 3-14.

Fuhrman, J. A., and Ferguson, R. (1986), *Mar. Ecol. Prog. Ser.*, **33**, 237-242.

Goldman, J. C., Caron, D. A., and Dennett, M. R. (1987), *Limnol. Oceanogr.*, **32**(6), 1239-1252.

Gommers, P. J. F., van Schie, B. J., van DijKen, J. P., and Kuenen, J. G. (1988), *Biotechnol. Bioeng.*, **32**, 86-94.

Hansen, L., Krog, G. F., and Sondergaard, M. (1986), *Oikos*, **46**, 37-44.

Hopkinson, C. S. Jr., Sherr, B., and Wiebe, W. J. (1989), *Mar. Ecol. Prog. Ser.*, **51**, 155-166.

Ikeda, T. (1985), *Mar. Biol.*, **85**, 1-11.

Iturriaga, R., and Hoppe, H.-G. (1977), *Mar. Biol.*, **40**, 101-108.

Jannasch, H. W. (1965), *Anreicherungskultur und Mutantenauslese. Zentr. Bakteriol.* [*Suppl.*], **1**, 498-502.

Jannasch, H. W. (1967), *Limnol. Oceanog.*, **12**, 264-271.

Jannasch, H. W. (1968), *J. Bacteriol.*, **95**(2), 722-723.

Jensen, L. M. (1985), *Oikos*, **45**, 311-322.

Jorgensen, N. O. G., and Sondergaard, M. (1984), *Microb. Ecol.*, **10**, 301-316.

Jumars, P. A., Penry, D. L., Baross, J. A., Perry, M. J., and Frost, B. W. (1989), *Deep-Sea Res.*, **36**, 483-495.

Karl, D. M., Knauer, G. A., and Martin, J. H. (1988), *Nature*, **332**, 438-441.

Kroer, N., Coffin, R. B., and Jorgensen, N. O. G. (1991), in preparation.

Lambert, W. (1978), *Limnol. Oceanogr.*, **23**(4), 831-834.

Lancelot, C. (1984), *Mar. Ecol. Prog. Ser.*, **12**, 115-121.

Landeck, R. E. (1990), *Development of a Time-Variable Phytoplankton-Bacteria Model and Its Application to Carbon Dynamics in a Microcosm*, Masters Thesis, Manhattan College, 71 pp.

Larrson, U., and Hagstrom, A. (1979), *Mar. Biol.*, **52**, 199–206.

Larrson, U., and Hagstrom, A. (1982), *Mar. Biol.*, **67**, 57–70.

Lau, A. O., Strom, P. F., and Jenkins, D. (1984), *J. Water Pollut. Control Fed.*, **56**, 41–51.

Linley, E. A. S., Newell, R. C., and Lucas, M. I. (1983), *Mar. Ecol. Prog. Ser.*, **12**, 77–89.

Linton, J. D., and Stephenson, R. J. (1978), *FEMS Microbiol. Lett.*, **3**, 95–98.

Mague, T. H., Friberg, E., Hughes, D. J., and Morris, I. (1980), *Limnol. Oceanogr.*, **25**(2), 262–279.

McGee, R. D., Drake, J. F., Fredrickson, A. G., and Tsuchiya, H. M. (1972), *Can. J. Microbiol.*, **18**, 1733–1742.

Moloney, C. L., and Field, J. G. (1989), *Limnol. Oceanogr.*, **34**, 1290–1299.

Nyholm, N. (1976), *Biotechnol. Bioeng.*, **18**, 1043–1056.

Pengerud, B., Skjoldal, E. F., and Thingstad, T. F. (1987), *Mar. Ecol. Prog. Ser.*, **35**, 111–117.

Pett, R. J. (1989), *Mar. Ecol. Prog. Ser.*, **52**, 123–128.

Powell, E. O. (1967), in *Microbial Physiology and Continuous Culture* (E. O. Powell, C. G. T. Evans, R. E. Strange, and D. W. Tempest, Eds.), Her Majesty's Stationary Office, London, U.K., pp. 34–56.

Rhee, G.-Y. (1978), *Limnol. Oceanogr.*, **23**, 10–25.

Riemann, B., and Sondergaard, M. (1984), in *Heterotrophic Activity in the Sea* (J. E. Hobbie and P. LeB. Williams, Eds.), Plenum Press, New York, pp. 233–248.

Riemann, B., Jorgensen, N. O. G., Lampert, W., and Fuhrman, J. A. (1986), *Microb. Ecol.*, **12**, 247–258.

Robarts, R. D. (1988), *Hydrobiologia*, **182**, 137–148.

Robarts, R. D., and Sephton, L. M. (1989), *Hydrobiologia*, **182**, 137–148.

Ryder, D. N., and Sinclair, C. G. (1972), *Biotechnol. Bioeng.*, **14**, 787–798.

Scavia, D., Eadie, B. J., and Robertson, A. (1976), *NOAA Tech. Rept. ERL 371-GLERL 12*, National Oceanic and Atmospheric Administration, Boulder, CO.

Seto, M., and Alexander, M. (1985), *Appl. Environ. Microbiol.*, **50**, 1132–1136.

Shailaja, M. S., and Pant, A. (1986), *Mar. Ecol. Prog. Ser.*, **32**, 161–167.

Sharp, J. H. (1977), *Limnol. Oceanogr.*, **22**(3), 381–399.

Simkins, S., and Alexander, M. (1985), *Appl. Environ. Microbiol.*, **50**, 816–824.

Stamer, J., et al. (1979), *J. Water Pollut. Control Fed.*, **51**, 918–925.

Sushchenya, L. M., and Khmeleva, N. M. (1967), *Dokl. Biol. Sci.*, **176**, 559–562.

Sykes, R. M. (1972), *J. Water Pollut. Control Fed.*, **45**, 888–895.

Taylor, A. H., and Joint, I. (1990), *Mar. Ecol. Prog. Ser.*, **59**, 1–17.

Tetra Tech, Inc. (1979), *Methodology for Evaluation of Multiple Power Plant Cooling System Effects, Volume II. Technical Basis for Computations*, Electric Power Research Institute Report EPRI EA-1111.

Thingstad, T. F., and Pengerud, B. (1985), *Mar Ecol. Prog. Ser.*, **21**, 47–62.

Vaccaro, R. F., and Jannasch, H. J. (1966), *Limnol. Oceanog.*, **11**, 596–607.

Watson, S. W. (1978), in *Upwelling Ecosystems* (R. Boje and M. Tomczak, Eds.), Springer-Verlag, New York, pp. 139–154.

Wiebe, W. J., and Smith, D. F. (1977), *Mar. Biol.*, **42**, 212–223.

Williams, F. M. (1980), in *Contemporary Microbial Ecology* (D. C. Ellwood, J. N. Hedger, M. J. Latham, J. M. Lynch, and J. H. Slater, Eds.), Academic Press, London, pp. 349–375.

Williams, P. J. LeB. (1984), in *Flows of Energy and Materials in Marine Ecosystems* (M. J. Fasham, Ed.), Plenum Press, New York, pp. 271–299.

Wolter, K. (1982), *Mar. Ecol. Prog. Ser.*, **7**, 287–295.

Wright, R. T. (1988), *Hydrobiologia*, **159**, 111–117.

Wright, R. T., Coffin, R. B., and Lebo, M. E. (1987), *Cont. Shelf Res.*, **7**, 1383–1397.

Zlotnik, I., and Dubinsky, Z. (1989), *Limnol. Oceanogr.*, **34**(5), 831–839.

POSTFACE

"Messieurs, les microbes auront le dernier mot."

—*attributed to Louis Pasteur*

INDEX

Ability to support bacterial growth, 155, 156
Accuracy:
   of equation formats, 149
   numerical test, 171, 173
   visual test, 157
Acidothermus cellulolyticus, 107
ACV formation and conversion, kinetics of, 137
Additive error equation format, 150
Additive error model, 151
Additive error structure, 154
Aerobic metabolism, 8
Aerobic treatment, 8
Alt equation, 185
Aminoadipyl-cysteine-valine synthetase (ACVS), 116, 131
Ammonia, 225
Antibiotic production model, 115
Antibiotic synthesis, regulation of, 118
Arabinose, 48
Articulation, 242
Ascendancy, 221, 242, 243
Attractants, 179
Autocatalytic equation, 34
Autodigestion, 10

Bacterial chemotaxis, 177
Bacterial growth, ability to support, 155, 156
Bacterial motility, 177
Bacterial population genetics, 61
Bacterial transport coefficients, 198
Bacteria-substrate interactions, 253
Bacteriophage, 70
Biochemical oxygen demand, 9, 258
Biodegradation, 61
Biological growth model, 18
Biomass, 16
Biomass yield, 11
Biosynthesis, 116
Blackman kinetics, 35
Blackman model, 36

Capillary assay, 193, 199, 201
Carbon–oxygen cycle, 6
Carrying capacity, 34
Catabolite repression, 125, 134
Cell balance equations, 184
Cell recycle, 19
Cell transport coefficients, 192
Cellulase, 90
Cellulase producing organisms, 89
Cellulose, 90
Cell yield, 12, 18
Cephamycin C, 116, 117
Cephamycin C synthesis, kinetics of, 135
Chemical oxygen demand, 8, 11
Chemosensory mechanism, 181
Chemotactic coefficient, 191
Chemotactic sensitivity coefficient(s), 183, 192, 200
Chemotactic velocity, 207
Chemotaxis, 178, 180, 205
   general description, 179
Chlorobenzoate, 53
Coefficient:
   bacterial transport, 198
   cell transport, 192
   chemotactic, 191
   chemotactic sensitivity, 183, 192, 200
   growth yield, 253
   maintenance, 37
   random motility, 183, 192, 197
   yield, 253
Comparison of equation formats, 156
Compartment design, 72
Compartment model(s), 61, 64, 74
Computer simulation, 82
Conductivity, 155, 156
Conjugation, 69
Contamination simulation, 77
Cycle, carbon–oxygen, 6

Death, 250

280 INDEX

Death phase, 96
Death rate, 96, 252
Decay model, 16
Decay rate, 11, 132
Desacetoxycephalosporin C hydroxylase (DACS), 117
Desacetoxycephalosporin C synthetase (DAOCS), 116, 129
Development capacity, 242, 243
Diaminopimelic acid pathway, 119
Diazo nitrogen, 225
Diffusion, 53
  Fickian, 196
Dilution rate, 17
Dinitrophenol, 51
Dissipation, 243
Dissolved inorganic nitrogen, 219
Dissolved organic carbon, 260
Dissolved organic nitrogen, 219, 225
Dissolved oxygen, 9
Dual-substrate kinetics, 53
Dual-substrate model(s), 38, 47

Enterobacter cloacae, 155
Enzymatic reduction, 74
Enzyme activities, 143
Enzyme inactivation kinetics, 125
Enzyme inactivation rate, 125
Enzyme synthesis rate, 125
Equation:
  additive error format, 150
  Alt, 185
  autocatalytic, 34
  cell balance, 184
  exponential decay, 153
  exponential format, 153
  Haldane, 14, 21, 37
  logistic, 33, 40
  logistic format, 152
  Monod, 12, 13, 21, 35, 37, 40, 43, 250
  multiplicative error format, 151
  population balance, 184
  rectangular hyperbolic, 12
  Segel, 187
  Stroock, 185
  Volterra, 98
Equation formats:
  accuracy of, 149
  comparison of, 156
Escherichia coli, 182, 198, 204
Excess biomass, 22
Excision, 69
Exponential decay equation, 153

Exponential equation format, 153
Exponential growth function, 46
Exponential growth phase, 93
Exponential models, 153
Export, 71, 243
Extracellular organic carbon, 266

Feedback loops, 240
Fickian diffusion, 196
First order kinetics, 51
Flows, 226, 227
Flux, 229

Gene expression, 79
General systems modeling, 63
Genetic configuration, 66
Genetic ecology, 62
Genetics, bacterial population, 61
Glucose, 48
Grazing rate, 252
Growth, 91, 250
Growth curve, 10
Growth kinetics, 9, 47, 51
Growth modeling, 89
Growth rate, 10, 17, 33, 252, 257
Growth rate hysteresis, 24
Growth yield, 263
Growth yield coefficient, 253

Haldane equation, 14, 21, 37
Haldane model, 36
Haldane relationship, 21
Half-saturation constant(s), 257, 263
  Michaelis, 251
Hydraulic retention time, 17

Import, 243
Inactivation rate values, 155
Incubation temperature, 155
  incorporation in models, 156
Incubation time, 155
Induction, 79
Inhibition constant, 14
Inhibitory substrates, 37
Inputs, 232
Interactions, matrix of, 227
Internal interactions, 232
Isopenicillin N synthetase (IPNS), 116, 127

Klebsiella oxytoca, 155

Labile substrate, 258
Lag phase, 92
Laser densitometry assay, 194, 205
Linear regression, 41, 151, 159
Logistic equation, 33, 40
Logistic equation format, 152
Logistic model, 99, 152
Logit transformation, 152, 170

Maintenance coefficient, 37
Maintenance rate, 37
Malthusian parameter, 33
Maximum growth rate, 263
mer-Operon, 73
mer-Operon system, 64, 73
Metabolic blockage, 25
Michaelis half-saturation constant, 251
Michaelis–Menten kinetics, 135
Michaelis–Menten model, 35
Microbial growth, 9
Microbial growth modeling, 92
Microcystis aeruginosa, 222
Mineralization, 47
Mineralization curves, 51
Mineralization of phenol, 43
Model:
    additive error, 151
    antibiotic production, 115
    biological growth, 18
    Blackman, 36
    compartment, 61, 64, 74
    decay, 16
    dual substrate, 38, 47
    exponential, 153
    failure of, 170
    framework for, 124
    general systems, 63
    growth, 89
    Haldane, 36
    incorporation of incubation
       temperature in, 156
    logistic, 99, 152
    Michaelis–Menten, 35
    microbial growth, 92
    misdirection of, 170
    Monod, 36
    multiplicative error, 151
    NETWRK4, 235
    physiological response, 49
    population compartmental, 65
    population decay (die-off), 149
    population dynamics, 33
    population equation, 31
    product formation, 49
    RTBL, 188
    simulation, 49, 62
    Stella, 76, 77
    structured, 91
    unstructured, 91
    Volterra, 98
    wastewater treatment, 1
Model relatedness, 40
Model simulations, 64
Model validation, 42, 82, 211
Model verification, 103
Modulating function, 111
Monod equation, 12, 13, 21, 35, 37, 40, 43, 250
Monod growth kinetics, 39
Monod kinetics, 40, 196, 251
Monod model, 36
Multiphasic kinetics, 52
Multiple linear regression, 150
Multiplicative error equation format, 151
Multiplicative error model, 151
Multiplicative error structure, 154
Multisystem kinetics, 52

Natural substrate, 258
Network analysis, 219
Network theory, 218
NETWRK4, 235
Nexus, 221, 240
Nitrate, 225
Nitrite, 225
Nitrogen cycling, 217
Nonlinear regression, 42, 51
Non-living seston nitrogen, 225
Nontoxic substrates, 21

Operon, 65
Optimization methods, 80
Organization of concepts, 64
Organization of data, 64
Organomercuric compounds, 73, 74
Organomercury, 77
Organotropic microorganisms, 7
Outputs, 232
Overhead, 242, 243
Oxidative assimilation, 26

Parameter estimation, 41
Paranitrophenol, 45, 54
Parathion, 49
Particulate nitrogen, 223
Particulate organic carbon, 253, 268
Partitioning of substrate, 53

Phage (bacteriophage), 70
Phenol, 44
Physiological response models, 49
Phytoplankton, 265
Phytoplankton nitrogen, 225
Plasmid replication, 69
Plasmids, 69
Population balance equation, 184
Population compartmental model, 65
Population decay rates, 149, 150
Population dynamics, 47, 72
Population dynamics models, 33
Population ecology models, 31
Population simulation, 76
Predation, 252
Prediction, 64
Product formation model, 49
Pseudo-first order mineralization, 54
Pseudomonad bacteria, 64
Pseudomonas alcaligenes, 53
Pseudomonas fluorescens, 155, 198
Pseudomonas species, 55

Random motility, 180, 205
Random motility coefficient, 183, 192, 197
Reactor engineering, 16
Rectangular hyperbolic equation, 12
Recycle flow, 19
Redundancy, 243
Refractory substrate, 258
Reliability, degree of, 231
Repellents, 180
Respiration, 7, 15
Respiration rate, 252, 263
Respirometric concept, 16
RTBL model, 188

Salmonella typhimurium, 47, 198, 205
Sediment nitrogen, 225
Segel equation, 187
Segregation, 70
Shock loadings, 24
Simulation model(s), 49, 62
Soil, 49
Sorption, 53
State variable(s), 225, 229
Stationary phase, 96
Statistical design, 150
Stella model, 76, 77
Stochastic analysis, 81
Stopped-flow diffusion chamber assay, 194, 200, 209
Streptomyces clavuligerus, 116, 127, 129, 143
Stroock equation, 185

Substrate(s), 7, 258
 labile, 258
 natural, 258
 nontoxic, 21
 phytoplankton generated, 265
 refractory, 258
 toxic, 21
Substrate availability, 53
Substrate concentration, influent, 23
Substrate flux, 53
Substrate inhibition, 55
Substrate utilization, 257
Substrate utilization rate, 18
Survival, of viruses, 150
Survival ratio, 152
Survival ratio values, 156
Synthesis, 7
System throughput, 243

Temperature, 155, 156
Testing:
 of concepts, 64
 of data, 64
Theory of continuous culture, 18, 23
Thermomonospora fusca, 108
Time, incorporation into models, 155
Total organic carbon, 9, 260
Toxic substrates, 21
Transcription, regulation of, 132
Transcription kinetics, 125
Transcription rate(s), 79, 132
Transduction, 70
Transfection, 70
Transformation, 71
Translation rate, 132
Transposable elements, 68
Transposition, 68
Transposon, 66
Trichoderma viride (*Trichoderma reesei*), 103
Turbidity, 155, 156

Upstream regulation, 120

Virus stability, 150, 174, 175
Virus survival, 150
Volatilization, 74
Volterra equation, 98
Volterra model, 98

Waste substrate, 8
Weak arc(s), 221, 240

Yield coefficients, 253